FLAMMABILITY OF CELLULOSIC MATERIALS, Part II

VOLUME ELEVEN
FIRE AND FLAMMABILITY SERIES

Edited by

DR. CARLOS J. HILADO

a TECHNOMIC® publication
TECHNOMIC Publishing Co., Inc.
265 Post Road West, Westport, CT. 06880

FLAMMABILITY OF CELLULOSIC MATERIALS, Part II

VOLUME ELEVEN
FIRE AND FLAMMABILITY SERIES

265 Post Road West, Westport, CT 06880

a TECHNOMIC® publication

Printed in U.S.A.
Library of Congress Card No. 73-82115
I.S.B.N. 0-87762-171-3

To Melody

INTRODUCTION

One of the goals of the Journal of Fire and Flammability and its supplements is to so guide their content that within five years a complete collection of issues would be the basic literature source in this field of technology. Compilations of Journal articles in areas of specific interest thus become a valuable literature resource, and the Fire and Flammability Series of books embodies this concept.

Cellulosic materials represent a major source of materials for mankind, and appear primarily in the form of paper, wood, and cotton. They are prominent in both building construction and building contents, and widely used in consumer products. Because cellulosic materials are combustible, their flammability characteristics have been the subject of considerable study because of the need for defining criteria for their safe use.

The first compilation of articles on cellulosic materials, the 242-page Volume One on Flammability of Cellulosic Materials, contained sixteen articles on this subject. Since Volume One went to press, eighteen articles relevant to cellulosic materials have appeared in the Journal and its supplements, and these articles are assembled in this volume for the benefit of those interested in this subject.

This continuing interest in cellulosic materials is well justified by volume of use. While the United States uses some 12 million tons of plastics in one year, in this same period it uses some 50 million tons of paper and paperboard alone, together with 35 billion board feet of lumber, over 15 billion square feet of plywood, and about 3 million tons of cotton. Cellulosic materials are also renewable resources, and will be very much with us for many years to come.

CONTENTS

PUBLISHER'S NOTE

SECTION ONE

Paper

ASHLEY S. CAMPBELL

Department of Mechanical Engineering
University of Maine
Orono, Maine

FIRE SPREAD OVER PAPER

(Received December 10, 1973)

ABSTRACT: An experimental study of the influences of sheet thickness and initial temperature on the steady state rate of spread of a fire moving downward over filter paper. The data indicates that a simple relationship exists between rate of spread and heat flux from the flame.

INTRODUCTION

THE THEORETICAL TREATMENT of the rate of spread of a free-burning fire over a solid fuel has been reported by de Ris [1], Lastrina *et al*, [2] and Parker [3]. For the case of a thin sheet, the analysis predicts the following relationship

$$\sqrt{2}\,\rho c \tau V\,(T_p - T_o) = k_g\,(T_f - T_p) \quad (1)$$

where ρ, c and τ are the density, heat capacity and thickness of the sheet, V is the steady state rate of spread, k_g the thermal conductivity of the gas in the region between flame and sheet surface, T_f the flame temperature, T_p the pyrolysis temperature of the fuel, and T_o the fuel temperature at a large distance from the burning region. The purpose of the work reported here has been to examine the validity of Equation (1), for the case of a free-burning fire moving down a sheet of filter paper. To do this we have studied the separate influences of changes in sheet thickness τ and initial temperature T_o on the rate of spread V.

EXPERIMENTAL APPARATUS

The experimental arrangement is simple. A sheet of filter paper 3″ wide and 10″ or more in length is placed in a vertical plane, with the long dimension vertical. Both long edges are held between pairs of metal straps, which support the sheet and prevent the burning zone from reaching the edges. The paper is ignited at the top edge. A burning zone about 2½″ wide is established quickly, and is symmetrical on the two paper surfaces. The lower edge of the burning zone is horizontal, except for slight curvature where it touches the straps. The burning zone creeps down the sheet at a uniform rate, which is measured by noting the elapsed time between pencil marks drawn on the surface. Downward propagation is easy to study, because it proceeds at a steady rate, and the analysis can ignore radiation.

Sheets of varying thickness were formed by glueing a number of sheets, using a 2% solution of methyl cellulose in water, following Murty [4]. The composite sheet is placed between metal plates and oven dried for 24 hours at 100°C. The basic sheet has the following properties;

τ, thickness = .015 cm
c, heat capacity = .2 cal/g°C
k, thermal conductivity = .00021 cal/cm sec°C
ρ, density = .62 g/cm^3
α, thermal diffusivity = $k/\rho c$ = .0017 cm^2/sec

The limit for thickness is 12 sheets. For thicker sheets the burning zones are not symmetrical on the two surfaces, and the flames wander erratically.

This method of forming thick sheets provides for a systematic study of the influence of sheet thickness, keeping density constant. The method has two drawbacks. First, in the burning zone, only the outermost sheets burn; the inner sheets separate and char. This prevents any study of the burning zone itself. The separation occurs well above the lower edge of the burning zone, however, and we believe that there is no influence on events in the heating region ahead of the burning zone. The second drawback is the presence of glue between sheets, and the possibility that the thermal conductivity normal to the sheet surface is altered. When a glued sheet is torn apart, the tear does not always follow a glued surface, indicating a good bond. A cursory look at thermal conductivity showed that two sheets glued together offer twice the resistance to heat flow.

Hand sheets were formed by draining a slurry of pulp stock (cotton linters, supplied by the filter paper manufacturer, Schleicher and Schuell, Keene, New Hampshire) on a fine screen. Sheets made in this way do not provide for independent variation of density and thickness; rather they provide for tests of the influence of $\rho\tau$. The hand sheets tested varied in thickness from .013 to .22 cm, and in surface density from .0055 to .12 g/cm^2.

Paper temperature is measured with .002″ chromel/alumel thermocouples. The weld head is flattened to a disc, and is placed between sheets when forming a composite sheet, or under a flap cut with a scalpel, as first suggested by Parker [3]. Usually the junction was placed at or near the mid-point of the thickness.

The initial paper temperature T_o was controlled by a pair of electric heaters which irradiated the sheet. T_o values up to 200°C can be obtained. No attempt was made to control ambient humidity. All paper samples were oven dried, and burned with a few minutes after removal from the oven. Presumably, local humidity should not influence the burning behavior for T_o greater than 100°C.

EXPERIMENTAL RESULTS

Figure 1 is a log-log plot of rate of spread against number of sheets in a glued sheet, at four T_o conditions. Equation (1) predicts a uniform slope of −1, assuming flame temperature T_f and pyrolysis temperature T_p are independent of T_o and thickness. The lines show some departure from predicted slope. The data given here, and elsewhere in this report, for glued sheets are an average of five or more runs.

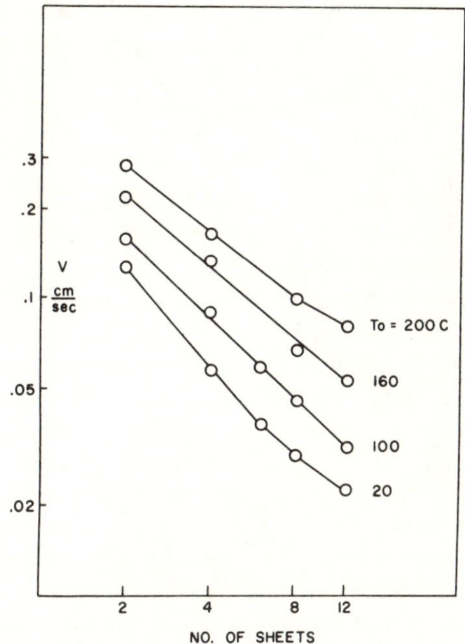

Figure 1. Glued sheets. Average V values. See Table

Figure 2 shows the same data plotted according to Equation (1) for glued sheets, and in Figure 3 a few data from hand sheets. For plotting purposes we have taken $T_f = 1500°C$, $T_p = 325°C$ and $k_g = 1.66 \times 10^{-4}$ cal/cm sec°C. The curves, identified by the open symbols, are influenced by thickness or surface density.

The theory describing steady state fire spread over a thin sheet of fuel ignores heat transfer by internal conduction from the burning zone forward to the unburned fuel. The fuel is assumed to be heated entirely by heat transfer from the flame to the paper surface. This is supported by the observation, first reported by Parker [3], that the lower (leading) edge of the flame in a downward propagating fire on paper extends a few millimeters below the blackened zone where pyrolysis is taking place. But heat transfer from the flame to the surface must produce a temperature gradient in the paper between surface and center, since by symmetry the temperature gradient at the center is zero. The temperature variation across the thickness of the sheet may be taken into account by rewriting [1] in the form

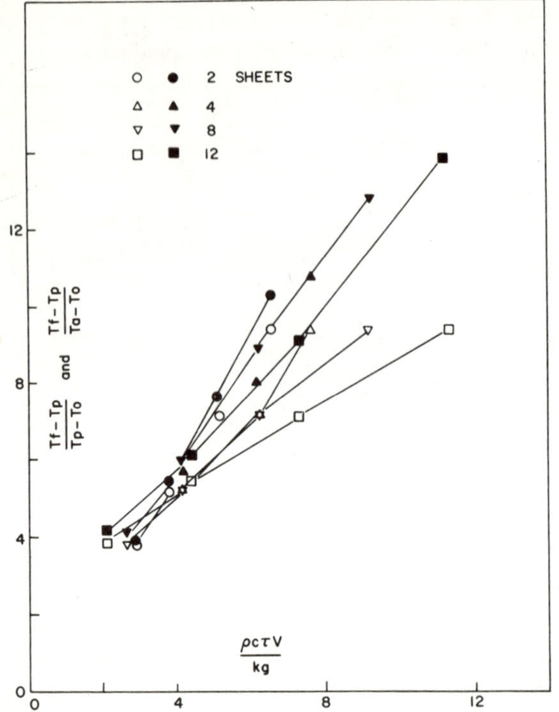

Figure 2. *Glued Sheets* $T_p = 325C$
Open Symbols: Equation (1) $T_f = 1500\ C$
Filled Symbols: Equation (2) $k_g = 1.66 \times 10^{-4}\ \frac{cal}{cm\ sec\ C}$

$$\sqrt{2}\,\rho c\tau V\,(T_a - T_o) = k_g\,(T_f - T_p) \qquad (2)$$

where T_a is the average paper temperature at the instant the surface temperature reaches T_p.

T_a can be inferred from the time history of the thermocouple record. Figure 4 is a typical recorder output. As the burning zone creeps downward there is a small temperature increase which ends abruptly at A, followed by a linear rate of rise between B and C, the heating period, followed by an abrupt decrease to a plateau between D and E, the pyrolysis period. Beyond E the thermocouple emerges from the burning zone and is exposed to direct flame heating.

We consistently find that the lower edge of the burning zone passes over the thermocouple at some point between B and C. We cannot determine the exact moment of passing. The fibre character of the paper surface blurs the edge of the burning zone, and the precise location of the embedded thermocouple is not

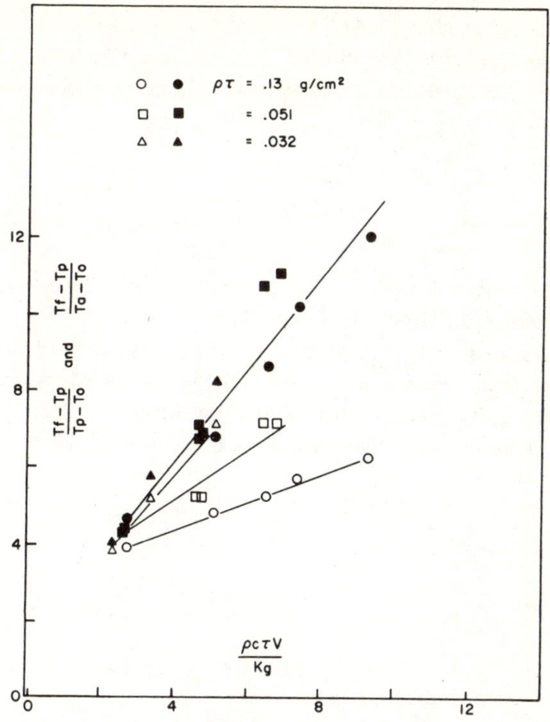

Figure 3. Hand Sheets
Open symbols: Equation (1)
Filled symbols: Equation (2)

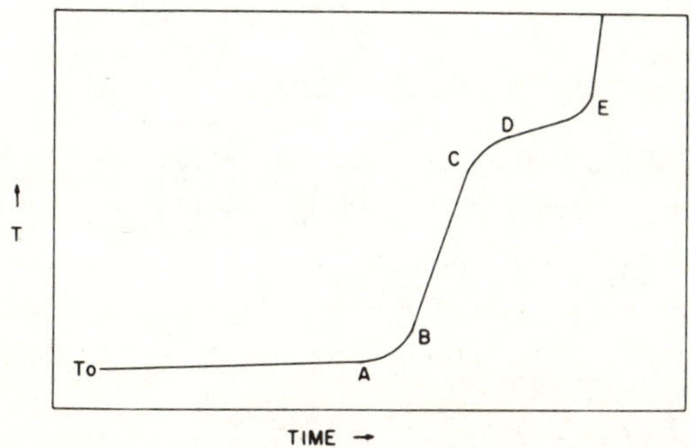

Figure 4. Typical temperature-time curve for a thermocouple placed inside filter paper.

known. But if the linear character of the temperature rise is taken as an indication that a temperature gradient is established and persists during the heating period, T_a is easily evaluated. Let q denote the heat flux (cal/cm^2 sec) absorbed at the paper surface, then

$$q = \frac{1}{2} \rho c \tau V \frac{\Delta T}{\Delta t} \tag{3}$$

where $\Delta T/\Delta t$ represents the linear rate of temperature rise with time. (The quantity $\Delta T/\Delta t$ is found to vary from 15°C/sec, for 12 sheets at $T_o = 20°C$, up to 260°C/sec, for 2 sheets at $T_o = 200°C$.) Let T (x,y) denote the temperature in the sheet. The origin of the coordinate system is at the center of the sheet. The x-axis passes through the surface of the sheet where the surface temperature is T_p, i.e. $T(\tau/2, 0) = T_p$. The y-axis is positive toward the burning zone. Then

$$T(x,o) = T_p - \frac{\tau q}{4k} \left[1 + \left(\frac{2x}{\tau}\right)^2\right]$$

and for T_a we have

$$T_a = \frac{2}{\tau} \int_0^{\tau/2} T(x,o)\, dx = T_p - \frac{\tau q}{6k}. \tag{4}$$

The filled symbols in Figures 2 and 3 are plots of Equation (2). The influence of thickness on the glued sheets, and surface density on the hand sheets, is substantially reduced, and Equation (2) is approximated rather closely.

Lastrina *et al*, [2] have shown that the criterion for a "thin" sheet is given by a dimensionless thickness which must be less than unity for a thin sheet,

$$\bar{\tau} = \frac{\tau}{2} \sqrt{\frac{V}{\alpha \delta}}. \tag{5}$$

Here δ is the length of the heating zone, and may be calculated by a simple energy consideration. In accordance with Equation (1), δ would be evaluated from

$$2\delta q = \rho c \tau V (T_p - T_o) \tag{6}$$

On the other hand, when internal heat conduction is taken into account, we define the length of the heating zone as δ', and from energy consideration

$$2\delta' q = \rho c \tau V (T_a - T_o) - 2 \int_0^{\tau/2} k \frac{\partial T(x,o)}{\partial y}$$

where

$$k\frac{\partial T\,(x,o)}{\partial y} = \frac{k}{V}\frac{\Delta T}{\Delta t} = \frac{2\alpha}{\tau V}\,q,$$

so that

$$\delta' = \delta - \frac{\alpha}{V} - \frac{\tau^2}{12\alpha}V \tag{7}$$

Table 1 lists some average data for the glued sheets, and calculated values for δ, δ' and $\overline{\tau}$. The difference between δ and δ' are substantial, and reflect the influence of internal conduction. The values of δ and δ' are the proper order of magnitude, as judged by visual observation of the extent of flame overhang below the lower edge of the burning zone. With the exception of the thickest sheets at T_o 200°C, all $\overline{\tau}$ are less than unity, so the data in this report represents the behavior of a thermally thin fuel.

Figure 5 is a plot of log V vs q, from average data for glued sheets. There is considerable scatter, and this prompted our study of burning in hand sheets; we were concerned that the thermal conductivity in the glued sheets might be a contributing factor. The plot for hand sheets is on Figure 6, and the scatter is similar to Figure 5. There is also considerable difference between the locations of the data points on these two figures, for which we have no ready explanation. The basic paper fibre used for the manufactured sheets and the hand sheets is the same; the difference between the two lies only in the manner of sheet formation.

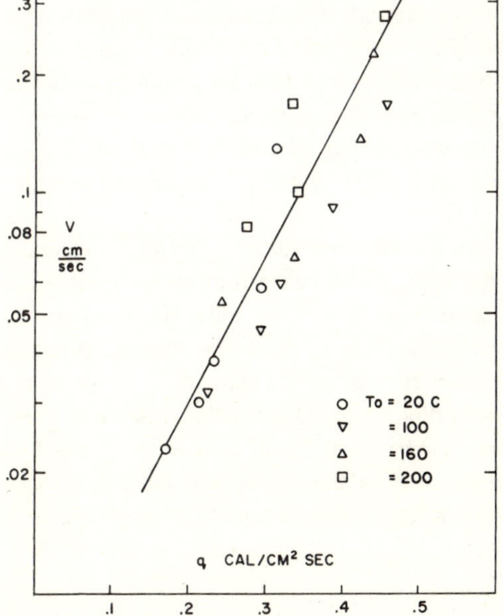

Figure 5. Glued Sheets. Averaged V and q values. See Table 1.

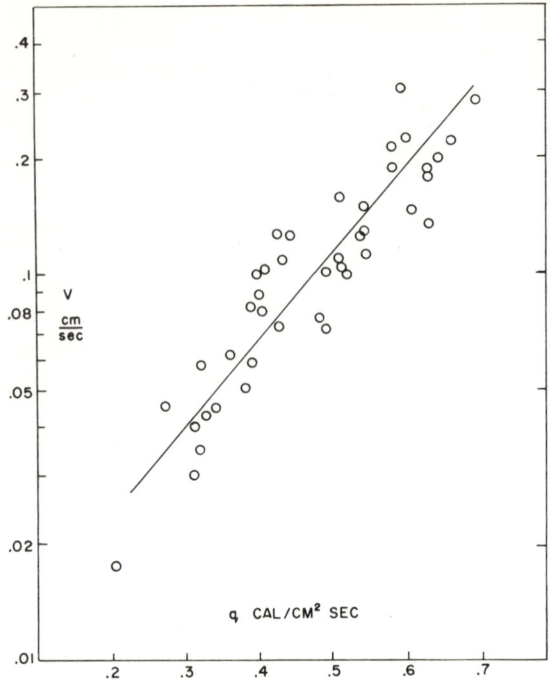

Figure 6. Hand Sheets

It seems reasonable to attribute the scatter to local variations in surface density, $\rho\tau$, for these give rise to variations in the calculated heat flux q, but not in the rate of spread V, since the latter is a gross measurement made over a half-inch of sheet. Now if this conjecture is correct, then it is legitimate to draw a faired line through the data points, and to conclude that a relationship exists between q and V which is independent of density, thickness and initial temperature of the sheet. Which is to say that the spreading behavior of the fire is controlled entirely by events at the paper surface.

To push the analysis a small step further, consider Figure 7, where the flame surface is fixed and the sheet moves upward at velocity V. As a result of pyrolysis in the burning zone, gas diffuses away from the surface and mixes with the ambient air to form a combustible mixture at the flame surface. In particular, gas diffuses downward along the paper surface with a velocity V_x (relative to the paper) which varies in magnitude according to the local partial pressure gradient. Since the flame is supposed stationary, V_x becomes equal to V at the flame surface. The heating region is assumed to lie beneath the flame, and extends a distance δ' below the lower edge of the burning region. A crude application of one-dimensional diffusion theory leads to

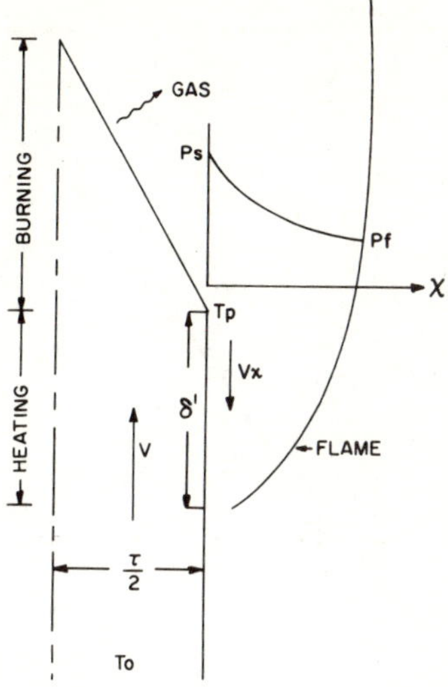

Figure 7. Cross-section of the heating and burning regions in a sheet of paper. The flame tip is maintained stationary by the diffusion of pyrolyzed gases originating at the lower edge of the burning region, while the paper sheet moves at constant velocity V. The flame tip extends distance δ' below the burning zone.

$$V\delta' = D \frac{P}{p_f} \ln \left(\frac{P - p_f}{P - p_s} \right) \tag{8}$$

where D is a diffusion coefficient, P the total pressure and the partial pressures p_s and p_f of the pyrolysis gas refer, respectively, to the burning surface and the flame surface. If one now plots $V\delta'$ against q, we get Figure 8, for the glued sheets. The open points use average q values, Table 1. The filled points use faired values from Figure 5, Table 2. The filled symbols lead us to the empirical relationship

$$\log (V\delta') = aq + b, \qquad a, b \text{ constants}$$

and through Equation (8)

$$q = f\,(p_s)$$

taking D, P and p_f as fixed parameters, which is to say that the partial pressure of the pyrolysis gas at the surface of the paper, and therefore the rate at which this gas is produced, is related in some way to the heat flux from the flame. This lends confirmation to the conclusion that the steady spread rate of downward fire propagation in a thin sheet of paper is a diffusion controlled phenomenum.

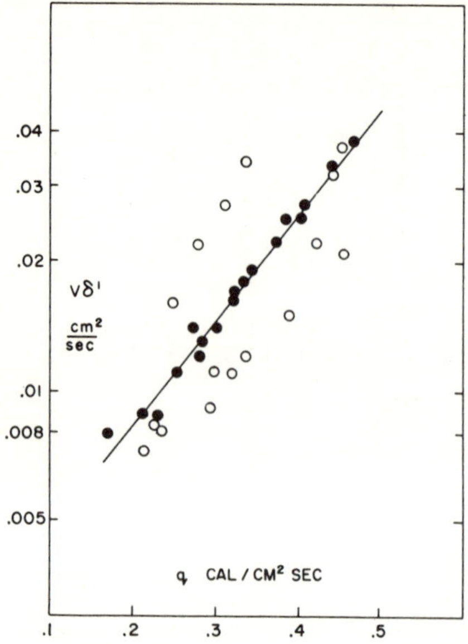

Figure 8. Glued Sheets. Distance δ' shown in Figure 7.
Open symbols: δ' calculated from average q values. See Table 1.
Filled symbols: δ' calculated from q values taken from the faired curve in Figure 5. See Table 2.

CONCLUSIONS

Analysis of experimental results on the rate of spread of a free-burning fire moving downward over thin sheets of filter paper supports the solution of the problem as a diffusion-controlled flame. The theory is correct if the sheet is sufficiently thin, i.e. if the temperature gradient across the half-thickness can be ignored. Over the full range of thermally thin sheets the theory is not correct, because the gradient is substantial. This study suggests that the relationship between rate of spread and heat flux from the flame is independent of density, thickness and initial temperature of the paper.

ACKNOWLEDGMENTS

The work was supported under the RANN program of NSF, grant GI-35509. The author thanks Thomas Eaton for his able assistance.

Table 1. Data and Calculated Values for Glued Sheets
V and q are average values from five or more runs

T_o C	No. sheets	V cm/sec	q cal/cm² sec	T_a C	δ cm	$\bar{\tau}$ —	δ' cm	Vδ' cm²/sec
20	2	.128	.313	317	.23	.28	.21	.027
	4	.058	.296	311	.23	.41	.19	.011
	6	.038	.235	308	.28	.46	.22	.0081
	8	.030	.212	305	.32	.52	.24	.0072
	12	.023	.172	300	.46	.56	.35	.0081
100	2	.164	.454	314	.15	.41	.13	.021
	4	.091	.388	306	.20	.56	.16	.015
	6	.059	.319	302	.24	.62	.19	.011
	8	.046	.292	297	.27	.71	.20	.0092
	12	.032	.224	293	.36	.78	.26	.0083
160	2	.228	.440	314	.16	.46	.14	.032
	4	.137	.421	305	.20	.68	.16	.022
	8	.069	.437	292	.25	.92	.18	.012
	12	.054	.247	289	.41	.95	.29	.016
200	2	.287	.453	314	.15	.54	.13	.037
	4	.168	.335	309	.24	.65	.20	.034
	8	.102	.342	292	.28	1.08	.19	.019
	12	.083	.277	285	.42	1.26	.26	.022

Note: T_a is calculated on the assumption $T_p = 325$ C.

Table 2. Glued Sheets
q Values are taken from the faired line in Figure 5

T_o C	No. sheets	V cm/sec	q cal/cm² sec	δ cm	δ' cm	Vδ' cm²/sec
20	2	.128	.373	.19	.17	.022
	4	.058	.280	.24	.20	.012
	6	.038	.230	.29	.23	.0087
	8	.030	.204	.33	.25	.0075
	12	.023	.170	.46	.35	.008
100	2	.164	.403	.17	.15	.025
	4	.091	.332	.23	.19	.018
	6	.059	.281	.27	.22	.013
	8	.046	.252	.31	.24	.011
	12	.032	.210	.38	.28	.0088
160	2	.228	.442	.16	.14	.033
	4	.137	.381	.22	.18	.025
	8	.069	.300	.28	.21	.014
	12	.054	.270	.38	.26	.014
200	2	.287	.468	.15	.13	.038
	4	.168	.407	.20	.16	.027
	8	.102	.345	.28	.19	.019
	12	.083	.322	.36	.20	.017

REFERENCES

1. John de Ris, The Spread of a Laminar Diffusion Flame, XII Symposium (International) on Combustion (1969), p. 241.
2. F. A. Lastrina, R. S. Magee and R. F. McAlevy III, Flame Spread Over Fuel Beds: Solid-Phase Energy Considerations, XIII Symposium (International) on Combustion (1971), p. 935.
3. W. J. Parker, Flame Spread Model for Cellulosic Material, *Journal Fire & Flammability*, Vol. 3, p. 254 (1972).
4. K. Murty, Thermal Decomposition Kinetics of Wood Pyrolysis, Combustion and Flame, Vol 18, p. 72 (1972).

Ashley S. Campbell

Ashley S. Campbell has been a Professor of Mechanical Engineering at the University of Maine since 1968. He received his B. S., S. M., and Sc. D. degrees from Harvard University. He was Dean at the University of Maine from 1950 to 1957, and Dean at Tufts University from 1957 to 1968.

KOGAKU KOMAMIYA

The Research Institute of Industrial Safety
Ministry of Labour, Japan
5-35-1 Shiba, Minato-ku, Tokyo, Japan

QUENCHING DISTANCE FOR A COMBUSTIBLE SOLID IN THE OXYGEN-ENRICHED ATMOSPHERE

(Received June 9, 1972)

ABSTRACT: In connection with fire prevention in oxygen-enriched atmosphere, "flame quenching distances" of various combustible solids are measured.

Membranous specimens such as plastic sheets and papers are held between two metal plates having a V- shaped notch, and are ignited at the open end in the combustion chamber in which the pressure is at 1–5 bar. The quenching distance is determined from the width of the notch where the flame ceases to spread.

The results obtained have a good reproducibility and show that the quenching distance can be used as an effective measure for combustibility of solid materials.

In this paper, the details of experimental apparatus and procedure are described, and the correlation between the quenching distance and the "oxygen-index" is discussed.

INTRODUCTION

AS IS WELL known, the combustion of solid materials such as plastic, wood and paper are quenched when they contact a cool body.

This phenomenon is considered to be caused by the heat loss from burning solid. It has fully studied for flame propagation in an explosive gas mixture [1], and the results are utilized in the design of flame arresters But, no or little study has ever been made for the flammable solids[1].

From such standpoint, the quenching distance of membranous solid in oxygen-enriched atmosphere is investigated here experimentally. The results show that the quenching distance becomes an effective measure for combustibility of solids as well as explosive gas mixtures.

EXPERIMENTAL

Apparatus

For supporting the combustible specimen, the brass holder shown in Figure 1 is

[1] Recently the author became aware of R. L. Durfee and J. M. Spurlock's report [2] for the plastics burning in oxygen under a reduced-pressure. But their test method is different from that of the author.

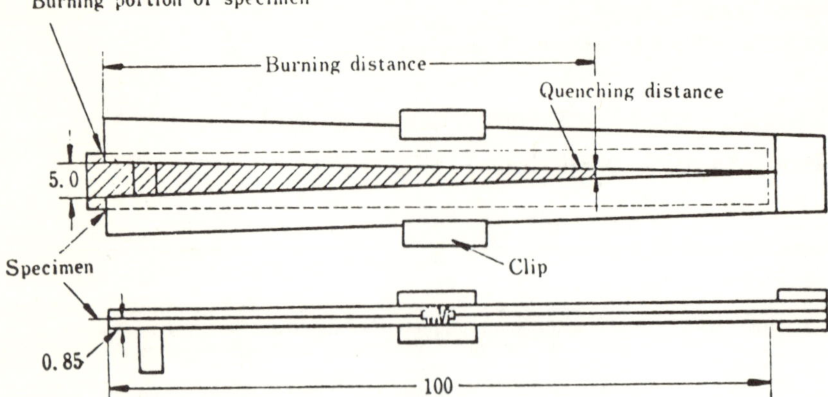

Figure 1. *Holder for measuring quenching distance.*

used. The holder consists of two parts of upper and lower plate with the cut area of "V" shape. The specimen is inserted between these plates, which are then fastened tightly together by means of clips. Thus the test specimen sandwiched between two metal plates has a combustible area of an equilateral triangle whose base is the starting lines of flame spread.

The combustion vessel used for the present experiment is shown in Figure 2. A cover with a glass window is provided for visual observations. A Bourdon type pressure gauge, a valve for supplying an oxidizing gas and an ignition heater were installed at the positions shown in the figure.

Experimental Procedure

A piece of Japanese paper (a sort of paper which is made peculiarly and traditionally in Japan and composed of almost pure cellulose about 10 X 5 mm) is used as ignition aid at the end of a test specimen. The holder prepared with a test specimen is fixed horizontally in the combustion vessel, and after the vessel is closed and purged, the oxidizing gas is introduced through a valve up to a preselected pressure. The following five oxidizing gas mixtures are used: (1) 21%O_2 79%N_2, (2) 42.5%O_2 57.5%N_2, (3) 58.1%O_2 41.9%N_2, (4) 76.3%O_2 23.7%N_2, (5) 100%O_2. Oxygen concentration is measured by gas-interferometer, but the decrease of it in the course of burning is negligible.

The paper is first ignited, the flame then propagates along the test specimen. The flame spread stops at a certain distance depending on the properties of specimen and atmosphere. The length beyond which the flame ceases to propagate denotes the "burning distance," and the minimum width of the burning area at this length is the "quenching distance." The former is measured by a vernier calipers and the latter is calculated from the burning distance.

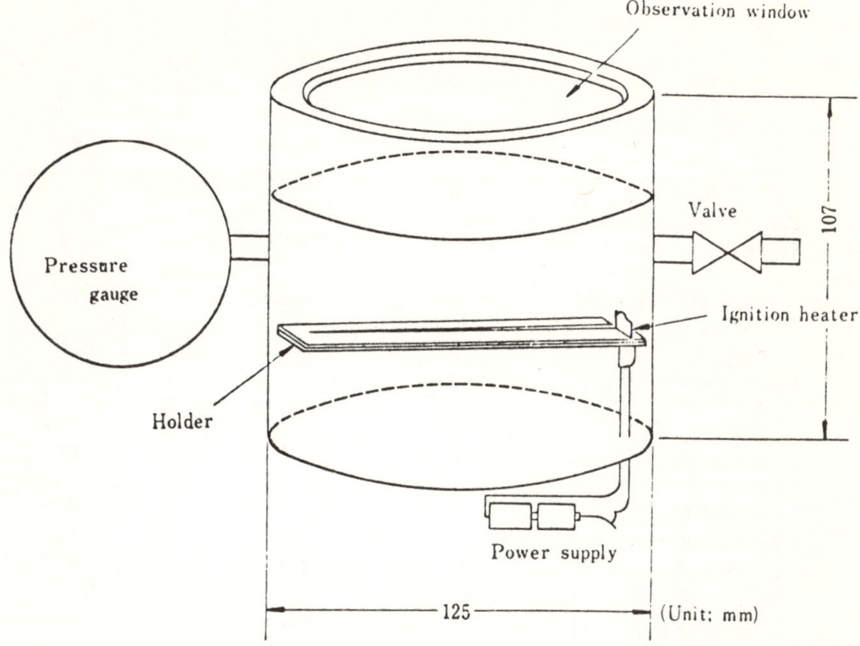

Figure 2. *Combustion vessel.*

Specimens

The characteristics of test specimens used are shown in Table 1. These are mostly commercial ones and the amounts of the additives and the plasticizer are not determined. The sizes of specimens are 25 X 200 mm for the case when air is used as the atmosphere, and 7 X 100 mm for other cases. The specimens are cut from rolled film in the direction perpendicular to the axis of the roll. Also, the test specimens were kept in a silica gel desiccator for over two days before experiment.

RESULTS

Quenching Distance of Various Materials

Quenching distance and burning distance are measured for various materials in an atmosphere of 100% oxygen at 1 bar. Results obtained are summarized in Table 2 (mean value of three trials). As seen in this table, the quenching distance of polyethylene or Japanese paper is very small in comparison with the hydrocarbon polymers containing halogen as polytetrafluoroethylene. Also polychlorotrifluoroethylene was difficult to measure in the oxygen of atmospheric pressure because of a high flame resistance.

It is an interesting fact that the quenching distances for these combustible materials are near those of explosive gases.

Table 1. Characteristics of Specimens

Specimen	Shape	Process	Weight per unit section (g/cm^2)	Thickness (mm)
Japanese paper	Paper		0.00105	0.05
Polyethylene	Film	Extrusion	0.0046	0.05
Nomex	Paper		0.0040	0.055
Polyethyleneterephthalate	Biaxial oriented film	Extrusion	0.0071	0.055
Cellulose triacetate	Film	Casting	0.0065	0.05
Polyvinylchloride	Film	Extrusion	0.0058	0.05
Polyvinylidenefluoride	Biaxial oriented film	Extrusion	0.0114	0.05
Polyvinylidenechloride	Tubular film	Extrusion	0.0065	0.04
Fluorinated ethylene propylene-FEP100	Tubular film	Extrusion	0.07	0.34
Polytetrafluoroethylene	Film	Slicing	0.010	0.05
Polychlorotrifluoroethylene	Film	Extrusion	0.025~0.05	0.15~0.3
Carbon fiber	Paper non-woven		0.013	0.25

Table 2. Burning and Quenching Distances of Various Materials ($1 kg/cm^2$, $100\% O_2$, Horizontal Spread)

Specimen	Mean burning distance (mm)	Mean quenching distance (mm)	State of flame
Japanese paper	86.8	0.66	Blue flame
Polyethylene	85.1	0.75	Blue flame
Nomex	85.0	0.75	Change into glow combustion on the way
Polyethyleneterephthalate	81 2	0.94	Bright flame
Cellulose triacetate	79.9	1.01	Bright flame
Polyvinylchloride	77 2	1 14	Bright flame accompanied by soot
Polyvinylidenefluoride	74.4	1 28	Bright flame
Polyvinylidenechloride	57.3	2.14	Red and short flame accompanied by white smoke
Carbon Fiber	17.2	4.14	Glow combustion
Fluorinated ethylene propylene-FEP100	13.0	4.35	Bright carbon particles in blue flame
Polytetrafluoroethylene	6.5	4.68	Bright carbon particles in blue flame
Polychlorotrifluoroethylene	(26.7)*	(3.67)*	A candle like flame

(*$2 kg/cm^2$)

Effect of the Holder Plate Thickness

In order to see the effect of the plate thickness, holders with four different thicknesses were tested. Results obtained from these measurements for polyethylene and Japanese paper are shown in Table 3. According to these results it is found that the plate thickness has considerable effect. In this study, a brass holder with a plate thickness of 0.85 mm was used.

Table 3. Effect of Plate Thickness on Burning Distance
(1kg/cm^2, Horizontal Spread, Unit mm)

Specimen	Oxygen (Vol %)	Plate thickness (mm)				
		0.85	1.0	3.0	9.0	16.0
		85.0	86.0	73.6	63.9	62.8
Polyethylene	100	85.2	85.5	72.2	62.5	63.4
(mean)		85.0	86.0	71.8	65.0	64.9
		(85.1)	(85.8)	(72.5)	(63.8)	(63.7)
		87.0	86.7	77.6	76.0	
Japanese paper	100	86.8	86.6	77.2	77.3	
(mean)		86.5	87.0	78.4	76.8	
		(86.8)	(86.8)	(77.7)	(76.7)	
		31.8	27.6	16.7	14.6	
Japanese paper	air	32.9	26.2	16.2	14.4	
(mean)		32.0	27.6	17.3	14.8	
		(32.2)	(27.1)	(16.7)	(14.6)	

Effect of the Specimen Thickness

The effects of specimen thickness on the quenching distance are examined by using multilayered sheets. The fuels tested were Japanese paper and polyethylene. The results shown in Table 4 indicate that the quenching distance is not greatly influenced by the specimen thickness. The burning time from ignition to quenching, however, increases with the increase in number of sheets.

Table 4. Effect of Specimen Thickness on Burning Distance
(1kg/cm^2, 100%O_2, Horizontal Spread, Unit mm)

Specimen	Number of piled sheets			
	1	2	4	6
Polyethylene	85.1	84.5	81.7	
Japanese paper	86.8	86.4	85.7	85.8

Effect of Orientation

The effect of orientation was studied by considering the following four cases: (1) upward vertical, (2) downward vertical, (3) longitudinal-horizontal and (4) horizontal directions (see Figure 3). The test fuels are once again Japanese paper and polyethylene and the atmosphere was an oxygen-nitrogen mixture with 40.7%

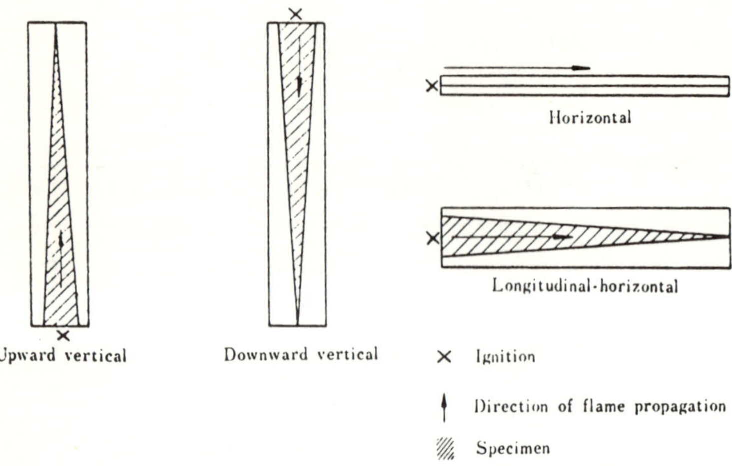

Figure 3. *Orientation of flame propagation (side-views).*

oxygen under 1 bar. The results obtained are shown in Table 5. For Japanese paper, the burning distance decreases in the order of upward vertical, longitudinal horizontal, downward vertical, and horizontal direction. This behavior is similar to the flame propagation in an explosive gas. For polyethylene, on the other hand, the burning distance decreases inthe order of downward vertical, longitudinal horizontal, upward vertical, and horizontal directions. This discrepant behavior is noted to be mostly due to the dripping melt.

Table 5. Relationship Between Orientation and Burning Distance (1kg/cm², Unit mm)

Specimen	Oxygen (Vol %)	Upward Vertical	Longitudinal Horizontal	Downward Vertical	Horizontal
Polyethylene	40.7	68.3	68.9	69.9	66.2
Japanese paper	air	39.2	30.6	28.7	25.6

Effects of the Oxygen Concentration and Pressure

The burning distance for different oxygen concentration and pressures is measured in the horizontal direction for polyethylene, polyvinylchloride, and polytetrafluoroethylene. The results are shown in Figures 4 and 5. In these tests, the

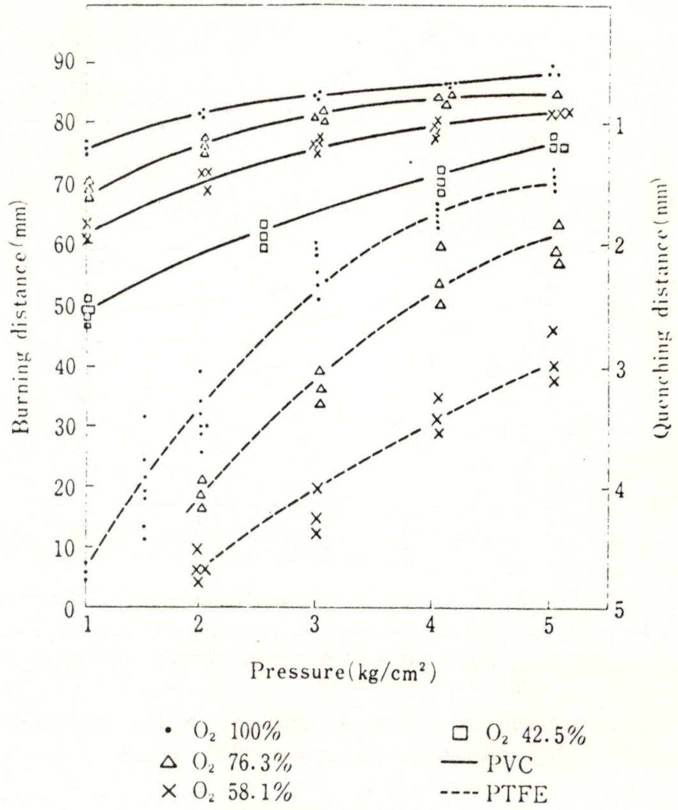

Figure 4. *Effect of oxygen concentration and pressure on quenching distance of polyethylene (horizontal spread).*

reproducibility was inferior for polytetrafluoroethylene due to the fact that the test specimen was prepared by slicing, so its surface was quite irregular microscopically. The quenching distance decreases with the pressure and the oxygen concentration. Usually the effect of the latter is far larger than that of the former. For example, the quenching distance corresponding to air at a pressure of 5 bar is about 1.87 mm. On the other hand, corresponding to pure oxygen at a total pressure of 1 bar, it is about 0.75 mm. If combustion is controlled by the partial pressure of oxygen, these two quenching distance measurements should be equal to one another. The

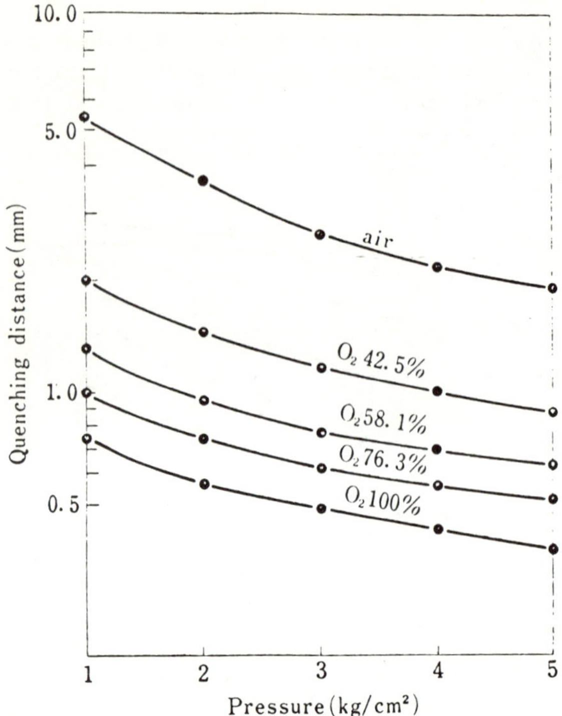

Figure 5. *Relation between oxygen concentration and/or pressure and burning distance of polyvinylchloride and polytetrafluoroethylene (horizontal spread).*

great difference between the two values suggests that the combustion of solid materials does not depend solely on the oxygen partial pressure.

Flame Separation Phenomenon in Polyvinylchloride

In the experiments on polyvinylchloride, the flame is observed to be separated into two stages in the range of oxygen concentration of 40–50%. The first flame is associated with the first step of the PVC pyrolysis whereas the second flame is connected with a burning of the residue. The second flame is quenched earlier than the first one. The detailed mechanism of this interesting phenomenon is not yet understood. A sketch obtained from the experiment is given in Figure 6.

COMPARISON WITH OXYGEN-INDEX METHOD

As a means for measuring the combustibility of solid materials, the limiting oxygen-index method developed by Fenimore and Martin [3] is very useful and many measured LOI values have been previously published [4, 5, 6].

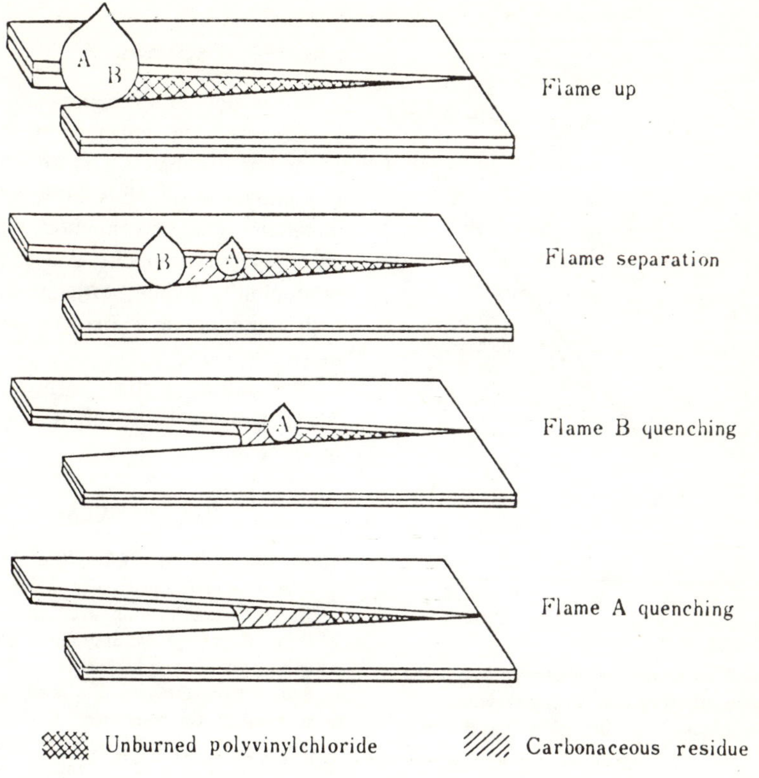

Figure 6. *Flame separation of polyvinylchloride (O_2 42.5%, 1kg/cm², horizontal spread).*

This method based on the critical oxygen concentration of the atmosphere at which the specimen ignited by a pilot flame can be kept to burn, has a disadvantage that it is impossible to obtain the index for highly flame resistant materials such as polychlorotrifluoroethylene because the oxygen-index becomes greater than 100.

On the other hand, the present method based on the quenching distance can be applied to any material by the control of ambient pressure, for example the quenching distance of above-mentioned flame resistant polymer is easily determined as 3.67 mm under an atmospheric pressure of 2 bar. Moreover, as shown in Figure 7, the results obtained by this experimental procedure correlate fully well with the oxygen-index.

In Figure 7, the oxygen-indices reported by Isaacs [4] are plotted with the present quenching distances on a semilogarithmic diagram. It is clear from this figure that the relationship between the two measures of flammability follows law.

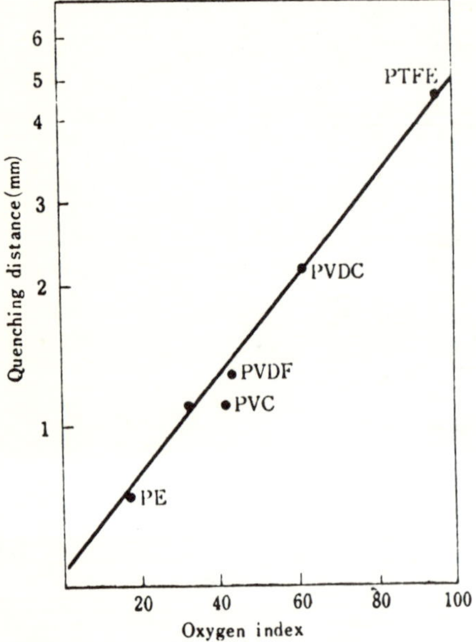

Figure 7. *A relationship between quenching distance and oxygen index (all these quenching distances are for 100% oxygen 1.0 kg/cm²).*

SUMMARY

From the above experimental results, we can conclude that the quenching distance is a useful measure to compare the combustibility of solid materials. The new method has the advantage of simplicity and good reproducibility while correlating well with Fenimore and Martin's oxygen-index method.

The author is thankful to Dr. K. Akita, University of Tokyo, for helpful discussions.

REFERENCES

1. B. Lewis, G. Von Elbe, "Combustion, Flames and Explosions of Gases," Academic Press (1961) pp. 228, 328.
2. R. L. Durfee, J. M. Spurlock, "Quenching and Extinguishment of Burning Solids in Oxygen Enriched Atmospheres," Contract No. NAS9–8470, Atlantic Research Corporation (1969).
3. C. P. Fenimore, F. J. Martin, "Flammability of Polymers," Combustion and Flame, *10*, 135, (1966).
4. J. L. Isaacs, "The Oxygen Index Flammability Test," J. Fire & Flammability, *1*, 36, (1970).
5. K. Komamiya, "Flammability of Fabrics," Kasai (Journal of Fire Prevention Society of Japan), *12*, 75, (1962) (in Japanese).
6. K. Komamiya, "Flammability Test of Plastics Films," Anzen Kogaku (Journal of Japan Society for Safety Engineering), *2*, 270, (1963) (in Japanese).

SECTION TWO

Wood

Michael M. O'Mara

B. F. Goodrich Chemical Co.
Avon Lake, Ohio

THE COMBUSTION PRODUCTS FROM SYNTHETIC AND NATURAL PRODUCTS – Part 1: Wood

(Received June 4, 1973)

ABSTRACT: Animal exposure studied and large scale fire data indicate that one of the early life hazards in a developing fire is from the generation of carbon monoxide and other combustion gases. A gas chromatograph has been interfaced to an NBS Smoke Chamber to study the rate of generation of carbon monoxide and other gases during the burning of materials. Rate data are presented in this initial report for four kinds of wood: white pine, white oak, plywood and Masonite. The effect of environmental oxygen concentration on the release of carbon monoxide was also studied.

The presence of other combustion and degradation products were qualitatively and semi-quantitatively determined.

INTRODUCTION

Over the past ten-year period, more than 12,000 people have died each year as a result of fire involvement [1]. The major sources of life hazard in a fire are, 1) flame and superheated air, 2) oxygen depletion, 3) toxic gases and 4) smoke. Many fire officials feel that the latter two may be responsible for the majority of deaths. The need for more knowledge relating to the toxicological aspects of fires was recently reviewed [2]. Two areas appear to be of most interest, from a research standpoint, in defining the magnitude of the toxic gas problem. These areas are the analytical problems of both qualitatively and quantitatively measuring gaseous combustion products, and animal testing studies. With the advent of new (and exotic) polymeric materials, a considerable amount of work has been carried out in both areas.

With respect to animal testing studies, the work of Pryor and co-workers is of fundamental importance in understanding the nature of the overall problem. In an early paper, [3] Pryor defined the physiological effects of elevated temperature, oxygen depletion, carbon monoxide and carbon dioxide on man. Limits of exposure to these four fire parameters are summarized in Table 1. In a subsequent study, these authors studied the effects of two and more combined fire parameters on animals [4]. The results of this work showed the presence of synergism for several of the investigated fire parameters (i.e., oxygen depletion, temperature, carbon monoxide and carbon dioxide).

Table 1. Limits of Exposure to Heat, Carbon Oxides Oxygen Depletion [3]

Exposure time (minutes)	5	30
Heat	140°C	100°C
Oxygen	9%	11%
Carbon monoxide	3000 ppm	1500 ppm
Carbon dioxide	50,000 ppm	40,000 ppm

Later studies by this same author involved the same type of assessment with respect to other noxious and toxic combustion gases [5, 6, 7]. As an example of toxicological synergism Pryor showed that when the combustion products from wood samples were diluted with air to a point where the carbon monoxide concentration was reduced to 1400–1600 ppm, seven out of ten test animals died when exposed to this atmosphere. Yet in a control experiment involving exposure to 1500 ppm of carbon monoxide alone, only two out of ten animals died [8].

More recent work by Higgins and co-workers [9] has shown no evidence of toxicological synergism when animals were exposed to hydrogen fluoride, hydrogen chloride, nitrogen dioxide and hydrogen cyanide in combination with carbon monoxide. The main difference between these two studies was the high concentration/short time exposure method of Higgins [9] in contrast to the low concentration/long time exposure method of Pryor [6]. Thus one may assume that these two studies are not necessarily in conflict but rather that they point to the need for additional research on the question of time exposure and toxicological synergism.

A considerable amount of analytical work has been done on the combustion products from polymeric materials. Information relating to the absence of phosgene from polyvinyl chloride is available [12, 14, 15] along with data on the relative concentration of hydrogen chloride and carbon monoxide [12, 14, 15, 16]. Studies on the combustion products from polyphenylene oxide [17], polysulfone [18], polyimide [18], and polycarbonate [18], have also been carried out. Studies are now underway on urethane polymers [19]. Analytical combustion studies have been reported on polyvinyl chloride, polyamide, urethane foam, urea resin foam, polyester resins and polyacrylonitrile [20, 21]. A broad class of end-use products have been evaluated with respect to smoke and toxic gas generation [22]. Wood and a number of other building products also have been studied in a combined toxicity study involving test animals [6].

Although the test methods employed in the above studies varied significantly in both the methodology of generating and analyzing combustion products, a wealth of useful information does exist on the toxicological aspects of natural and synthetic materials. The major source of life hazard in a fire involving these materials does appear to be generation of carbon monoxide. As more information becomes available on animal testing studies, the question of toxicological synergism can be answered.

Avoiding the issue of synergism until more research has been carried out, one is faced with obtaining a method for evaluating the toxicity of combustion products from natural and synthetic materials. It is well known that a number of performance tests exist for evaluating flammability (Oxygen Index Test, ASTM E-84 Tunnel Test, and ASTM-E119 Endurance Test, among others) and smoke (National Bureau of Standards Smoke Chamber, Rohm and Haas XP-2 Chamber and ASTM E-84 Tunnel Test). However similar performance methods do not exist for evaluating the toxicity of combustion products. In a recent article on this subject, Tsuchiya and Sumi discussed the use of a "toxicity factor" for evaluating the toxicity of combustion products [23]. These authors defined the toxicity factor for a single component as

$$T = Ce/C_f$$

where T is the toxicity index and Ce and C_f are the concentration of the component in air and the concentration which is lethal for a thirty-minute exposure, respectively. The overall toxicity would then be a sum of the toxicity indices for each component in the combustion gases of a material:

$$\Sigma T = \Sigma Ce/C_f$$

They suggested that the values be obtained under fire conditions which favor the formation of the most toxic gases or favor the formation of the highest concentration of toxic gases. In their experimental design, one gram of material would be decomposed in 1 m^3 of air. In this way, the maximum toxicity of a material could be defined on a mass basis.

A problem develops in attempting to couple this concept with reported combustion gas analysis. In many cases the experimental design of generating the combustion products from materials and the subsequent "concentration" of combustion products does not lend itself to this type of analysis. In other cases, the method of generating the combustion products is either not well defined or difficult to reproduce. Some of the reported literature on combustion gas analysis is questionable because of the use of non-specific analytical methods. For example in a preliminary study involving the combustion products from vinyl chloride monomer, we found that the Draeger colorimetric tube used for phosgene analysis was useless in this application because of the interferences from other combustion products [24]. Similar interferences have been noted in using a colorimetric tube for hydrogen cyanide analysis [25].

In light of this discussion there appears to be a need for a standard experimental method for generating and measuring combustion gases. This standard method should have the following requirements:

1. reproducible thermal exposure to a material,
2. analytical methods which are specific for the major and minor

combustion products from materials, and

3. multi-point rather than single point data.

Hopefully this approach would provide some of the answers to questions of whether or not synthetic materials increase the life hazard in a fire situation.

The purpose of this paper is to focus on the problem of carbon monoxide generation. From an ignition standpoint it cannot be denied that many synthetic materials are far superior to cellulosic materials. However from a standpoint of carbon monoxide and other toxic gas generation, a question arises. When synthetic materials, which have been made ignition resistant, are forced to burn, is the life hazard increased above that created by the burning of more conventional cellulosic materials? Or is the hazard the same as with cellulosic materials? Can the hazard which is created by the burning of these materials be diminished through new additive chemistry? In an attempt to answer some of these questions, an experimental program has been undertaken to study the combustion and degradation products from synthetic and cellulosic materials.

In this initial report, we have concentrated on measuring the generation of carbon monoxide and carbon dioxide from a variety of wood products. The concept of *rate* rather than single point analysis was an integral part of this study. A limited amount of work was carried out on the other combustion products from wood. The effect of oxygen depletion on the generation of carbon monoxide and carbon dioxide was also investigated.

EXPERIMENTAL APPROACH

In selecting an experimental approach to carry out this work it became obvious that the ability of others to duplicate our approach was essential. For this reason, an investigation of various methods for generating combustion gases was carried out. While a number of procedures were tried, one method was selected because of its simplicity of design, commercial availability, present use and acceptability as a fire test and the extensive research by others that has already gone into its development. The method chosen for this study was to interface a gas chromatograph with the National Bureau of Standards (NBS) smoke chamber. This method allows the exposure of materials under a variety of conditions and, in addition to toxic gas analyses, a simultaneous smoke analyses can be performed.

In investigating methods for quantifying the combustion products from materials, we rejected the use of colorimetric tubes because of the large number of interferences that are possible. Previous work by us on the thermal degradation of polymers relied heavily on the use of gas chromatography [26, 27]. The success by others in this area also influenced this decision [12, 13, 15]. Mass spectrometric techniques have also been reported in assessing the toxicity of combustion products but unless a high resolution instrument is used, differentiation between carbon monoxide and nitrogen (M/e=28) especially in a mixture spectrum, is impossible [28]. Detailed experimental information with respect to the chamber, chromatography, etc., is contained at the end of this paper.

All experiments involve the exposure of various wood products to the radiant heater in the NBS smoke chamber. Experiments involving exposure to the radiant panel and the flame jets (flaming exposure) were carried out but under these conditions very little carbon monoxide was found after exposure. For this reason, all exposures in this study were limited to the radiant heater. The output of the heater was maintained at an incident heat flux of 2.5 watts/cm^2.

COMBUSTION OF α-CELLULOSE AND MATERIAL BALANCE OF CARBON

Initial experiments involved the combustion of α-cellulose (National Bureau of Standards Standard Reference Material 1006). The purpose of these experiments was to determine if a material balance in carbon, i.e.

$$C_{(cellulose)} \longrightarrow CO_2 + CO$$

could be obtained and the time required to reach a steady state concentration of carbon oxides. As a result of this early work we discovered that a material balance was possible but that the time required to establish a steady state concentration of carbon oxides in the chamber was very long. A dynamic view of carbon oxide formation and smoke formation is presented in Figure 1. Based on the internal

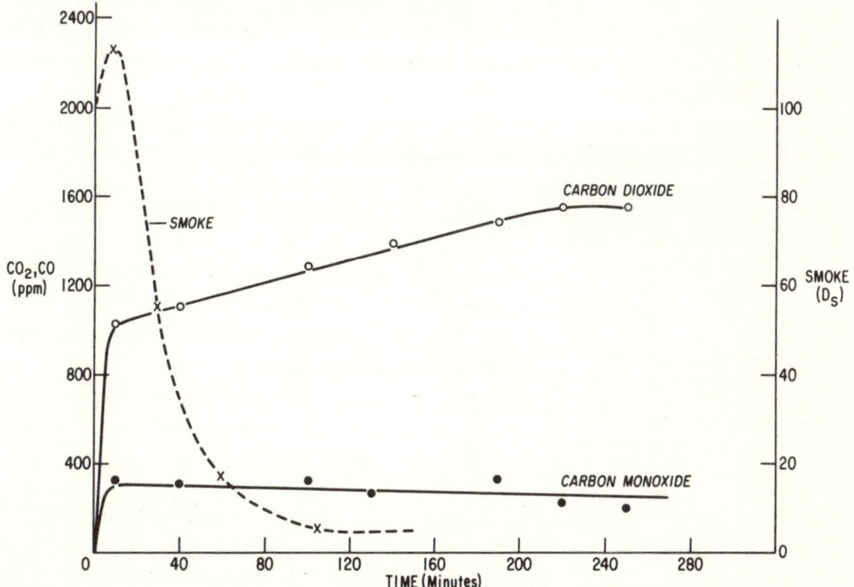

Figure 1. Generation of CO, CO_2 and smoke from 0.917g of cellulose in NBS chamber.

volume of the NBS chamber (518 liters) and the amount of cellulose exposed to the radiant heater in a typical experiment (0.917g), the theoretical concentration of carbon oxides, assuming complete oxidation of all degradation products, was 1466 ppm. The background concentration of carbon dioxide determined initially was 375 ppm. At the end of the three-hour exposure the concentration of carbon monoxide plus carbon dioxide was 1807 ppm. Thus, 98.1% (1807/1841 $\times$ 100) of the carbon in the original cellulose sample was accounted for as carbon monoxide and dioxide.

The most reasonable explanation for the long time to reach a steady state concentration of carbon oxides is the slow gas phase oxidation of degradation products from the cellulose which escaped initial combustion. From the curve in Figure 1, which depicts the buildup in carbon dioxide, one can calculate an average rate of increasing carbon dioxide concentration between 10 minutes and 190 minutes. This rate of buildup is 155 ppm/hr. This should also be the rate of degradation product decay in the chamber under the same conditions. To determine this value, methane was introduced into the chamber until a concentration of 870 ppm was reached. At this point the radiant heater was turned on and the decay of methane was monitored over a three-hour period. (The rate of methane leakage from the chamber is 7 ppm/hr.) The average rate of decay of methane under these conditions was 140 ± 1 ppm/hr.

This rate of methane decay, or rate of gas phase oxidation of methane, is in good agreement with the rate of carbon dioxide buildup in the chamber. This further substantiates the hypothesis that the long period of time required to reach a steady state concentration of carbon dioxide is due to a low rate of gas phase oxidation of the degradation products from cellulose.

COMBUSTION OF VARIOUS WOOD PRODUCTS

The major objective of this study is to ultimately investigate the carbon monoxide producing properties of a variety of fire-retarded and non-fire-retarded plastics relative to those of wood. In order to obtain an adequate "baseline," four kinds of wood products were chosen. This selection was based on the wide use of these materials in the construction industry. The four materials investigated were plywood, white oak, white pine and hardboard (Masonite).

Considerable information exists on the various aspects of wood chemistry. An excellent monograph on the structure, properties and chemistry of wood has been in existence for a number of years [29]. The variation in the chemical composition of various wood products has also been known for years [30].

There has been no attempt to relate the combustion profiles of the four wood products chosen in this study to their physical or chemical properties. The sole purpose of this study is to provide a baseline and starting point for more in-depth studies with polymeric systems.

In these combustion experiments the basic data collected were, 1) concentration of nitrogen and oxygen, 2) concentration of carbon monoxide and carbon dioxide, 3) weight loss, 4) exposure time and 5) light attenuation caused by smoke accumulation. A limited amount of information was also collected on the other combustion/degradation products from wood. In addition, the effects of the water content of wood and of oxygen depleted atmospheres on the carbon monoxide producing properties of wood, were also investigated.

A. Plywood

Commercial grade 1/4-inch (245 mils) plywood was obtained from a local lumberyard for evaluation. The widespread use of this material in the construction industry prompted this investigation. From a "purity" standpoint this was not a highly desired material. The presence of adhesives and other laminating components did introduce a number of unknowns. Nevertheless, from a standpoint of establishing a well defined baseline for carbon monoxide evolution from "wood," it was important to include this material in the evaluation.

Experimentally, this material was exposed to the radiant panel in the NBS chamber for periods up to 25 minutes. Sample weights of the 3" X 3" samples averaged 18.5 ± 0.4 grams. A summary of the combustion data from this material is contained in Table 2. Rates of carbon oxide and smoke formation are shown graphically in Figure 2.

Table 2. Combustion Profile of 1/4-inch Plywood[a]

Exposure Time (min)	Weight Loss (%)	CO_2, (ppm)	CO	CO_2/CO	O_2 %	Ds[b]
5	20.1	736	195	3.91	21.9	207
7	29.3	1,210	544	2.22	–	336
10	–	2,588	1728	1.50	–	535
13	49.7	3,666	2469	1.48	21.7	555
17	64.7	6,572	3581	1.84	–	528
21	72.4	8,854	4453	1.99	–	485
25	80.1	12,246	5461	2.24	20.5	427

[a]Sample weights: 18.5 ± .4 grams

[b]Optical density due to smoke generation

The data in Table 2 demonstrate a number of features unique to this type of combustion gas analysis. Since this is "fuel-lean" combustion the amount of oxygen consumed during the total exposure time is minimal. In this experiment only 6.4% of the available oxygen was consumed. It is interesting to note that the CO_2/CO ratio decreases then increases during the exposure. This change in ratio is due to a changing rate of carbon dioxide since the rate of carbon monoxide is constant

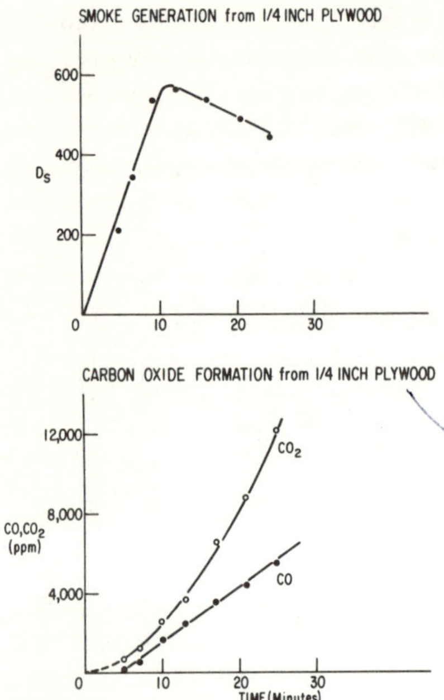

Figure 2. Smoke and carbon oxide generation from 1/4-inch plywood.

throughout the exposure time. The minimum CO_2/CO ratio occurred at the same time, about 13 minutes, as the maximum smoke value, $D_5 = 555$. The two phenomenon may be related.

In terms of life hazard, this analysis points up the fact that the effects of smoke production should not be minimized. At the 12-minute point, the smoke from the plywood had reached a maximum, yet the life hazard from carbon monoxide was not yet at the critical concentration of 3000 ppm.

B. White Oak

A commercial grade of oak of 3/4-inch thickness was obtained for this analysis. Samples were prepared by sawing along the longitudinal axis of the oak symmetrically so that the final thickness was 3/8 inch. Sample weights were in the range of 25.9 ± .7 grams.

Combustion data are summarized in Table 3 and graphically displayed in Figure 3. The results are similar to those obtained for plywood. The ratio of CO_2/CO minimized and the concentration of smoke maximized early in the combustion process. This phenomenon occurred at the 13.5-minute period during exposure. As with the plywood, the carbon monoxide concentration is not at the critical value of 3000 ppm at this point of maximum smoke obscuration.

Table 3. Combustion Profile of 3/8-inch White Oak[a]

Exposure Time (min)	Weight Loss (%)	CO_2, (ppm)	CO	CO_2/CO	O_2 (%)	Ds[b]
5	11.9	580	52	11.2	21.9	47
7	15.1	840	123	6.8	–	260
10	32.6	2,330	1235	2.3	–	566
13.5	52.4	4,705	2582	1.8	21.3	710
17	59.7	6,253	3220	1.9	–	
20	72.5	8,456	4120	2.1	–	584
25	78.1	12,204	5280	2.3	20.6	473

[a]Sample weights: 25.8 ± .7 grams

[b]See Table 2

C. White Pine

A commercial grade of 1/2-inch thick white pine was obtained for this analysis. This material was also cut along the longitudinal axis in order to obtain 1/4-inch thick specimens. Sample weights were in the range of 18.1 ± 1.4 grams.

In this study, analyses were also performed on pine samples which had been dried and on pine samples which had been exposed to 100% relative humidity. Drying was accomplished by placing the pine samples in a vacuum oven at 70°C for 12 hours followed by three days of storage over a dessicant. For the high humidity exposures, samples were placed in a dessicator containing water. Exposure to this atmosphere lasted 31 days. By this time a constant sample weight was achieved. A sample weight increase of approximately 16–17% was noted. Pertinent combustion data from these three experiments are summarized in Table 4.

These data show that the moisture content of the wood affects both the carbon oxide formation and the development of smoke. As the progression is made from "dry" to "wet" pine, carbon monoxide, carbon dioxide and smoke generation are decreased. This is especially noticeable during the early stages of combustion. For example, after seven minutes of exposure the smoke from the "dry" pine (D_s = 584) was significantly greater than the smoke from the "wet" pine (D_s = 122).

The data plotted in Figure 4 indicates that the water content of the white pine does not influence the *rate* but rather the *induction period* to smoke and carbon monoxide generation. Thus, there is no indication that the water in the pine is significantly alerting the combustion process except by delaying the onset of degradation and combustion.

D. Hardboard (Masonite)

A commercially available 1/8″ thick hardboard (Masonite) was chosen for this phase of the study. In addition to the usual analysis, this material was also subjected to combustion in oxygen depleted atmospheres in order to determine the effect of atmospheric oxygen content on carbon oxide formation. Smoke and carbon oxide data for combustion in 21% oxygen are summarized in Table 5. In this instance the CO_2/CO ratio minimized before smoke maximized. For some reason the maximum in smoke was delayed relative to the other wood materials. No attempt has been made to correlate this phenomenon with the material's composition.

In the second phase of this analysis samples of Masonite were exposed to the radiant heater in reduced oxygen atmospheres. Experimentally this was accomplished by sealing the sample in the chamber, and introducing nitrogen through one of the sample ports. Chromatographic analysis insured the correct oxygen content in the chamber before the sample was placed in front of the radiant heater. Pertinent data are tabulated in Table 6. The total exposure time in each case was 11 minutes. Since it is impossible to duplicate the same mass loss every time, a simple comparison of CO_2 and CO at the same exposure time would be misleading. For

Table 4. Combustion Profiles of Pine Stored Under Different Relative Humidities (RH)

Exposure Time (Minutes)	0% RH[a] CO_2 (ppm)	0% RH[a] CO (ppm)	0% RH[a] CO_2/CO	0% RH[a] D_s	55% RH[a] CO_2 (ppm)	55% RH[a] CO (ppm)	55% RH[a] CO_2/CO	55% RH[a] D_s	100% RH[a] CO_2 (ppm)	100% RH[a] CO (ppm)	100% RH[a] CO_2/CO	100% RH[a] D_s
5	1,074	431	2.49	281	766	179	4.28	156	663	59	11.24	71
7	2,403	1229	1.96	584	1,335	594	2.25	337	972	355	2.74	122
10	3,996	2114	1.89	652	2,801	1357	1.94	513	2,483	1414	1.83	542
13	6,885	3187	2.15	652	5,552	2458	2.24	560	3,989	1976	22.02	589
17	9,242	3493	2.64	508	7,597	3120	2.43	517	6,289	2849	2.21	549
21	11,731	3940	2.98	434	10,864	4084	2.64	467	9,420	3343	2.82	439
25	14,727	5009	2.94	425	12,478	4525	2.76	465	11,698	3987	2.93	373

[a] Samples defined as "0% RH" were vacuum dried and stored over a dessicant; those defined as "55% RH" were stored at 22°C/55 ± 5% relative humidity/1 week; those defined as "100 RH" were stored at 22°C/100% relative humidity/1 month.

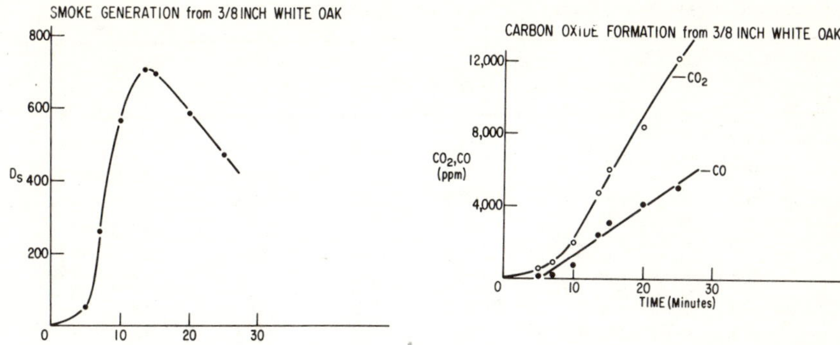

Figure 3. Smoke and carbon oxide formation from 3/4-inch white oak.

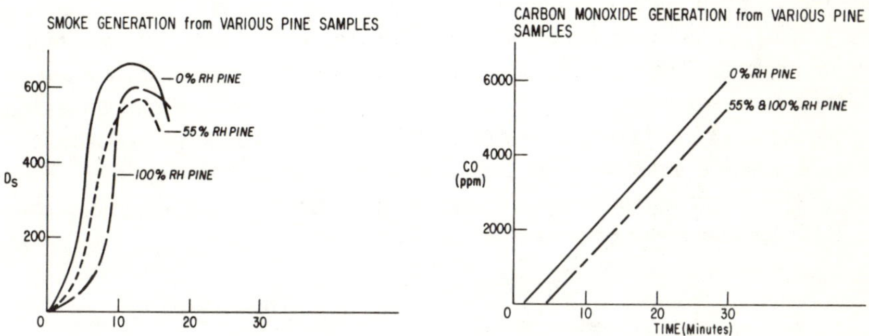

Figure 4. Smoke and carbon monoxide generation from various pine samples.

Table 5. Combustion Profile of Masonite[a]

Exposure Time (min)	Weight Loss (%)	CO_2 (ppm)	CO (ppm)	CO_2/CO	O_2 (%)	D_s[b]
5	18.4	721	109	6.6	21.9	127
7	27.8	1,033	347	3.0	–	233
10	40.1	1,691	823	2.0	–	442
13.5	70.7	3,694	1703	2.2	20.6	624
17	82.6	6,210	2336	2.7	–	638
21	88.3	11,048	3212	3.4	–	569
25	92.0	15,626	4039	3.9	20.1	–

[a] Sample weights: 19.3 ± .6

[b] See Table 3

Table 6. Combustion of Masonite in Reduced Oxygen Atmospheres for 11-Minute Exposure Time

Atmospheric O_2 (%)	Weight Loss (g)	CO_2 (ppm)	CO (ppm)	Average CO_2/g	Average CO/g
22.1	9.8	2410	1206	24.6	12.3
19.6	10.6	2803	1260	26.4	11.9
15.4	11.0	2842	1333	25.8	12.1
14.1	11.6	2615	1334	22.6	11.5
11.7	10.5	2408	1295	22.9	12.3
10.4	10.0	1979	1199	19.8	12.0
7.0	9.4	1891	1055	20.1	11.2
4.9	9.7	1485	990	15.4	10.3
1.7	9.3	1498	932	16.1	10.0

that reason these values are compared on a unit mass loss basis. The efficiency of Masonite to produce CO_2 and CO in reduced oxygen atmospheres is shown graphically in Figure 5. The quantity CO_2/unit weight and CO/unit weight are plotted against the percent oxygen in the atmosphere. It is obvious from the two curves that the production of carbon dioxide is much more dependent upon the concentration of oxygen in the atmosphere than is the production of carbon monoxide. Above 10% oxygen in the atmospheres, the amount of carbon monoxide which is released *and which escapes combustion* remains constant.

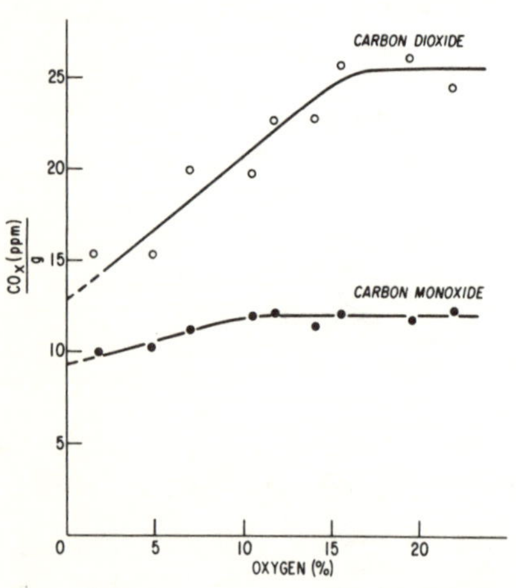

Figure 5. Production of carbon oxides as a function of atmospheric oxygen content.

DISCUSSION

It is clear from these analyses of the different wood products that all of the materials do not generate carbon monoxide at the same rate. For example, if the carbon monoxide concentrations from the various materials are compared at one exposure time, i.e. 15 minutes, the following is observed:

wood product	CO (ppm)	relative CO concentration (Masonite = 1.00)
plywood	3025	1.48
white oak	2900	1.51
white pine	2789	1.36
Masonite	2040	1.00

It might be anticipated that the reason for these differences is due to different mass loss rates for the various products. Figure 6 shows the relationship between weight loss and time for the four wood products. This does show that the rates of mass loss for the various samples are not identical. Unfortunately this is not sufficient to explain the observed difference in carbon monoxide concentrations. On the contrary, an apparent contradiction exists. Comparing the weight loss of white oak with that of plywood at an exposure time of 15 minutes (Figure 6), it can be seen that more of the oak is consumed than plywood, yet at this point, the carbon monoxide concentrations from these two materials are nearly identical. It is obvious that the presence of other combustion and degradation products must account for these weight differences. The other combustion products from white oak and plywood were analyzed quantitatively after the 15-minute exposure time. This analysis is only semi-quantitative (response factors were not used) but it demonstrates that the degradation profile of these two materials are quite different. A comparison of weight loss, carbon monoxide, carbon dioxide and "other" combustion gases, including water, for white oak and plywood at the 15-minute exposure time is shown in Table 7.

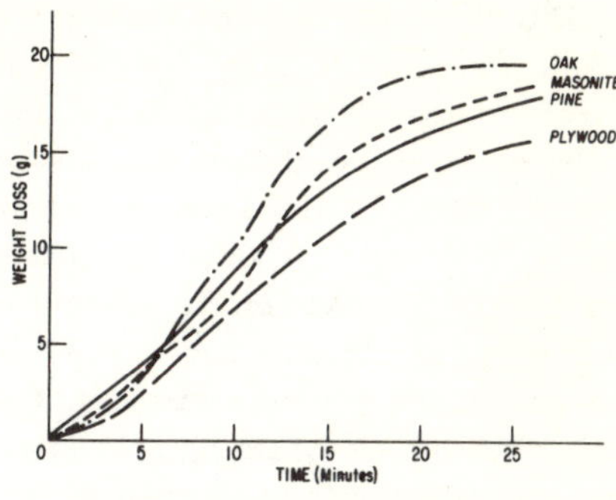

Figure 6. Weight loss of wood products as a function of exposure time.

Table 7. A Comparison of the Combustion Profiles of White Oak and Plywood after 15-Minute Exposure to 2.5 Watts/cm²

	Plywood	White Oak
Weight loss (g)	10.5	16.7
Carbon monoxide (ppm)	3,025	2,900
Carbon dioxide (ppm)	5,119	6,107
Water (ppm)[a]	20,500	21,500
Other combustion products (ppm)[b]	748	1,731

[a]Water includes water content of laboratory air, water of combustion and equilibrium water content of wood.

[b]These were qualitatively the same and are identified below.

This composite analysis demonstrates that a higher conversion of organic degradation products to carbon dioxide and carbon monoxide is realized for the slower burning plywood. White oak, which degrades at a faster rate, produces more than twice the concentration of these organic degradation products in the gas phase as compared to plywood.

This difference in carbon monoxide concentrations for the two materials at the 15-minute exposure time is not due solely to differences in rate of weight loss. Carbon monoxide is produced in two ways. One is by thermal breakdown of cellulose, lignin and the other components of wood, the other by the partial oxidation of carbon. Since the chemical composition of wood varies it might be expected that each of the four wood products gives varying amounts of carbon monoxide upon thermal breakdown. To determine the significance of this variance each of the four materials was subjected to inert pyrolysis at 600°C and the carbon monoxide released during pyrolysis was quantified. Experimental pyrolysis procedures have been published elsewhere [26]. The relative quantities of carbon monoxide released per unit mass of wood pyrolyzed are summarized in Table 8.

Table 8. A Comparison of the Relative Amounts of Carbon Monoxide Released from Various Wood Products During Inert Pyrolysis

Wood product	Relative Carbon Monoxide
White pine	1.00
Masonite	1.09
Plywood	1.12
White oak	1.22

These various analyses indicate that the generation of carbon monoxide from wood is dependent upon both the rate of thermal breakdown and combustion of the wood and upon the chemical composition of the particular wood. Apparently the physical factors which influence the rate of thermal degradation of the wood sample (i.e., heat transfer through the wood, type of char formation, diffusion rates of gaseous decomposition and degradation products through these chars, moisture content of the wood, among others) are more important in determining the overall rate of carbon monoxide generation than is the chemical composition of the wood.

From a life hazard standpoint, two concepts are important with respect to the generation of carbon monoxide from wood. These are the induction period to generation and the rate of generation. The basic carbon monoxide/time data in Tables 2–5 were subjected to a least squares analysis. The equations resulting from this analysis provided the following rate and induction period data for the generation of carbon monoxide:

Wood product	Rate of CO production (ppm/min)	Induction period (min)
Masonite	208	5.6
White pine	232	3.9
Plywood	267	4.1
White oak	276	5.3

With these four analyses completed, a suitable baseline for carbon monoxide generation has been set. Two of the materials, white oak and Masonite, establish the upper (276 ppm/min) and lower (208 ppm) limits of this baseline. This is shown graphically in Figure 7.

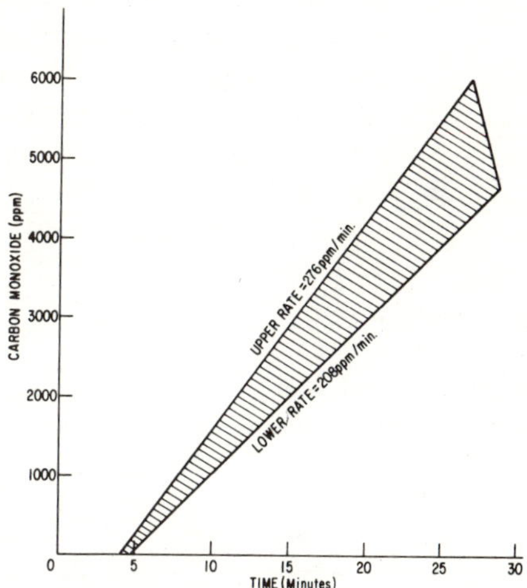

Figure 7. Limits of carbon monoxide generation from four wood products.

ANALYSES OF OTHER COMBUSTION PRODUCTS

In the previous section, mention was made of the presence of other combustion and degradation products from wood. While the major combustion products are carbon monoxide, carbon dioxide and water there are varying degrees of other products. Analyses of these products from the four wood samples indicate that the same combustion products result from each of the four materials.

Using mass spectrometry and/or gas chromatographic retention times of pure components, these other combustion and degradation products have been identified. An exposure time of 15 minutes has been chosen for generating these products. All separations have been carried out on one of the two chromatographic columns used throughout these analyses (the Porapak Q column). A typical separation showing the presence and identifications of these products is shown in Figure 8. A list of these products along with the chromatographic retention data are summarized in Table 9. Concentrations of these gases are also included but it should be pointed out that these are only approximate since response factors were not used.

There are other combustion products present but the concentration of these is so low that positive chemical identifications could not be made.

It should be pointed out that this is by no means an exhaustive examination of the combustion products from wood. There is no doubt that other products exist

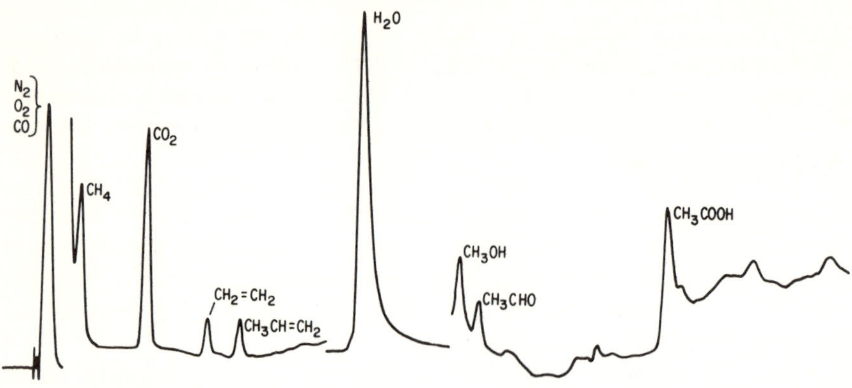

Figure 8. Combustion products from white pine after 15-minute exposure.

Table 9. Chemical Identification, Retention Time Data and Approximate Concentrations of Other Combustion Products from White Pine

Retention Time (minutes)	Component	Concentration (ppm)
3.6	methane	260
9.6	ethylene	50
11.0	propylene	50
20.4	methyl alcohol	90
21.3	acetaldehyde	44
30.1	acetic acid	320

which were too dilute to identify. Many of the degradation products from cellulose which have been identified by others [31, 32] undergo combustion under the experimental conditions of this study. Nevertheless the above analysis indicates the nature of other combustion products with respect to chemical classes, i.e., alcohols, aldehydes, acids, aliphatic and olefinic hydrocarbons.

EXPERIMENTAL

A. NBS Smoke Chamber

This chamber has been described by a number of authors in studies involving the generation of smoke from materials. Our instrument was purchased from Aminco and except for the addition of a sample-removing stainless steel arm, the instrument has not been modified. Basically, the instrument is 18.3 cubic feet in volume and contains a radiant heater and flame jets (on sample holder) for carrying out flaming and smoldering combustion. A vertical light path allows the monitoring of smoke buildup. Percent transmission and time are recorded on an x-y recorder. During exposure of a sample, pressure in the chamber builds up to 3–6 in. H_2O. Additional pressure is then bled off. The output from the heater is maintained at 2.5 watts/cm^2.

B. Gas Chromatograph

An F and M 810 Research Gas Chromatograph was coupled to the NBS smoke chamber through the use of a LOWENCO six part gas sampling valve. A vacuum pump was connected to one arm of the valve so that a continuous stream of combustion products could be removed from the NBS chamber. The valve-chromatograph-NBS chamber link was made via two 1/4″ stainless steel tubes. One runs from the fitting on the top of the NBS chamber to the gas sampling valve and the other from the valve to the injection port of the chromatograph. We discovered that direct coupling of this line to the chromatographic column was necessary to avoid excessive peak tailing. To complete the cycle, the carrier gas line was disconnected from the injection port and reconnected to the sampling valve. A 5 ml sampling loop was used on the sampling valve.

The valve can be operated in two positions. In the closed position, carrier gas from the chromatograph travels to the sampling valve and back to the chromatographic column. At the same time the vacuum pump is removing a continuous stream of gaseous products through the other stainless steel line to the valve and through the sampling loop. When the valve is opened, the line to the pump and the line to the NBS chamber are closed and the sampling loop with sample is opened to the chromatograph. All lines involving the transport of combustion products are maintained at 100°C.

C. Chromatographic Column

In order to separate the products of interest (i.e., N_2, O_2, CO, CO_2, H_2O, etc) a dual column system was used. The two columns chosen for this analysis were an 8′ X 1/4″ Porapak Q (100–120 mesh) column and a 2′ X 1/4″ Molecular Sieve 5A (60–80 mesh) column. Another LOWENCO gas sampling valve was necessary to operate these columns in such a way as to obtain the desired analysis. The molecular sieve column was attached to the valve. This column was wrapped with heating wire and could be heated to 180°C. A typical separation of O_2, N_2, CO and CO_2 was obtained by putting both columns in series. Once the O_2 eluted, the valve was closed thereby isolating the molecular sieve column. After a while the CO_2 eluted from the Porapak column and the valve was then opened. The molecular sieve column was heated and the N_2 and CO then eluted. A chromatogram showing this separation is contained in Figure 9.

D. Chromatographic Details

Helium flow in the chromatograph is maintained at 35 cc/min. The detector bridge is maintained at 160 ma and heated to 150°C. Gas chromatographic areas are

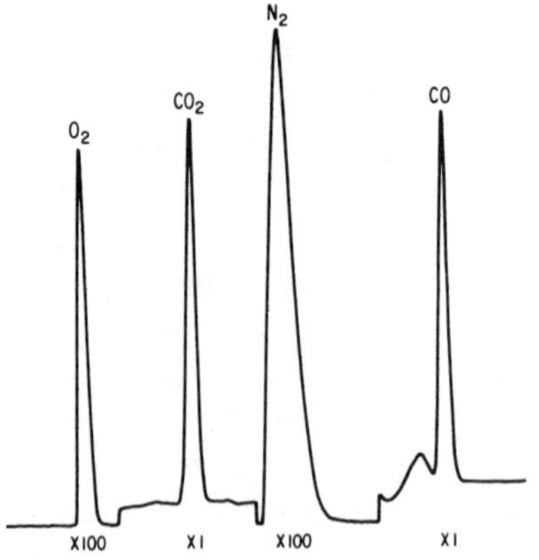

Figure 9. Gas chromatographic separation of O_2, N_2, CO, CO_2 with dual column system.

recorded by means of a Hewlett Packard Printing Integrator. Response factors for the various gases were experimentally determined by using the pure component, or ambient dry air, or synthetic gas mixtures. These response factors, all relative to $N_2 = 1.000$ are checked periodically.

E. Sampling Inside the Chamber

A micro valve was placed on the line extending from the sampling valve to the vacuum pump in order to control the rate of sample removal from the chamber. Presently we are sampling at a rate of 885 cc/min. Considering the internal volume of the sampling apparatus this rate corresponds to a delay time of about 3 seconds. In some early experiments we probed at various places within the smoke chamber by mounting a 1/4″ stainless steel probe from the sampling ports at the top of the box. There was no indication of gas stratification inside the box. The only time a problem of reproducibility (typically within 8%) occurred was when no internal probe was used. If gas sampling was done right at the exit port, reproducibility was not good. It should be pointed out that excess pressure in the smoke chamber is bled off at a point near this exit port and the turbulence created by this operation could explain the lack of reproducibility. Once sampling was performed away from the exit ports, reproducibility improved markedly.

F. Typical Procedure

The following is a typical procedure for determining the combustion products from a material.

An approximate 3″ by 3″ sample of the material is weighed and placed in the sample holder. After the radiant panel has reached thermal equilibrium (checked with a radiometer), the sample holder is placed in position in front of the panel. The door to the chamber is closed and zero time is noted. At a point one minute before a sample of the atmosphere in the chamber is taken, the vacuum pump is turned on. This procedure allows the system to become conditioned with combustion products. After this one minute purge time, the sampling valve is opened. At this point the sample is removed from the path of the radiant panel by means of the sample removal arm. While the chromatographic analysis is taking place, the chamber is evacuated and the remains of the sample are weighed.

In addition to the gas analysis, the optical density (D_S) can also be obtained by noting the % transmission at the time of sampling.

ACKNOWLEDGMENT

The author acknowledges the experimental work and thoughtful comments of R. L. Jackson along with the helpful criticisms of L. B. Crider. The mathematical analyses performed by H. S. Haller were also greatly appreciated.

REFERENCES

1. "Multiple-Death Fires, 1971," Fire Journal (May 1972), 65.
2. John Autian, *J. Fire & Flammability*, 1, (July 1970), 239.
3. A. J. Pryor, et al, "Mass Fire Life Hazard," Clearinghouse for Federal Scientific and Technical Information, AD 642 790, September 1966.
4. A. J. Pryor, et al, "Mass Fire Life Hazard," Clearinghouse for Federal Scientific and Technical Information, AD 672035, March 1968.
5. Ibid, p. 13.
6. A. J. Pryor, et al, "Hazards of Smoke and Toxic Gas Produced in Urban Fires," National Technical Information Service, AD 697839, September 1969.
7. Ibid, p. 42.
8. Ibid, p. 43.
9. E. A. Higgins, et al, "The Acute Toxicity of Brief Exposures to HF, HCL, NO_2 and HCN Singly and In Combination with CO," Federal Aviation Administration, Office of Aviation Medicine, Washington, D.C., 20590, July 1971.
10. J. A. Zapp, Arch. of Environ. Health, 4, 335 (1962).
11. H. H. Cornish, E. L. Abar, Arch. of Environ. Health, 19, 15 (1969).
12. E. A. Boettner, G. Ball, B. Weiss, J. Appl. Polymer Sci., 13, 377 (February 1969).
13. J. D. Seader, I. N. Einhorn, W. O. Drake, C. M. Mihlfeith, "Analysis of Volatile Combustion Products and A Study of Their Effects on Laboratory Animals," presented at Polymers and Materials Conference Series, U. of Utah, June 21–26, 1971.
14. Y. Kobayashi, Mo Hori, H. Murata, Japan Plastics, 5, #1, 40 (January 1971).
15. W. D. Woolley, Br. Polym. J., 3, 186 (July 1971).
16. G. W. V. Stark, Rubber and Plastics Age, 50, #4, 283 (April 1969).
17. G. Ball, B. Weiss, E. A. Boettner, Amer. Indus. Hyg. Assoc. J., 31, 572 (September-October, 1970).
18. E. A. Boettner, University of Michigan, School of Public Health, Ann Arbor, Michigan 48104, prepublication release.
19. E. A. Boettner, private communication.
20. E. Schroeder, "The Composition of the Combustion Gases of Plastics," paper presented at 3rd INTAB, Leipzig, October 22–24, 1968. Institute of Organic High Polymers, Leipzig, Germany.
21. E. Hagen, Plaste u. Kant., 15, #10, 711 (October 1968).
22. D. Gross, "Smoke and Toxic Gases Produced by Burning Aircraft Interior Materials," Clearinghouse for Federal Scientific and Technical Information, AD 675513, June 1968.
23. Y. Tsuchiya, K. Sumi, *J. Fire & Flammability*, 3, 46 (January 1972).
24. M. M. O'Mara, L. B. Crider, R. L. Daniel, Amer. Ind. Hyg. Assoc. J., 32, 153 (1971).
25. M. M. O'Mara, B. F. Goodrich Chemical Co., unpublished work.
26. M. M. O'Mara, J. Polym. Sci., A-1, 8, 1887–1899 (1970).
27. M. M. O'Mara, J. Polym. Sci., A-1, 9, 1398–1400 (1971).
28. G. A. Kleinberg, D. L. Geiger, "Simultaneous Thermogravimetry (TG) and Evolved Gas Analysis (EGA)," paper #91 presented at the 1972 Pittsburgh Conference on Analytical Chemistry and Applied Spectroscopy, Inc., Cleveland, Ohio, March 6–10, 1972.

29. L. E. Wise, E. C. Jahn, *Wood Chemistry*, Volume 1, 2nd ed., Reinhold, New York, 1952.
30. G. J. Ritter, L. C. Fleck, Industrial and Engineering Chemistry, 18, 6, 1926.
31. S. L. Madorsky, V. E. Hart, S. Straus, J. Research NBS, 60, 343 (1958) RP 2853.
32. A. E. Lipska, F. A. Wodley, J. Appl. Polym. Sci., 13, 851 (1969).

Michael M. O'Mara

Michael M. O'Mara is a Development Scientist with the B. F. Goodrich Chemical Company. He joined B. F. Goodrich in 1968 after receiving his Ph.D. degree in chemistry at the University of Cincinnati. Publications have been in the area of the flammability and pyrolysis of vinyl polymers and the analysis of combustion products from burning materials. Research interests include high temperature pyrolysis, gas analysis, mass spectrometry and mechanistic chemistry.

K. L. Paciorek, R. H. Kratzer, J. Kaufman, J. Nakahara
and A. M. Hartstein*

Ulatrasystems, Inc.
2400 Michelson Drive
Irvine, CA 92664

THERMAL OXIDATIVE DEGRADATION STUDIES OF WOODS

(Received May 2, 1974)

ABSTRACT: Untreated and variously impregnated samples of standard southern yellow pine were subjected to thermal oxidative degradations under static and dynamic conditions. The volatiles formed were identified and quantitated on a mg/g basis. Correlations between the impregnation treatments and the thermal oxidative behavior as reflected by thermogravimetric and differential thermal analyses and the products formed are discussed.

INTRODUCTION

Wood is used extensively in mines as structural material. For this type of an application the wood has to be made flame retardant and must be treated against rot and decay. The objective of the present study was to determine the effects of these treatments on the nature and relative concentration of the products formed on thermal oxidative decomposition.

Cellulose, lignin, hemicellulose, resins, tannins, fats, waxes, and some coloring matter [1–3] are constituents of wood and it is the degradation of all these components that leads to the variety of chemicals observed on its pyrolysis. A large number of investigations have been published and reviewed pertaining to wood degradation [4, 5]. Of these probably the most useful and most inclusive is the compilation by the Canadian Wood Council [5]. The work done by Lipska [6, 7] and others [8], to mention just a few, although pertaining only to cellulose is nevertheless of direct application to wood studies since wood consists largely of cellulose. Regarding the effect of preservatives and flame retardants on wood's thermal oxidative behavior and the nature and relative concentration of volatiles the data in the literature are more limited. The extent of penetration is usually of the order of ¾–1″ depending both on the type of wood, its water content and the method of impregnation. A number of theories have been advanced for the action of flame retardants in woods; to summarize briefly the current belief is that a flame

*Bureau of Mines, Pittsburgh Mining and Safety Research Center.

retardant influences the mechanism of decomposition, lowers the temperature of decomposition onset and increases the char yield. Byrne et al [8] state that an effective flame retardant must act at temperatures below that at which cellulose would normally pyrolyze. Our results support these stipulations, as will become apparent later. The action of the preservatives on the thermal oxidative behavior of wood would depend on the nature of the agent or agents and the method of application (aqueous or oil). In actual practice the wood is treated both with preservatives and with fire retardants. To determine the effect of both the preservatives and flame retardants on the thermal oxidative behavior of woods, tests were performed on untreated and variously impregnated samples of pine both under static and dynamic conditions and the volatiles were identified and quantitated.

EXPERIMENTAL

The quiescent investigations were conducted using a 2000 ml calibrated sealed volume which included a bulb and a fingerlike projection wherein the sample was placed. The system was filled with air at pressure somewhat below one atmosphere (usually 400 mm) and the sample was heated at 370°C for 30 min by immersing the finger in a preheated metal bath. After cooling to room temperature the liquid nitrogen volatiles (O_2, N_2, CO, CH_4) were collected in a Sprengel pump, measured and determined by mass spectrometry. The room temperature volatile −196°C condensibles were fractionated in vacuo through traps held at −23, −78, and −196°C. Each of these fractions was measured and weighed and then subjected to infrared and mass spectroscopy, gas chromatography (using 8′ × 1/8″ stainless steel Porapak Q column programmed from 50–220°C at 8°C/min employing thermal conductivity detector) and wet analysis (for hydrogen chloride). The solid carbonaceous residue left in the finger of the bulb was removed and weighed whereas the involatile oils and tars deposited on the wall were examined by infrared spectroscopy.

For the dynamic studies a stagnation burner arrangement previously described [9, 10] was employed. In the past the burner was used as an open system; in the current work all the outlets but one were closed. The open outlet was connected to three traps: the first at −78°C, the other two at −196°C. The cooling was sufficiently effective to condense most of the air flowing over the sample. Both the air and heating block were at 400°C; the sample's residence in the stagnation burner was 15 min. After the completion of the experiment the traps were evacuated while cooled in liquid nitrogen. Subsequently the room temperature volatiles were fractionated and analyzed in the same manner as the products from the quiescent degradation, whereas the involatiles (oils) were examined by infrared spectroscopy.

Differential thermal and thermogravimetric analyses were performed using a DuPont 951/990 thermal analysis system; heating rates of 10°C/min were employed. The sample size in the thermogravimetric analyses was 10–11 mg.

RESULTS AND DISCUSSION

Five pine samples, namely the virgin standard southern yellow pine and four treated samples of this material were studied. The samples are listed in Table 1 together with the details of the impregnations performed by Koppers Company, Pittsburgh, Pennsylvania which are standard procedures described in Wood Preservers Handbook.

Table 1. List of Pine Samples Studied

Material	Material Description
Untreated pine - 6D	Standard southern yellow pine
Treated pine - 2D	Standard southern yellow pine treated with oil-borne preservative (pentachlorophenol); 0.37 lbs/ft^3
Treated pine - 3D	Standard southern yellow pine treated with creosote (coal tar distillate); 10.6 lbs/ft^3
Treated pine - 4D	Standard southern yellow pine treated with fire retardant type-C, Minalith ($(NH_4)_2HPO_4$, 10%; $(NH_4)_2SO_4$, 60% $Na_2B_4O_7$, 10%; H_3BO_3, 20%): 0.4 lbs/ft^3
Treated pine - 5D	Standard southern yellow pine treated with water-borne preservative-CCA (CrO_3, CuO, As_2O_5); 0.40 lbs/ft^3

The DTA curves presented in Figure 1 show in general three processes in which heat is either consumed or produced by the sample a) an endotherm at ~ 120°C present in all five samples which is most likely caused by evaporation of water (drying) and/or formation of water via dehydration (e.g., $-CHOH \longrightarrow -C \leqq + H_2O$), b) a very gentle reaction exotherm, which may indicate slow oxidation and/or simultaneous oxidation and dehydration, and c) a final sharp exotherm, which definitely points to oxidative decomposition. It is apparent that untreated pine, creosote treated pine, and CCA treated pine exhibit very similar DTA curves, with the exception that the final steep reaction exotherm for CCA treated pine occurs ~ 22°C lower (~ 333°C) than for the other two materials (~ 355°C). It is also apparent that pentachlorophenol treated pine exhibits a DTA curve significantly different from that of untreated pine since there appears an additional endotherm at ~280°C which is most likely associated with the sublimiation (evaporation) of some of the impregnant. The fire retardant, Minalith, treatment of pine obviously has the most profound effect on the wood decomposition processes. This DTA curve shows two additional reaction endotherms which could very well be due to impregnant decomposition and/or evaporation; furthermore, the onset of the final steep reaction exotherm starts here as low as 250°C compared to 355°C for untreated pine. Such a lowering of the decomposition temperature would be expected for an efficient flame retarding action according to Byrne et al [8].

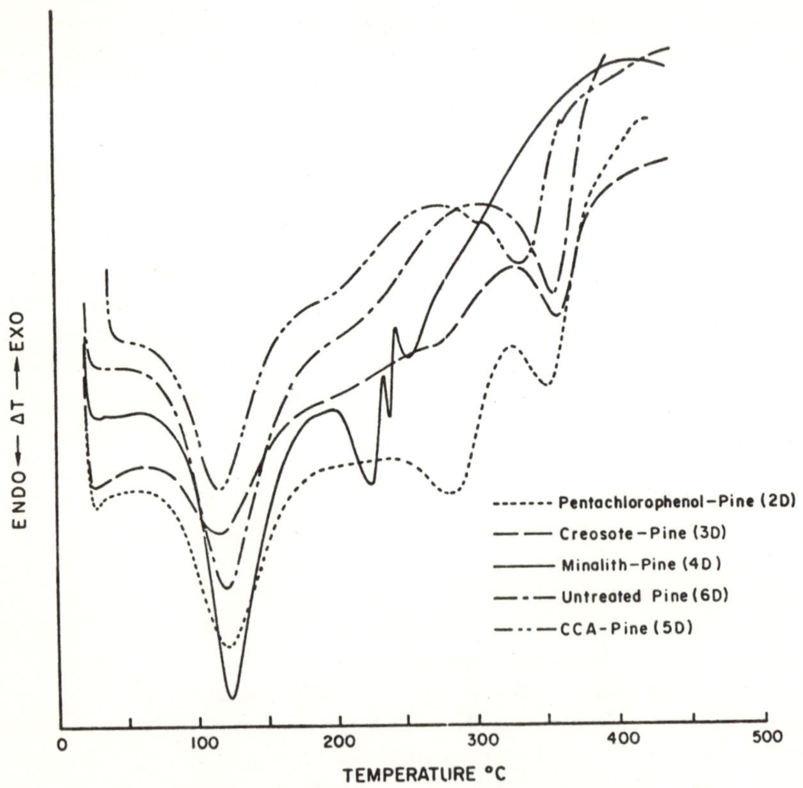

Figure 1. Differential thermal analyses of pine samples.

DTA curves tend to show exotherms much lower than would be expected from heat of combustion values and from calculations of standard bond energies. This is due to the failure of the DTA method to measure accurately the heat produced by oxidation of gaseous degradation products and also due to incomplete combustion.

Yet, based on the DTA curves one can at least get an idea of the general effect of a particular additive on the thermal behavior of a given material such as wood.

In Figure 2 are presented the TGA curves of the various pine samples. Comparison of the five curves shows clearly that in the case of Minalith treated pine no rapid material loss occurs beyond 270°C which points to the absence of combustion. Furthermore, at 400°C the char yield is about 44% which is in a rough agreement with the 40% obtained in the stagnation burner treatment. The lower stagnation burner value can be explained by the longer residence time there at the temperature as compared to the TGA conditions. The almost complete material loss in the case of the CCA treated pine, found in the stagnation burner test, again could have been predicted from the TGA curve. The sharp break or actually hook at *ca* 350°C observed in the TGA of the untreated pine and at 460°C in cresote treated pine is due to strong exothermal reaction which makes the sample reach a higher temperature than that programmed by the instrument. This is in agreement with the DTA curve where a sharp exotherm was evident in this region. The absence of glow in the case of creosote impregnated pine and the low residue obtained from the pentachlorophenol treated sample is indicated neither by TGA nor the DTA data. Thus it is obvious that no one method can provide all the necessary information to predict a thermal behavior of a composition. On the other hand the DTA/TGA investigations provide an insight how a given impregnation affects the thermal behavior of a material and thus allow conclusions to be drawn regarding the advantages and disadvantages of the selected treatment insofar as the mining environment is concerned. Yet to obtain a complete picture one also needs to know the nature and relative concentrations of the volatiles produced on exposure to heat since as the TGA curves show the danger of toxic off-gassing exists even in the absence of fire.

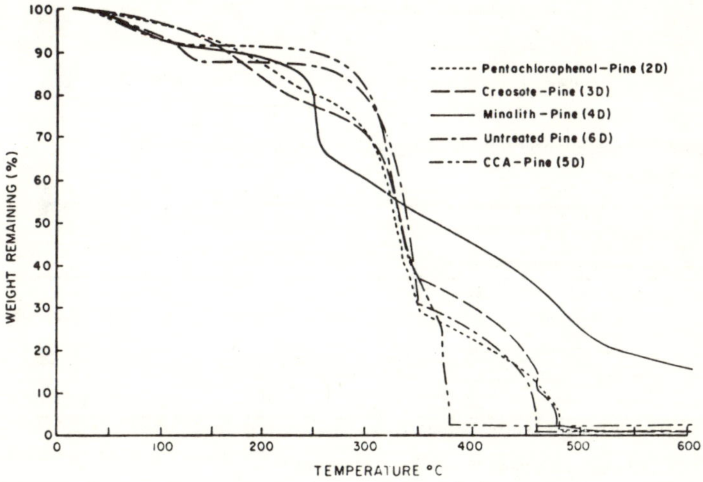

Figure 2. Thermogravimetric analyses of pine samples.

The actual thermal oxidative tests were performed using a quiescent (sealed tube) and a dynamic (stagnation burner) system. The experimental details of the sealed tube studies are summarized in Table 2 whereas the data for the stagnation burner are compiled in Table 3. Comparing the experimental data for the two procedures it is apparent that the stagnation burner, although operated at only 30°C higher temperatures than the sealed system, provides a much more severe environment which is reflected both in the lower residue yields and in the high carbon dioxide content in the volatiles. It could be argued that the sealed system is more related to the TGA environment since the char yields, based on TGA data, correspond closely with those found for the sealed tube studies (i.e., material 6D 34.0 versus 31.2%; material 2D 32.9 versus 26%; material 3D 32.8 versus 33.8%; material 4D 47.0 versus 47.2%; material 5D 35 versus 27%).

Table 2. Experimental Data for Sealed Tube Degradations of Pine Samples

Pine Sample	Sample Wt mg	Residue %[a]	Oxygen Consumed mg	Oxygen Consumed %[b]	Total Volatiles %[c]
6D	597	34.0	35.5	9.5	67.6
2D	625	32.9	43.2	11.6	60.6
3D	630	32.8	33.3	9.0	57.7
4D	603	47.0	54.0	14.7	79.4
5D	605	35.0	49.0	13.3	72.8

a Percent of the weight of the starting material; this is only the solid removable portion of the residue and does not include the tars and oils deposited on the side of the tube.

b Percent of oxygen available.

c Percent of total products expected based on sample weight loss and oxygen consumed.

Table 3. Experimental Data for Stagnation Burner Degradations of Pine Samples

Pine Sample	Glow min	Sample Wt mg	Residue mg	Residue %[a]	Condensible Volatiles mg	Condensible Volatiles %[a]	RT[b] Frac. mg	RT[b] Frac. %[a]
6D	8-15	1020	132	12.9	905	88.8	148	14.5
2D	6-15	1007	34	3.4	798	79.2	202	20.1
3D	no glow	1008	264	26.2	630	62.5	116	11.5
4D	no glow	1018	404	39.7	983	96.5	17	1.7
5D	4-6	1003	14	1.4	1220	121.6	144	14.4

a Percent of the weight of the starting material.

b Room temperature; this material was volatile at 400°C but involatile at room temperature.

The main effort under this program was to identify and quantitate all volatile products formed upon oxidative thermal decomposition which possibly could enter and remain in a mine atmosphere. No attempt was made to determine materials such as phenols, cresols, highly substituted furan derivatives and related species. However mass spectral evidence indicated the presence of compounds such as 2-furyl ethyl ketone (m/e 95, 124), 2-furyl methyl ketone (m/e 95, 110), 5-methyl-2-furaldehyde (m/e 110, 109), dimethylphenol (m/e 122, 107), benzofuran (m/e 118, 90), and furfuryl alcohol (m/e 97). The materials of direct interest to this investigation were the toxic species such as acrolein, formaldehyde, formic acid, and furfural. Unfortunately the values given for the first three compounds in our compilations (Tables 4 and 5) are probably much lower than the actual values. This is due to strong retention of these products in water and tars and their apparent tendency to associate during condensation, which results in a greatly reduced volatility. Thus it is quite possible that a significant fraction of these chemicals remained in the room temperature involatile oils and tars. This unexpected behavior has been also reported by Byrne et al [8].

The product distribution and the nature of the materials formed is fairly close for the stagnation burner and the sealed system, if one neglects the higher carbon dioxide yields in the stagnation burner studies. Yet it is apparent on close examination that the stagnation burner did provide a more drastic environment, as evidenced by, e.g., the increased production of chloride ion from the pentachlorophenol treated pine as compared to the results of the sealed system. The chloride ion values obtained for the stagnation burner tests are always significantly lower than the true or actual chloride ion production due to the reaction of the chloride ion formed with the burner walls and metal connections. Since no carbon monoxide determinations were carried out in the stagnation burner studies no comparison between the two methods can be made insofar as this species is concerned. From the sealed system studies it can be said that the creosote treated pine (material 3D) afforded a significantly lower quantity of CO than the untreated woods, whereas the Minalith treated pine (material 4D) formed definitely higher amounts of CO. The small room temperature involatile fraction in the case of Minalith treated pine is in agreement with the theories advanced above that flame retardants do act by changing the mechanism of decomposition, i.e., increase in yields of volatiles such as water and CO_2 and decrease in production of tars, an important constituent of which (levoglucosan) is believed to be one of the flammable species [11].

Insofar as toxic product formation is concerned the Minalith treated pine is, on the basis of the products identified and quantitated, the most dangerous of the samples investigated in view of SO_2 formation, definite hydrogen cyanide presence, and the increased CO production. As given in Table 1 the Minalith treatment includes, among others, ammonium sulfate, yet no ammonia was detected. Since a substantial quantity of SO_2 was found it can be deduced that the sulfate is reduced to sulfur dioxide by ammonia. The pentachlorophenol and the creosote impregnations are carried out in organic (oil) media; this accounts for the relatively high content of toluene in the volatiles formed.

Table 4. Products Obtained on Thermal Oxidative Decomposition of Pine Samples (Sealed Tube Studies)

Products	Untreated Pine-6D mg/g	Treated Pine-2D mg/g	Treated Pine-3D mg/g	Treated Pine-4D mg/g	Treated Pine-5D mg/g
H_2	0.07	0.13	0.01	0.01	0.13
CO	48.3	45.1	20.7	58.1	43.4
CH_4	0.78	1.18	2.79	0.98	2.30
H_2O	276.6	268.1	243.1	306.3	320.2
HCl	-	0.51	-	-	-
HCN	-	-	-	0.51	-
CO_2	115.8	73.8	77.7	88.7	135.4
SO_2	-	-	-	15.0	-
COS	-	-	-	0.08	-
C_xH_y[a]	0.85	1.22	1.03	1.04	0.36
C_6H_6	0.28	0.82	0.92	1.43	0.50
$C_6H_5CH_3$	0.54	5.64	17.7	0.29	0.25
Xylenes	0.02	0.18	0.04	-	Trace
CH_3Cl	-	3.41	0.11	-	0.05
C_2H_5Cl	-	-	0.03	-	-
CH_3OH	7.74	5.15	6.48	9.42	7.97
HCHO	2.38	1.84	1.68	3.64	2.40
CH_3CHO	1.54	3.57	0.43	0.07	1.21
Furfural	2.18	3.04	4.32	1.39	4.28
5-Methyl-2-furaldehyde	0.03	0.01	0.02	-	0.02
Acrolein	0.62	1.21	0.43	?	0.47
$(CH_3)_2CO$	3.15	1.77	2.72	2.01	1.49
$CH_3COC_2H_5$	1.34	0.36	1.19	1.24	0.86
$CH_3COC_3H_5$	1.83	0.56	2.75	0.02	0.35
2,3-Pentanedione	0.10	0.29	0.30	-	-
HCOOH	?	?	?	0.23	?
CH_3COOH	7.27	7.68	10.9	-	4.57
Methyl formate	2.78	5.44	2.32	0.93	0.84
Methyl acetate	1.84	2.08	1.95	3.96	0.88
Vinyl acetate	2.28	2.02	1.37	0.08	0.18
Furan	1.46	2.83	1.13	0.36	1.60
2-Methylfuran	2.13	1.23	0.32	0.35	2.13
2,5-Dimethylfuran	0.40	0.27	3.38	0.13	0.84

a C_xH_y denotes hydrocarbons C_2 through C_4.

Table 5. Products Obtained on Thermal Oxidative Decomposition of Pine Samples (Stagnation Burner Studies)

Products	Untreated Pine-6D mg/g	Treated Pine-2D mg/g	Treated Pine-3D mg/g	Treated Pine-4D mg/g	Treated Pine-5D mg/g
H_2O	486.2	334.3	224.3	468.5	410.3
HCl	-	8.57	-	-	-
CO_2	199.6	226.8	171.8	274.0	590.9
SO_2	-	-	-	22.3	-
COS	-	-	-	0.12	-
HCN	-	-	-	1.32	-
C_xH_y[a]	0.85	1.37	0.59	0.87	1.53
C_6H_6	0.86	0.85	1.61	1.08	0.19
$C_6H_5CH_3$	0.07	20.0	22.0	0.03	0.19
Xylenes	Trace	Trace	-	0.03	0.01
CH_3Cl	0.22	0.46	0.30	-	-
CH_3OH	2.78	4.15	3.68	8.81	5.17
HCHO	2.09	0.37	0.71	0.50	9.97
CH_3CHO	2.53	0.84	0.81	0.76	1.12
Furfural	1.09	1.15	2.23	4.85	3.18
5-Methyl-2-furaldehyde	Trace	Trace	-	Trace	-
Acrolein	0.21	0.70	0.59	0.22	0.68
$(CH_3)_2CO$	2.81	2.02	1.80	0.45	1.45
$CH_3COC_2H_5$	0.69	0.15	0.62	0.48	0.70
$CH_3COC_3H_5$	0.71	1.74	1.55	-	0.52
2,3-Pentanedione	Trace	Trace	-	-	0.04
HCOOH	?	?	?	0.13	?
CH_3COOH	0.01	0.62	3.04	0.30	2.83
Methyl formate	0.28	1.23	1.12	0.04	0.23
Methyl acetate	0.76	0.96	0.85	0.45	0.18
Vinyl acetate	1.48	1.76	1.90	0.06	0.16
Furan	1.41	0.88	1.26	0.37	0.79
2-Methylfuran	1.05	1.77	1.06	0.06	0.76
2,5-Dimethylfuran	0.79	1.62	0.96	0.03	0.18

a C_xH_y denotes hydrocarbons C_2 through C_4.

To conclude, it can be said that the Minalith treatment did affect most significantly the thermal oxidative behavior of the pine. The most important aspect of this treatment is the lowered decomposition onset, the increased production of carbon monoxide (in the sealed system; no CO data have been obtained for the stagnation burner work), formation of substantial quantities of sulfur dioxide, and some hydrogen cyanide, absence of glow, increased char yield and strongly decreased tar production. The pentachlorophenol and creosote treatments resulted, among others, in an increased toluene content of the volatiles, most likely caused by the oil impregnation medium. The CCA treatment was reflected in the highest carbon dioxide production of all the samples both in the static and dynamic investigations; this is most apparent in the latter case.

ACKNOWLEDGMENT

This investigation was supported by the Bureau of Mines, Department of Interior, under Contract HO133004. The authors are indebted to Drs. Forshey and Hofer for helpful discussions and suggestions.

REFERENCES

1. T. C. Braus, "Lignin and Other Noncarbohydrates," High Polymers, *480* Cellulose and Cellulose Derivatives, 2nd ed., Part I., E. Ott, H. M. Spurlin, and M. W. Grafflin, Eds., Interscience Publishers, Inc., New York, 1954.
2. C. B. Purves, "Chain Structure," High Polymers, V. *54*, Cellulose and Cellulose Derivatives, 2nd ed., Part I., E. Ott, H. M. Spurlin, and M. W. Grafflin, Eds., Interscience Publishers, Inc., New York, 1954.
3. J. Marton, "Lignin Structure and Reactions," Advances in Chemistry Series *59*, R. F. Gould, Ed., American Chemical Society Publications, Washington, 1966.
4. A. E. Lipska and N. J. Alvares, *J. Fire and Flammability, 3*, 204 (1972).
5. J. B. Fang, C. D. M. MacKay and G. Bramhall, "Wood Fire Behavior and Fire Retardant Treatment," a review of literature, November 1966, published by Canadian Wood Council, Ottawa, Canada (1966).
6. A. E. Lipska and F. A. Wodley, "Isothermal Pyrolysis of Cellulose Kinetics and Gas Chromatographic Mass Spectrometric Analysis of the Degradation Products," paper presented at the Spring Meeting of the Combustion Institute, Los Angeles, 1968.
7. A. E. Lipska, "The Fire Retardance Effectiveness of High Molecular Weight, High Oxygen Containing Inorganic Additives in Cellulose and Synthetic Materials," paper presented at the Spring Meeting of the Combustion Institute, Arizona, 1973.
8. G. A. Byrne, D. Gardiner, and F. H. Holmes, J. Appl. Chem., *16*, 81 (1966).
9. K. L. Paciorek, et al., J. Polymer Sci., *11* (7), 1455 (1973).
10. K. L. Paciorek, et al., Amer. Ind. Hyg. Assoc. J. *35* (3), 175 (1974).
11. A. Basch and M. Lewin, *J. Fire and Flammability, 4*, 92 (1973).

Kay L. Paciorek

Kay L. Paciorek received her Ph.D. in Organic Chemistry from the University of Western Australia in 1956. This was followed by a year post-doctoral study at Wayne State University. At Wyandotte Chemical Corporation (1957 to 1961) she was involved mainly in the synthesis and mechanistic studies involving fluorinated organic materials. Subsequently at the U.S. Naval Ordnance Laboratory, Corona, California (1961–1964) she pursued research in organometallic chemistry with emphasis on thermally stable organometallic polymers. This work was continued at MHD Research, Inc. (1964–1966) and at The Marquardt Company (1966–1970). At that time she also branched out into glow discharge investigations and thermal oxidative studies. The latter studies, as well as the organometallic polymer synthesis, have been further pursued at Dynamic Science and Ultrasystems, Inc.

(1970–present). Dr. Paciorek is a member of the American Chemical Society and has 31 scientific literature publications and patents.

Reinhold H. Kratzer

Reinhold H. Kratzer received his Dr. rer. nat (D.Sc) in Inorganic Chemistry at Ludwig-Maximilian University in Munich, Germany in 1960; this was followed by a year of post-doctoral fellowship at the University of Southern California. At the U.S. Naval Ordnance Laboratory, Corona, California (1962–1964) he pursued research in organometallic chemistry including among others, metal hydride and organometallic polymer synthesis. This work was further pursued at MHD Research, Inc. (1964–1966) and at The Marquardt Company (1966–1970). In addition he became involved in electric discharge reactions and thermal and thermal oxidative degradation investigations. He continued the organometallic polymer synthesis and thermal oxidative studies including combustion at Dynamic Science (1970–1972) and is currently involved in the above areas at Ultrasystems, Inc. (1972–present). Dr. Kratzer is a member of the New York Academy of Sciences, the American Chemical Society, the Chemical Society (London), the German Chemical Society, the American Association for the Advancement of Science, and Phi Lambda Epsilon. He has 20 scientific literature publications and patents.

Jacquelyn Kaufman

Jacquelyn Kaufman obtained her B.S. degree in Chemistry at the St. Lawrence University, Canton, New York in 1962. From 1962–1965 she was employed at Hunt Food Industries in an analytical capacity. At MHD Research, Inc. (1965–1966) and at The Marquardt Company (1966–1970) she was involved mainly in instrumental analytical chemistry specifically infrared spectroscopy, mass spectrometry, and gas chromatography. She continued this work at Dynamic Science (1970–1972) and is currently associated with the thermal and thermal oxidative studies of materials at Ultrasystems, Inc. (1972–present). Miss Kaufman is a member of American Chemical Society and has co-authored two papers in scientific literature.

James H. Nakahara

James H. Nakahara obtained his B.A. degree in Chemistry at the University of California, Irvine in 1972. He joined Ultrasystems, Inc. after graduation. His experience to date has mainly been in instrumental analytical chemistry, specifically gas chromatography, vacuum line separations, infrared spectroscopy, and mass spectrometry.

Arthur M. Hartstein

Arthur M. Hartstein received his Ph.D. in Physical Chemistry from Adelphi

University in 1970. He was employed as Research Chemist at Gulton Industries from 1970–1971. In 1971 he joined U.S. Bureau of Mines where he is currently among others a technical project officer on programs dealing with thermal oxidative degradation of materials used in underground mines. Dr. Hartstein has co-authored 7 papers published in scientific literature and is a member of the American Chemical Society.

C. A. Holmes

Forest Products Laboratory *
Madison, Wisconsin 53705

FLAMMABILITY OF SELECTED WOOD PRODUCTS UNDER MOTOR VEHICLE SAFETY STANDARD

(Received April 30, 1973)

ABSTRACT: Motor Vehicle Safety Standard No. 302 specifies the burn-resistance requirement and the test procedure for materials used in the occupant compartments of motor vehicles. In this study, the fire performance of some selected wood and wood-based products, including 1/2-inch lumber, veneers, plywood, hardboard, corrugated fiberboard, and kraft paper, were determined under this standard. Only the 0.012-inch-thick kraft paper burned at a rate in excess of the 4 inches per minute limitation of the standard. The other materials had zero or very low burn rates. Enamel and clear lacquer did not add any flammability by this test method to 1/8-inch birch plywood or hardboard. This study strongly indicated that wood and wood-fiber products in general will have burn rates less than the 4 inches per minute limitation of Standard No. 302.

INTRODUCTION

The National Highway Safety Bureau has issued Motor Vehicle Safety Standard No. 302 [1] effective September 1, 1972. This standard specifies resistance requirements for materials used in the occupant compartments of motor vehicles. The justification for the safety standard is the occurrence of thousands of fires per year that have their origin in vehicle interiors.

Fires in the United States have caused an annual 12,000 deaths in recent years. Of this number, the National Fire Protection Association estimates the average annual number of vehicular fire deaths at about 4,000 [2]. For 1971, the NFPA estimated 482,400 fires in pleasure and transportation vehicles with a total loss of $96,540,000 [3].

The scope of the Standard limits the applicable materials to those used in the occupant compartments of motor vehicles. Some of the components that may be made of wood or wood-fiber products, and must meet the test requirements, are trim panels, compartment shelves, sun visors, engine compartment covers, seats, and arm rests. The wood or wood-product item may be a surface material and would therefore be tested with its surface exposed to the test flame. If the wood product is part of a composite to which another surface material is bonded or attached, then the composite is tested with the surface material exposed to the test flame.

---

*Maintained at Madison, Wis., in cooperation with the University of Wisconsin.

The test method described in Standard No. 302 is similar in most respects to Federal Test Method Standard No. 191, Method 5906, Burning Rate of Cloth; Horizontal (Dec. 1968). Method 5906 is used for cloth that has not been flameproofed and provides a test condition giving a slower burning rate than other Federal methods for cloth.

Standard No. 302 may be considered a somewhat lenient test method in that the specimen is placed horizontally, the igniting burner is left on for only 15 seconds, and a 4 inches per minute burning rate is permitted. In the discussion regarding this standard in the Federal Register [1], Director D. W. Toms of the Highway Safety Bureau stated that the requirements were not too stringent in order that suppliers will be able to meet the requirements by the effective date. However, he states that further study will be made of the "feasibility of, and justification for, imposing more stringent requirements with a later effective date."

Of course, many different products and materials are used in the passenger compartments of motor vehicles. Flammable materials may be classed into two groups: Plastics including synthetic fiber fabrics, and materials of natural fibers including wood and wood-base products. Storrs and Lindemann reviewed the application of Standard No. 302 to plastic materials [4], but no similar review has been made for natural fibers.

The performance of wood products under this Standard is of concern to manufacturers who supply these products for use in vehicles. Data have not been available on the performance of wood products in the prescribed test because the new standard employs a test method not heretofore used for wood products. The Forest Products Laboratory, therefore, investigated the performance of the basic wood products by this standard test.

SELECTION OF TEST MATERIAL

Most often, if a wood or wood-based product is used, it has a decorative coating or is covered with a fabric or plastic material. According to the Standard, the portions of the vehicle interior components that shall meet the requirements are:

(a) The surface material taken separately if it is not bonded, sewed, or mechanically attached to underlying material.

(b) A composite consisting of the surface material bonded, sewed, or mechanically attached to underlying material, if such a composite is used in the component.

(c) Padding and cushioning materials taken separately, if those materials are not bonded, sewed, or mechanically attached to surface materials.

A variety of wood materials was selected for this study to obtain an indication of the test performance of representative wood products. Samples of lumber, thin veneers, plywood, corrugated board, and kraft paper were included. In addition, coated hardboard and plywood were tested and also plywood that had been pressure-impregnated with a fire-retardant chemical, monoammonium phosphate.

TEST APPARATUS AND PROCEDURE

The test apparatus is described in the Standard No. 302, which is reproduced in the appendix. Figure 1 shows the metal testing cabinet with the horizontal

Figure 1. *Cabinet for experimental testing according to Motor Vehicle Safety Standard No. 302. (M 140 558).*

U-shaped specimen support frame and bunsen burner in place. As all test specimens in this study were self-supporting, the additional U-shaped frame with support wires as recommended in the Standard was not needed.

The Standard specifies conditioning of the test specimens prior to fire testing at 70°F. and 50 percent relative humidity for 24 hours. The fire tests are also conducted at these conditions. Wood eventually reaches an equilibrium moisture content of about 9.2 percent based on its ovendry weight, under these conditions of temperature and relative humidity.

All specimens used in this study were held at the standard conditions until they reached a constant weight, indicating that they had come to an equilibrium moisture content. Fire testing was conducted thereafter.

Specimens were tested with a natural gas bunsen burner that had a temperature of 1,800°F. (at the center of a 1½-in.-long flame) when measured by an iron-constantan thermocouple. When a 1/4-inch plywood specimen was introduced into the flame in test position, the temperature at the point of flame impingement on the specimen was reduced to 1,675°F. The time of flame exposure for all tests was 15 seconds.

RESULTS AND DISCUSSION

Burn rates for the materials tested are given in Table 1. Figure 2 shows the

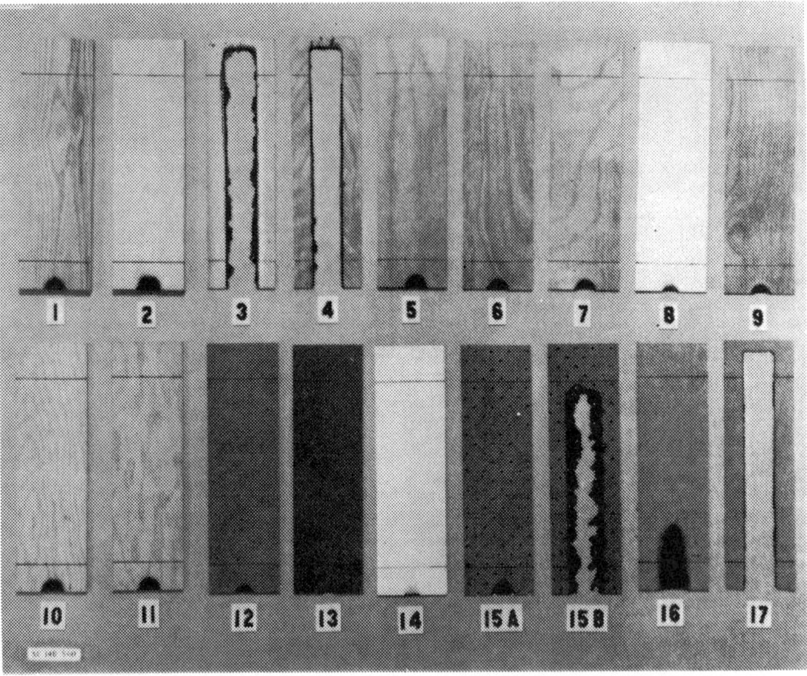

Figure 2. *Typical specimens of wood and wood products after fire testing according to Motor Vehicle Safety Standard No. 302. See Table 1 for the description of the wood specimens under the appropriate product number. (M 140 560).*

appearance after fire test of one typical specimen of each of the wood product types tested.

Thickness of the wood material determines its performance by this fire test method. The 1/2-inch lumber, and also the plywood specimens down to 1/8 inch thick, resisted ignition during the 15-second exposure to the bunsen burner flame. The thinner 1/16-inch veneers, however, charred completely through and reached ignition condition during the 15-second exposure. After the igniting burner was

Table 1. Fire Test Results[a] for Selected Wood Products Under the Motor Vehicle Safety Standard No. 302

Product No.	Description of Wood Product	Moisture content Pct.	Flame travel[b] Distance In.	Flame travel[b] Time Min.	Burn rate[c] In. per min.
1	1/2-inch red oak	5.7	0	–	0
2	1/2-inch white pine	6.3	0	–	0
3	1/16-inch ponderosa pine veneer	8.3	10	5.3	1.9
4	1/16-inch yellow birch veneer	7.5	10	5.1	2.0
5	1/4-inch yellow birch plywood	6.7	0	–	0
6	3/16-inch yellow birch plywood	6.6	0	–	0
7	1/8-inch yellow birch plywood	6.6	0	–	0
8	1/8-inch yellow birch plywood, coated[d]	6.1	0	–	0
9	1/8-inch yellow birch plywood, coated[e, f]	6.6	0	–	0
10	1/4-inch Douglas-fir plywood, exterior grade	5.9	0	–	0
11	1/4-inch Douglas-fir plywood, exterior grade, fire-retardant-treated[g]	8.1	0	–	0
12	1/4-inch standard hardboard	4.7	0	–	0
13	1/8-inch standard hardboard	5.6	0	–	0
14	1/8-inch standard hardboard, coated[h]	5.0	0	–	0
15A	1/8-inch standard hardboard, perforated[i]	6.2	0	–	0
15B	1/8-inch standard hardboard, perforated[f]	6.5	9.7	21.0	0.5
16	V3c corrugated fiberboard[j]	7.3	2.1	3.0	0.7
17	0.012-inch kraft paper	6.2	10	2.1	4.7

[a]Average of 5 tests unless indicated otherwise.
[b]Total measuring distance on specimen surface is 10 in.: From a starting point 1½ in. from burner end of specimen to 1½ in. from a clamped end.
[c]Acceptable burn rate under Standard No. 302 is 4 in. per min. or less.
[d]1 coat enamel undercoat, Federal Specification TT-E-543a, and 1 coat of gloss enamel Federal Specification TT-E-489d. Total dry coating weight, 15.6 g. per sq. ft.
[e]Clear lacquer, Type II, Federal Specification TT-L-50F. Dry coating weight, 5.2 g. per sq. ft.
[f]Average of 3 tests.
[g]Pressure treated with water solution of monoammonium phosphate to 3.1 lb. of dry chemical per cu. ft. of wood.
[h]Same enamel coating system as specimen No. 8 but with total dry coating weight of 14.5 g. per sq. ft.
[i]Average of 2 tests.
[j]As described in Federal Specification PPP-F-320D: Type CF (corrugated fiberboard), weather-resistant class, corrugating medium 0.010 in., outer facings 0.023 in.

turned off, flaming continued on both the upper and lower surface of the specimens.

The burn rate obtained on the 1/2-inch-thick red oak and white pine boards was zero, but on the pieces of 1/16-inch ponderosa pine and yellow birch veneers the flames traveled the full length of the specimens and burn rates of 1.9 and 2.0 inches per minute were obtained.

Two specimens of perforated hardboard are shown in Figure 2 to depict the two kinds of results obtained with this material. Two of the five specimens, identified as 15A in the table and photograph, did not burn any appreciable distance, and not to the 1½-inch point from which timing was started. However, three of these perforated hardboard specimens, identified as 15B, burned slowly for about their full length. The average burn rate was very low at 0.5 inch per minute. (The calculated burn rates for the three specimens were 0.45, 0.46, and 0.48 in. per min.) No complete explanation was determined why three of the perforated hardboard specimens should burn to about their full length while two did not. It is possibly a threshold condition having to do with the character of the specimen and the mildness of the test – short-time exposure to igniting flame and horizontal position of the specimen.

Hardboard will have variations in density over a small area due to clumping of fibers at some points and lack of fibers at other points. This could affect the ease of ignition. Also, the small amount of a flammable sizing agent in the hardboard may have a variation in distribution on the surface of the board. A heavier film of the sizing on the specimen at the point where the test flame impinges may provide a more easily ignitable fuel to start the burning process.

The corrugated fiberboard burned slowly at an average rate of 0.7 inch per minute for only 2.1 inches. The small blue flame was almost indiscernible. The fiberboard could not sustain combustion much beyond the heating influence of the bunsen burner flame.

Representative coatings that were not fire-retardant, a pigmented enamel and a clear acrylic lacquer, did not contribute any flammability by this test method on 1/8-inch birch plywood or 1/8-inch hardboard.

Heavy kraft paper, 0.012-inch thick liner board having a density of 0.7 gram per cubic centimeter, was the only wood product selected for this study which did not meet the acceptance requirement of the test standard. The average burn rate obtained for the five specimens of this material was 4.7 inches per minute.[1]

The results of this study have indicated that when wood products, except paper, are tested according to Standard No. 302, they will have burn rates below 4 inches per minute.

APPENDIX
MOTOR VEHICLE SAFETY STANDARD NO. 302[2]

Flammability of Interior Materials – Passenger Cars, Multipurpose Passenger Vehicles, Trucks, and Buses

S1. Scope

This standard specifies burn resistance requirements for materials used in the

[1] In an exploratory experiment, ordinary newspaper (0.0032 in. thick, and density 0.55 g. per cc.) burned at an average rate of 24.9 in. per min.

[2] Extracted from *Federal Register*, Vol. 36, No. 5, Jan. 8, 1971.

occupant compartments of motor vehicles.

S2. Purpose

The purpose of this standard is to reduce the deaths and injuries to motor vehicle occupants caused by vehicle fires, especially those originating in the interior of the vehicle from sources such as matches or cigarettes.

S3. Application

This standard applies to passenger cars, multi-purpose passenger vehicles, trucks, and buses.

S4. Requirements

S4.1 —The portions described in S1.2 of the following components of vehicle occupant compartments shall meet the requirements of S4.3: Seat cushions, seat backs, seat belts, headlining, convertible tops, arm rests, all trim panels including door, front, rear, and side panels, compartment shelves, head restraints, floor coverings, sun visors, curtains, shades, wheel housing covers, engine compartment covers, mattress covers, and any other interior materials, including padding and crash-deployed elements, that are designed to absorb energy on contact by occupants in the event of a crash.

S4.2 —The portions of the components that shall meet the requirements of S4.3 are all of the following:

(a) The surface material taken separately if it is not bonded, sewed, or mechanically attached to underlying material.

(b) A composite consisting of the surface material bonded, sewed, or mechanically attached to underlying material, if such a composite is used in the component.

(c) Padding and cushioning materials taken separately, if those materials are not bonded, sewed, or mechanically attached to surface materials.

S4.3 —Material described in S4.1 and S4.2 shall not burn, or transmit a flame front across its surface, at a rate of more than 4 inches per minute. However, if a material stops burning before it has burned for 60 seconds from the start of timing, and has not burned more than 2 inches from the point where timing was started, it shall be considered to meet this requirement.

S5. Test Procedure

S5.1 Conditions

S5.1.1 The test is conducted in a metal cabinet for protecting the test specimens from drafts. The interior of the cabinet is 15 inches long, 8 inches deep, and 14 inches high. It has a glass observation window in the front, a closable opening to permit insertion of the specimen holder, and a hole to accommodate tubing for a gas burner. For ventilation, it has a 1/2-inch clearance space around the top of the cabinet, ten 3/4-inch-diameter holes in the base of the cabinet, and legs to elevate the bottom of the cabinet by 3/8 of an inch.

S5.1.2 Prior to testing, each specimen is conditioned for 24 hours at a temperature of 70°F and a relative humidity of 50 percent, and the test is conducted under those ambient conditions.

S5.1.3 The test specimen is inserted between two matching U-shaped frames of metal stock 1 inch wide and 3/8 inch high. The interior dimensions of the U-shaped frames are 2 inches wide by 13 inches long. A specimen that softens and bends at the flaming end so as to cause erratic burning is kept horizontal by supports consisting of thin, heat-resistant wires, spanning the width of the U-shaped frame under the specimen at 1-inch intervals. A device that may be used for supporting this type of material is an additional U-shaped frame, wider than the U-shaped frame containing the specimen, spanned by 10-mil wires of heat-resistant composition at 1-inch intervals, inserted over the bottom U-shaped frame.

S5.1.4 A bunsen burner with a tube of 3/8-inch inside diameter is used. The gas adjusting valve is set to provide a flame, with the tube vertical, of 1½ inches in height. The air inlet to the burner is closed.

S5.1.5 The gas supplied to the burner has a flame temperature equivalent to that of natural gas.

S5.2 Preparation of Specimens

S5.2.1 Each specimen of material to be tested is a rectangle 4 inches wide by 14 inches long, wherever possible. The thickness of the specimen is that of the material as used in the vehicle, except that where the material's thickness exceeds 1/2 inch the specimen is cut down to that thickness. Where it is not possible to obtain a flat specimen, because of component configuration, the specimen is cut to not more than 1/2 inch in thickness at any point, from the area with the least curvature, and in such a manner as to include the face side. The maximum available length or width of a specimen is used where either dimension is less than 14 inches or 4 inches respectively.

S5.2.2 Material with directional effects is oriented so as to provide the most adverse results.

S5.2.3 Material with a napped or tufted surface is placed on a flat surface and combed twice against the nap with a comb having seven to eight smooth, rounded teeth per inch.

S5.3 Procedure

(a) Mount the specimen so that both sides and one end are held by the U-shaped frame, and one end is even with the open end of the frame. Where the maximum available width of a specimen is not more than 2 inches, so that the sides of the specimen cannot be held in the U-shaped frame, place the specimen in position on wire supports as described in S5.1.3; with one end held by the closed end of the U-shaped frame.

(b) Place the mounted specimen in a horizontal position, in the center of the cabinet.

(c) With the flame adjusted according to S5.1.4, position the bunsen burner and specimen so that the center of the burner tip is 3/4 inch below the center of the bottom edge of the open end of the specimen.

(d) Expose the specimen to the flame for 15 seconds.

(e) Begin timing (without reference to the period of application of the burner flame) when the flame from the burning specimen reaches a point 1½ inches from the open end of the specimen.

(f) Measure the time that it takes the flame to progress to a point 1½ inches from the clamped end of the specimen. If the flame does not reach the specified end point, time its progress to the point where flaming stops.

(g) Calculate the burn rate from the formula:

$$B = 60 \times \frac{D}{T}$$

where

B = burn rate in inches per minute,
D = length the flame travels in inches, and
T = time in seconds for the flame to travel D inches.

REFERENCES

1. National Highway Safety Bureau 1971. Flammability of interior materials in passenger vehicles, trucks, and buses. Title 49, Part 571 – Motor vehicle safety standards. Fed. Reg. 36(5). Jan. 8.
2. National Fire Protection Association 1969. Fire protection handbook, 13th ed.
3. National Fire Protection Association 1972. Fires and fire losses classified,1971. Fire J. 66(5).
4. Charles D. Storrs and Otto H. Lindemann 1972. Federal flammability standards for interiors of motor vehicles. Fire J. 66(4).

Carlton A. Holmes

Carlton A. Holmes is in the Fire Research Work Unit, U.S.D.A., Forest Products Laboratory, located at Madison, Wisconsin. He received a B.S. degree in wood technology from the University of Minnesota in 1948. His past experience was in wood utilization research. He has been in his present position since 1965 conducting research and evaluation on the fire performance properties of wood and wood products.

E. L. SCHAFFER, RESEARCH ENGINEER

Forest Products Laboratory *
Forest Service
U.S. Department of Agriculture

EFFECT OF FIRE-RETARDANT IMPREGNATIONS ON WOOD CHARRING RATE

(Received October 1, 1973)

ABSTRACT: Fire retardant and other chemicals impregnated in southern pine wood are examined for their influence on the charring rate as compared to untreated wood. Chemicals included at retentions of 15 percent of dry wood weight either did not influence or retarded the rate of char formation on fire-exposed sections. Polyethylene glycol (1000) reduced the rate of char 25 percent.

The following chemicals or resins reduce the rate of charring about 20 percent: boric acid, borax, ammonium sulfate, monosodium phosphate, potassium carbonate, sodium hydroxide.

Rates of charring for other chemically treated woods were equivalent to those untreated and included: THPC (tetrakis-hydroxymethyl phosphonium chloride: urea), dicyandiamide phosphoric acid, monoammonium phosphate, zinc chloride, sodium chloride.

INTRODUCTION

RESIN- AND fire-retardant-impregnated woods are utilized as components in housing and building construction, for models and dies, for furniture and cabinets. If a fire-retardant chemical is present, it reduces the flaming volatile contribution of the wood during fire exposure. As a result, a reduction in surface flammability is achieved.

Basic studies employing thermogravimetric analysis of fire-retardant wood, such as those of Browne and Tang [3, 4], have indicated that some fire-retardant chemicals promote wood decomposition at lower temperatures and thus direct the decomposition to carbon dioxide and water instead of intermediate flammable tar formation. Because of the lowering of the decomposition temperature, there is a possibility that the charring rate of wood treated with certain fire-retardant chemicals might be greater than for untreated wood. Certain other fire-retardant chemicals do not lower the decomposition temperature significantly.

A compensating factor which may reduce the rate of charring of fire-retardant-treated wood, is that the decrease in tar formation is accompanied by an increase in

*Maintained at Madison, WI 53705, in cooperation with the University of Wisconsin.

mass of the char formed. This is seen in data developed by Brenden [2] in Table 1. The percent char for treated ponderosa pine is double that of untreated in the more effective fire-retardant treatments such as borax and diammonium phosphate. An increased amount of char could provide additional surface insulation and also inhibit volatilization of the wood beneath during fire exposure. Others, e.g. [5], have also concluded that the rapid formation of a heavy layer of charcoal on fire-retardant-impregnated wood is induced (with a like reduction of volatile gases) by such retardants as ammonium phosphate when the timber is exposed to fire. The charcoal layer was described as denser than the charcoal of untreated wood, and did not smolder to any dangerous extent.

The Division of Forest Products, CSIRO [5], has recognized that though impregnated fire retardants have as their main effect the reduction of flammability of timber, there is also a significant increase in resistance to burnthrough. They further stated that a structural component made from fire-retardant-impregnated timber will continue to support its load longer than untreated components, and cited tests on wooden load-bearing walls and stressed-skin panels where times to failure were increased by 20 to 33 percent by fire-retardant treatment. In similar tests at the National Bureau of Standards, Mitchell [6] tested the fire endurance (ASTM E 119) [1], of fire-retardant (monoammonium phosphate)-impregnated solid wood walls in both the loaded and unloaded state. For the walls tested without load, the treatments increased the time to failure by 29 to 33 percent over that for untreated wood. The treatment also added 20 to 24 percent to the failure time of walls tested under load compared with walls of untreated wood.

This research was intended to help clarify whether impregnated chemicals significantly affect the charring rate of wood, and to determine the magnitude of the effect. The report describes the experimental results of charring rate determination for chemically impregnated southern pine slabs, and compares these results with those previously determined for untreated southern pine [7]. Included are 12 common fire-retardant and chemical treatments.

EXPERIMENTAL PROCEDURE

Specimens

Nominal 4- by 4-inch by 21¼-inch lengths of southern pine were cut from longer lengths and marked. A 1-inch density block was cut from each for dry specific gravity determination. The samples were prepared from the same stock of southern pine employed in a previous study on the charring rate of untreated wood [7].

Chemical Treatments

Chemicals evaluated in the past for fire retardancy have been classified into four groups according to their effect on the volatilization of wood in the early and late stages of pyrolysis. Browne and Tang [4] explain that at temperatures below the "exothermic point," or onset of fast pyrolysis for untreated wood (about 270°C). some chemicals greatly hasten volatilization in the early stage of pyrolysis independent of whether or not they decrease the ultimate volatilization significantly at completion of the pyrolysis process. Other chemicals are found to have little effect, if any, on early volatilization. Thermogravimetric analysis curves have been employed to reveal the effect of the chemicals on the rate of pyrolysis at moderate temperatures (such as 250°C).

The four class groups of chemicals referred to above may be described as follows [3] and typical ones are listed in Table 2:

I. Chemicals that reduce the volatilization at 400°C on the thermogravimetric curve with little increase in volatilization at 250°C.
II. Chemicals that reduce volatilization at 400°C but significantly increase volatilization at 250°C.
III. Chemicals that exert relatively little effect on volatilization at either 400° or 250°C.
IV. Chemicals that decrease volatilization very little at 400°C but increase volatilization significantly at 250°C.

The separation between chemicals in the four classes is rather arbitrary as the four classes gradually merge into one another. Class II chemicals are shown to produce char at temperatures as much as 100°C below that of untreated wood.

Based on this background and other sources, the following chemicals were selected for this study to represent each of the classes.

Class	Chemical
I	Boric acid
I	THPC, tetrakis-(hydroxymethyl) phosphonium chloride: urea (THPC (80 pct.) = 57.5 pct.; NaOH (50 pct.) = 6.2 pct.; urea = 11.4 pct.; and methylol-melamine resin = 24.9 pct.)
I	Sodium tetraborate (borax)
I	Dicyandiamide-phosphoric acid (pre-reacted to resin) (Composition on anhydrous weight basis: dicyandiamide = 45.3 pct.; phosphoric acid = 53.0 pct.; and formaldehyde = 1.6 pct.)
II	Ammonium sulfate
II	Monoammonium phosphate
II	Zinc chloride
III	Sodium chloride
III	Monosodium phosphate
IV	Potassium carbonate
—	Sodium hydroxide
—	Polyethylene glycol (molecular weight 1,000)

Southern pine was employed as the material for impregnation with fire retardant or resin because of its high permeability. Hence, when impregnated with chemical solution and then dried, the wood samples have an assumed homogeneous distribution of chemical throughout.

Twenty-inch lengths were epoxy tar end-coated and then treated with an appropriate chemical to desired concentration by vacuum-pressure impregnation. Treatment levels of all chemicals, except polyethylene glycol, were as close to 15 percent of ovendry weight of the wood as possible to maximize the effect of the chemical on char formation. Following impregnation, specimens were low-temperature kiln-dried, or air-dried and then low-temperature kiln-dried, and then brought to equilibrium at 80°F, 30 percent relative humidity.

The anhydrous salt added to a specimen was calculated by measuring the specimen block weight before and after treatment and by multiplying the difference by the known anhydrous salt concentration in solution. Untreated southern pine approached 7 percent moisture content when dried from a higher moisture content in an 80°F, 30 percent relative humidity environment. The apparent moisture content of chemically treated specimens will differ from untreated in the same environment in proportion to the hygroscopicity of the impregnated chemical. For purposes of this study, the percent of moisture available to alter the charring rate was calculated:

$$M = \frac{W_2 - W_1}{W_1} \times 100 \quad \text{(pct.)}$$

where

W_2 = conditioned weight of salt treated specimen, and
W_1 = ovendry weight of untreated specimen and anhydrous salt weight in specimen.

After reaching equilibrium, the sections were weighed and then bonded together into a 10- by 20-inch by 3½-inch-thick slab using thermosetting phenol-resorcinol adhesives. Surfaces to be bonded were planed just prior to glue application. Three specimens per chemical were then available for fire-exposure tests.

Fire Exposure

Slab specimens were weighed prior to fire exposure and then placed in position in a 20- by 20-inch gas furnace (Figure 1). The furnace was controlled to follow the ASTM E 119 time-temperature curve [1], employing a wrought-iron capped thermocouple 2 inches from the fire-exposed surface. Char developments to given depth were determined by taking increment core borings from the unexposed side at various times during exposure and calculating length of the charred portion from uncharred length measurements. The distance between cores was 2.3 inches, and no

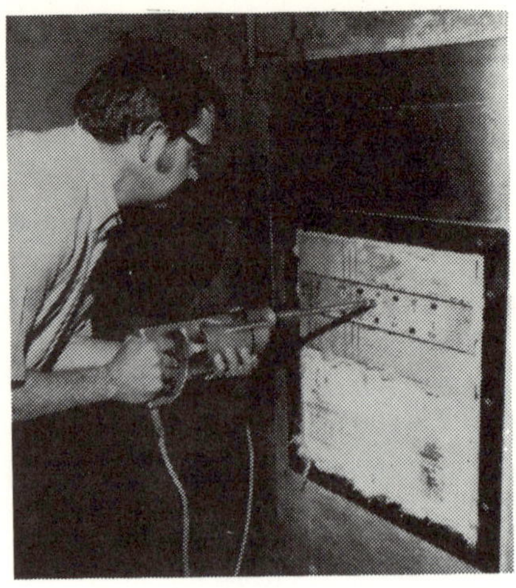

Figure 1. Treated southern pine specimen emplaced in 20- by 20-inch vertical gas-fired furnace, and showing boring at a selected location.

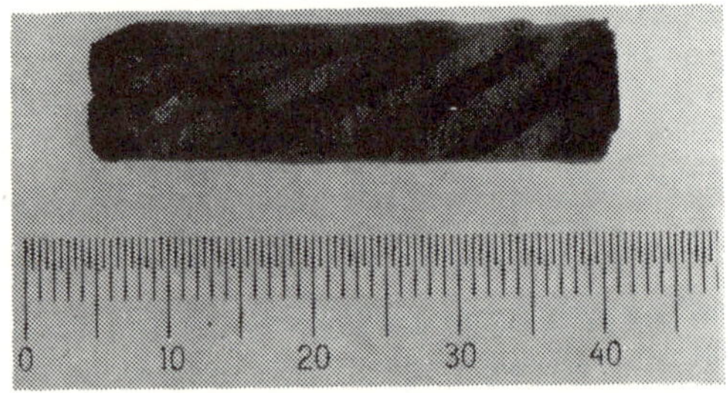

Figure 2. Charred core extracted from treated specimen after about 1 hour of fire exposure. (Scale in photo marked in millimeters.)

cores were extracted closer than 4 inches to the ends. Immediately after the specimen core was extracted, a birch dowel was inserted into the specimen to reduce localized effects. A core is shown in Figure 2. The uncharred length was recorded.

When char had formed at a depth of 1½ inches from the initial fire-exposed surface, the specimen was removed from the furnace opening and placed in an airtight box to extinguish glowing.

RESULTS

Schaffer [7] had previously examined the progress of char development as a function of time of exposure to ASTM E 119 fire conditions for untreated southern pine. He found that the char progression was linearly related to time of exposure and influenced mainly by dry specific gravity, *G*, and calculated moisture content (percent), *M*, at time of test:

$$t = \beta x$$

where

t = fire-exposure time (minutes),
x = char progression from initial fire-exposed surface (inches), and
$\beta = 2[(5.832 + 0.12M) G + 12.862]$

A plot of the predicted time, βx, as against time observed to form char at a given depth from the initial fire-exposed surface for untreated southern pine specimens is shown in Figure 3 [7]. Also indicated by dashed lines in Figure 3 are the estimated deviations of the observed time, *t*, as compared to predicted, βx.[2] The positions of these were simply determined as the minimum and maximum times observed as compared to predicted.

Shown in Table 3 are the dry (salt-free) specific gravities, calculated average moisture content at time of test, and anhydrous salt concentration of the center 4- by 4-inch section of each treated specimen.

The charring of untreated southern pine is little influenced by the moisture present. The expression for β, which includes the wood physical property effect, shows that charring rate is relatively insensitive to the moisture present. A change of only 1 minute in time for char to develop at a 1-inch depth occurs if a difference in moisture content of 9 percent exists between specimens. Water molecules forming a hydrated salt would be liberated during the drying due to the combustion of wood and should be considered as included in the initial moisture content. Calculated moisture contents for treated wood were therefore obtained by using anhydrous salt as a basis.

The observed times for char to form at a given depth are plotted against predicted times, βx, for resin and chemically treated specimens in Figures 4 through 7. Employed to compute β were the wood dry specific gravity (salt-free) and calculated average moisture content at time of test. Superimposed on these plots are the total deviations shown for untreated specimens (dashed lines) and the solid line where $t = \beta x$ as taken from Figure 3.

ANALYSIS

To determine if a given chemical treatment significantly affects the rate of charring (or time for char to form at a given depth) as normally found for untreated wood, observed times to progress to a given depth as plotted in Figures 4 through 7

[2] Statistical analysis to quantify the confidence level of this deviation is not possible due to the fact that a set of 8 data points represent observations from a single specimen. Hence the data are not independent.

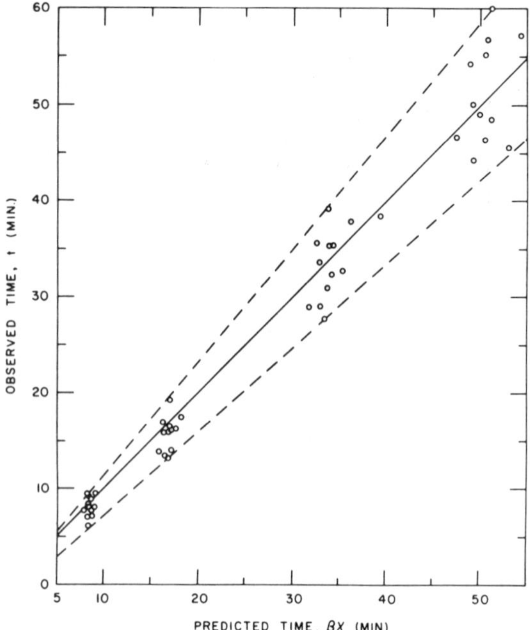

Figure 3. Observed time for char to form at a given depth from initially fire exposed surface versus predicted time for untreated southern pine (7).

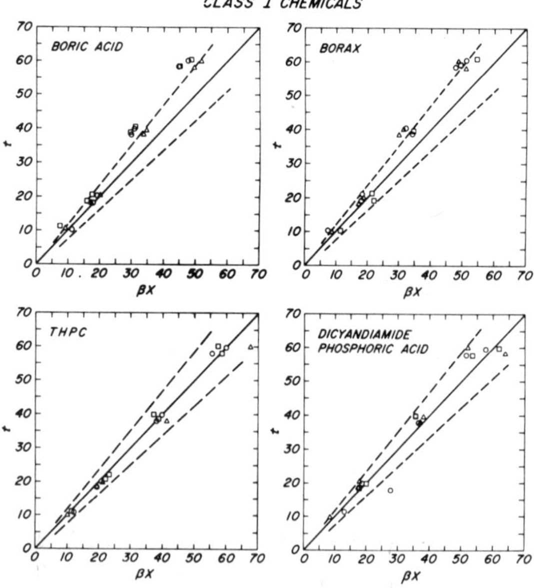

Figure 4. Observed times, t, to given char formation depth versus predicted, βx, for class I chemical treatments.

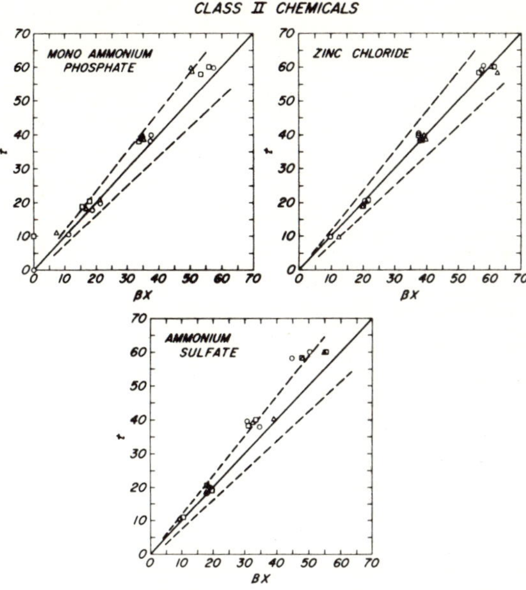

Figure 5. Observed times, t, at given char formation depth versus predicted, βx, for class II chemical treatments.

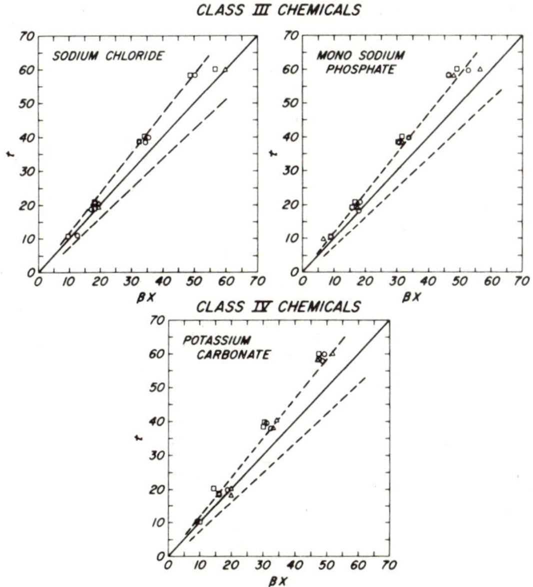

Figure 6. Observed times, t, at given char formation depth versus predicted, βx, for classes III and IV chemical treatments.

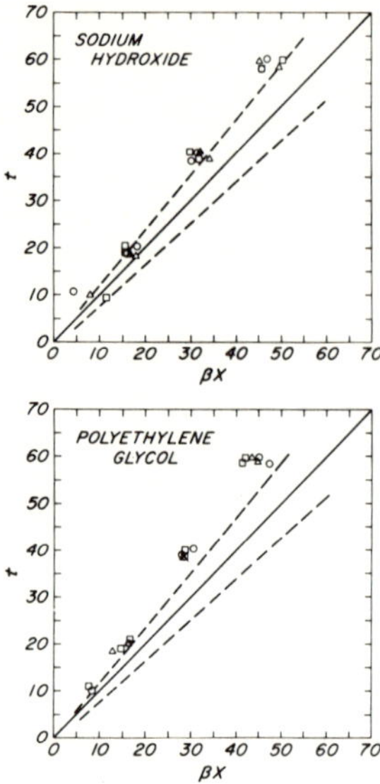

Figure 7. Observed times, t, at given char formation depth versus predicted, βx, for sodium hydroxide- and polyethylene-glycol-treated specimens.

for chemical treated wood should fall outside the deviation limit lines obtained from untreated wood tests shown in Figure 3. This, in essence, suggests that a charring rate is only significantly different from untreated wood if the majority of data points (observed times to produce char at a given depth) lie outside the deviation limit lines. If the majority of points are above (greater observed times), the charring is slower. If below, the charring is more rapid than untreated wood.

An examination of the data indicates the following:

1. None of the chemicals examined increased the charring rate of wood as compared to untreated.
2. PEG significantly reduced the mean rate of charring over the whole period of fire exposure as compared to untreated wood. Improvement: About 25 percent.
3. The following chemicals or resin treatments were found to decrease the mean rate of charring about 20 percent.

Class	Treatment
I	Boric acid
I	Borax
II	Ammonium sulfate
III	Monosodium phosphate
IV	Potassium carbonate
—	Sodium hydroxide

4. Charring rates for the following treated woods were equivalent to that of untreated:

Class	Treatment
I	THPC
I	Dicyandiamide phosphoric acid
II	Monoammonium phosphate
II	Zinc chloride
III	Sodium chloride

CONCLUSIONS

The results and analysis indicate that none of the included resin or chemical treatments of southern pine significantly increased the rate of charring. Neither was volatility activity class as given by Browne and Tang [4] indicative of charring rate alteration. The following resin or chemical treatments decrease the mean charring rate about 20 percent:

Boric acid	Monosodium phosphate
Borax	Potassium carbonate
Ammonium sulfate	Sodium hydroxide

Polyethylene glycol reduces the mean rate of charring over the entire period of fire exposure by about 25 percent.

Surprisingly, within the limits of this analysis, monoammonium phosphate did not significantly reduce the rate of charring as compared to untreated wood. While this is contrary to the report on fire endurance of monoammonium phosphate-treated wood walls (Mitchell [6]), it should be pointed out that in the evaluation of fire endurance of wall assemblies, there are other limiting criteria than penetration of fire which may govern performance.

Generally one may state that fire-retardant-impregnated woods do not increase charring rate during fire. Some chemical treatments even impede charring as compared to untreated wood.

Table 1. Major Products of 350°C Pyrolysis of Ponderosa Pine Shavings Treated with Inorganic Salts [2]

Salt	Composition Pct.	Treatment level Pct.	Products,[1,2] Char Pct.	Tar Pct.	Water Pct.	Gas Pct.
$Na_2B_4O_7$	15	24.12	43.06	14.90	31.92	10.11
	5	4.28	48.39	11.75	30.42	9.43
	1	[3]	44.13	20.39	29.33	6.15
$CaCl_2$	15	54.75	56.52	14.35	3.57	25.55
	5	23.55	46.03	22.23	20.38	11.36
	1	4.21	28.77	38.68	23.60	8.95
$(NH_4)_2HPO_4$	15	24.13	45.14	14.30	37.53	3.02
	5	6.69	45.49	16.76	32.01	5.73
	1	.51	33.85	35.21	25.07	5.87
NH_4Cl	15	26.31	29.84	35.16	27.00	8.00
	5	7.95	29.78	33.83	24.43	11.95
	1	4.23	27.33	43.17	23.63	5.87
NaCl	15	27.79	26.20	35.21	24.53	14.06
	5	10.34	25.91	35.69	25.18	13.21
	1	4.55	26.77	34.94	29.99	8.31
KBr	15	32.85	29.12	27.84	32.09	10.94
	5	17.06	28.10	29.72	31.97	10.21
	1	2.99	27.26	30.17	30.06	12.51
Na_3PO_4	15	28.78	42.89	14.92	24.49	17.70
	5	9.41	40.24	15.82	30.72	13.22
	1	[3]	36.55	21.20	31.41	10.85
K_2CO_3	15	30.49	43.46	5.59	32.07	18.87
	5	5.54	44.43	9.51	31.82	14.23
	1	1.03	41.13	13.54	30.32	14.99
Untreated wood (water extracted)	–	–	19.83	54.90	20.87	4.41

[1] Each value is the mean for 3 determinations.

[2] The proportion of products in each fraction is given on a salt-free weight basis.

[3] Less than can accurately be determined from gain in weight of specimens.

Table 2. Classification of Chemical Effect on The Volatilization of Wood

Impregnating Chemical
CLASS I
Boric acid
THPC, tetrakis-(hydroxymethyl)-phosphonium chloride
Sodium tetraborate
Ammonium pentaborate
Calcium chloride
CLASS II
Magnesium chloride
Aluminum chloride
Diammonium phosphate
Lithium chloride
Ammonium sulfate
Ammonium bromide
Monoammonium phosphate
Zinc chloride
Ammonium iodide
Ammonium chloride
Ammonium sulfamate
CLASS III
Ammonium fluoride
Ammonium formate
Sodium sulfate
Monosodium phosphate
Potassium chloride
Barium chloride
Sodium chloride
Potassium nitrate
Cesium chloride
Potassium bromide
Potassium iodide
Strontium chloride
CLASS IV
Trisodium phosphate
Potassium fluoride
Tetramethyl ammonium chloride
Sodium formate
Potassium carbonate

Table 3. Physical Data of Center Nominal 4- by 4-inch Sections of Chemically Treated Specimens

Chemical Treatment	Class	Salt-Free Dry Specific Gravity	Moisture Content[1] Pct.	Anhydrous[2] Salt Concentrations Pct.
Boric acid	I	0.525	4.2	14.5
		.515	5.2	14.7
		.526	4.5	14.8
Borax	I	.560	13.7	13.0
		.549	11.2	13.9
		.570	12.4	12.3
THPC	I	.483	5.9	15.4
		.480	4.3	14.5
		.491	4.7	14.5
Dicyandiamide-phosphoric acid	I	.532	5.3	14.6
		.518	6.6	14.7
		.506	6.3	15.2
Monoammonium phosphate	II	.505	7.7	16.3
		.530	6.8	15.6
		.536	6.7	15.4
Zinc chloride	II	.569	6.3	13.9
		.526	10.7	15.0
		.569	6.3	14.0
Ammonium sulfate	II	.543	4.8	16.3
		.560	4.7	15.5
		.548	5.4	16.1
Sodium chloride	III	.533	6.7	15.9
		.536	8.2	16.2
		.544	8.2	15.5
Monosodium phosphate	III	.538	8.0	16.5
		.541	10.0	16.4
		.551	8.6	15.7
Potassium carbonate	IV	.549	1.8	16.5
		.542	4.2	17.0
		.531	6.1	–
Sodium hydroxide	–	.525	15.3	11.9
		.517	14.5	10.1
		.520	16.8	13.3
Polyethylene glycol	–	.523	2.8	–
		.541	1.6	–
		.551	2.8	–

[1] Values are calculated from the initial sample weight, solution concentration, treatment retention, and weight after conditioning, and are subject to some error.

[2] Weight basis.

REFERENCES

1. American Society for Testing and Materials, 1971. Standard methods of fire tests of building construction and materials. ASTM E 119, pp. 508–525. Philadelphia, PA.
2. J. J. Brenden, 1965. Effect of fire retardant and other inorganic salts on pyrolysis products of ponderosa pine. Forest Prod. J. 15 (2): 69–72. Feb.
3. F. L. Browne, and W. K. Tang, 1962. Thermogravimetric and differential thermal analysis of wood and of wood treated with inorganic salts during pyrolysis. Fire Res. Abs. and Rev. NAS-NRC 4 (1) and (2). Jan. and May.
4. F. L. Browne, and W. K. Tang, 1963. Effect of various chemicals on thermogravimetric analysis of ponderosa pine. U.S. Forest Serv. Res. Pap. FPL 6. Forest Prod. Lab., Madison, WI.
5. Commonwealth of Scientific and Industrial Research Organizations, 1962. The fire resistance of timber. Part IV. Fire-retardant treated timber. CSIRO Forest Prod. Newsletter No. 283, pp. 3, 4. Melbourne, Australia. March.
6. Nolan D. Mitchell, 1947. Fire tests of treated and untreated wood walls. AWPA Proc. pp. 353–360.
7. E. L. Schaffer, 1967. Charring rate of selected woods – transverse to grain. U.S. Forest Serv. Res. Pap. FPL 69. Forest Prod. Lab., Madison, WI.
8. W. K. Tang, 1964. Effect of inorganic salts on the pyrolysis, ignition, and combustion of wood, cellulose, and lignin. Ph.D. Thesis. University of Wisconsin, Madison, WI.

Erwin L. Schaffer

Dr. Schaffer, a research engineer, and Project Leader of the Wood Product and Process Development Research Work Unit, Timber Utilization Research, at the U.S. Forest Products Laboratory, Madison, WI. Previously he studied the behavior of wood and wood structures when exposed to fire.

He obtained a Ph.D. in Engineering Mechanics (Minor in Chemical Engineering) in 1971 from the University of Wisconsin and holds BS and MS degrees in Structural Engineering from Wisconsin as well.

He is a registered Professional Engineer in the State of Wisconsin and is currently Executive Secretary of the Society of Wood Science and Technology. Other society affiliations are FPRS, ASCE, Sigma Xi, and NSPE.

SECTION THREE

Cotton

Michael J. Drews and Robert H. Barker

Textile Department, Clemson University
Clemson, South Carolina 29631

PYROLYSIS AND COMBUSTION OF CELLULOSE PART VI. THE CHEMICAL NATURE OF THE CHAR

(Received August 20, 1973)

ABSTRACT: An investigation has been carried out to characterize the chars produced from the burning of a series of cotton fabrics treated with different types of phosphorus compounds. The retention of the phosphorus during burning was determined. These retention values were shown to be closely related to the flame-retardant efficiencies of the phosphorus compounds as determined by static oxygen bomb calorimetry. In addition, these data have been shown to have implications concerning the overall mechanism of flame retardation. Although different types of organophosphorus compounds undergo different modes of interaction with the cellulose, the nature of the chars which are formed seems to be independent of the chemical structure of the phosphorus compound.

INTRODUCTION

Recent federal regulation has provided a tremendous impetus for textile flame-retardancy research. This has been especially true in regard to cellulosic materials. Basic research in this area of flame retardancy has been in progress at Clemson University for some time. The goal of this research is the elucidation of the various mechanisms by which phosphorus-containing flame retardants act.

To accomplish this goal it was first necessary to develop some quantitative means of measuring the efficiency of a given retardant. Since the heat evolved during the burning of a fabric is one measure of the flammability of the system, it was possible to devise a thermochemical method for determining these efficiencies. The method involves a series of static oxygen bomb calorimetry measurements and has been previously reported [1]. It is based on the combustion scheme shown in Figure 1 where the heat of combustion of cellulose and the heat of combustion of the char are determined calorimetrically. From these data the portion of the sample pyrolyzing to produce flammable gases (designated as Y) can be calculated from the relationship in Figure 2. Heat values are collected for samples containing varying weight fractions of flame-retardant additives (designated as X). The dependence of $Y/(1-X)$, the fraction of cellulose converted to flammable volatiles, upon the amount of flame retardant present in the fabric on a molar basis is then plotted to indicate the relative molar efficiency of the treatment [1].

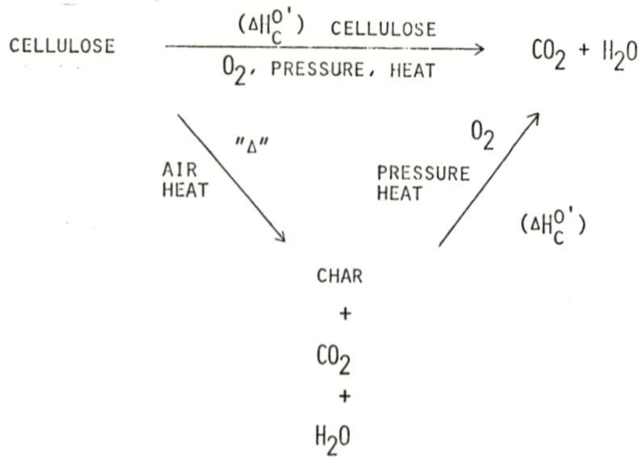

Figure 1.

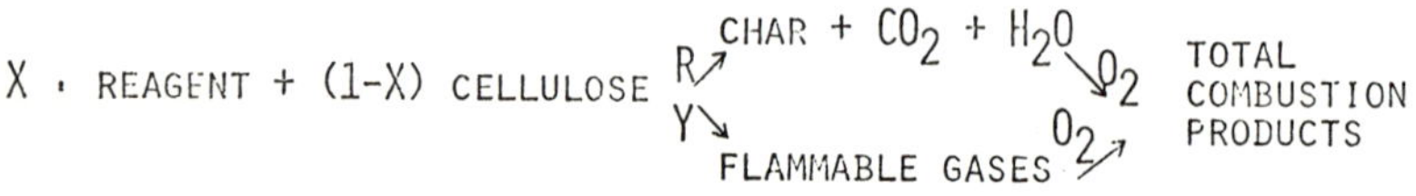

$$X \cdot \text{REAGENT} + (1-X)\ \text{CELLULOSE}\ \begin{array}{l} \xrightarrow{R} \text{CHAR} + CO_2 + H_2O \xrightarrow{O_2} \\ \xrightarrow{Y} \text{FLAMMABLE GASES} \xrightarrow{O_2} \end{array} \begin{array}{l}\text{TOTAL} \\ \text{COMBUSTION} \\ \text{PRODUCTS}\end{array}$$

$$\Delta = Y(\Delta H_c^{o'})_{\text{CELLULOSE}} + X(\Delta H_c^{o'})_{\text{REAGENT}} = (\Delta H^{o'}_c)\ \text{SAMPLE} - R(\Delta H^{o'}_c)\ \text{CHAR}$$

Figure 2.

This methodology has recently been applied to the study of compounds I, II and III [2].

I

$$(CH_3O)_2\overset{\overset{O}{\|}}{P}HC_2CH_2R$$

(a) R = $-C(O)NHCH_2OH$
(b) = $-C(O)NH_2$
(c) = $-CN$
(d) = $-C(O)OMe$

II

$$\overset{\overset{O}{\|}}{P}(NHCH_3)_3$$

III

$$(C_6H_5)_2PCH_2CH_2\overset{\overset{O}{\|}}{C}NH_2$$

For compounds of type I, the relative efficiencies on cotton fabric were found to be in the order Ia>Ib>Ic>Id. Fixation of Ia with a melamine resin resulted in only a very slight enhancement of the treatment's effectiveness. Conversely, the phosphoramide II, which has a highly electrophilic phosphorus and would be expected to react readily with cellulose, exhibited a surprisingly low efficiency when used alone. However, it showed marked enhancement when fixed with a melamine resin. The phosphine III, in which the phosphorus should be quite inert toward cellulose since it bears no displaceable leaving group, proved to be more predictable. It was found to be a very inefficient flame retardant for cotton.

When these efficiencies were plotted in the form Y/1—X (the fraction of cellulose producing flammable gases) vs log (%P) (the active species in the flame retardants), a set of straight lines resulted as shown in Figures 3 and 4. The slopes of these lines were shown to have considerable significance in studying the mode of interaction between the retardant and the substrate [2]. The steepest slope was interpreted as indicating the presence of a phosphoramide moiety in the active flame-retardant species. As expected, compound II alone or in combination with the melamine resin gave a line of this slope. In addition, compounds Ia and Ib also gave lines of similar slope indicating the formation of P—N bonds prior to reaction with the cellulose. Compounds Ic and Id, having no nucleophilic nitrogen in the molecule, acted like conventional phosphonates. Similarly, compound III which has a nucleophilic nitrogen but has no leaving group attached to the phosphorus, acted like a conventional phosphine. Its efficiency plot showed a line with a very gentle slope characteristic of almost complete lack of interaction with the cellulose.

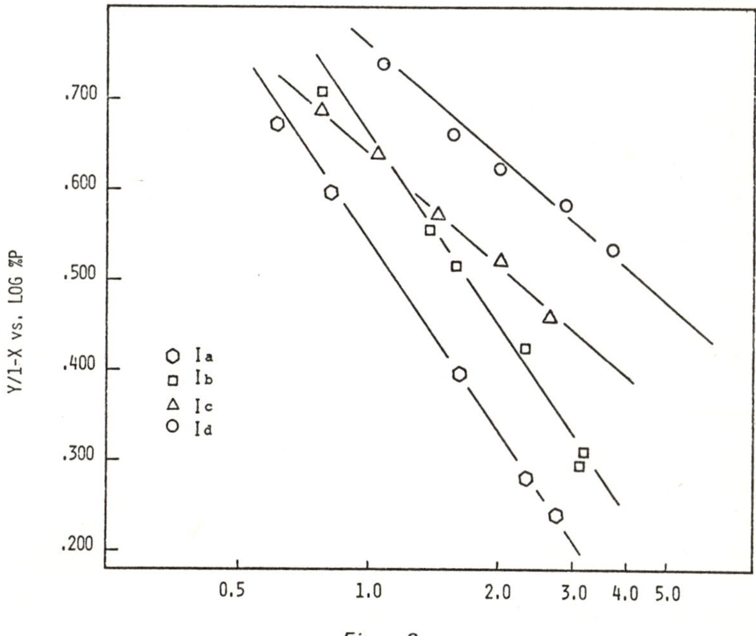

Figure 3.

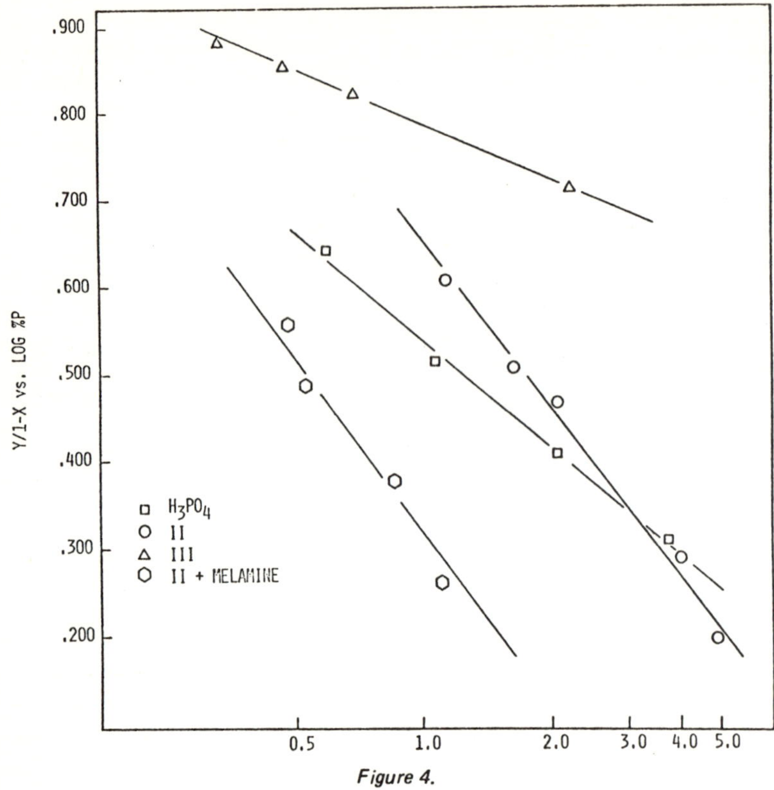

Figure 4.

EXPERIMENTAL

The fabric used was 80 X 80 bleached and mercerized 100% cotton print cloth from Test Fabrics, Inc., Middlesex, New Jersey. All of the compounds were applied to the fabric by either a pad-dry or a pad-dry-cure procedure as in the case of the two melamine treatments.

Compound Ia was obtained in a specially prepared pure form from the Ciba-Geigy Research Laboratories, Basel, Switzerland. Compounds Ib, Ic and Id were prepared by addition of dimethylphosphite to the appropriate acrylic acid derivative according to the method of Zahir [3]. Compound III was prepared by the addition of diphenylphosphine (obtained from the $LiAlH_4$ reduction of diphenyl chlorophosphine according to the method of Horner and Hofmann [4]) to acrylamide in ethanol using sodium ethoxide as a catalyst. All other chemicals were normal commercial laboratory grade reagents.

The samples of treated fabric were cut into 2 X 8 in. strips, weighed and mounted vertically. A paper match was used to ignite each sample at the bottom edge. Those samples which were self-extinguishing under these conditions were reignited until the entire length of the sample had been charred. The chars were then collected and weighed. All heats of combustion were determined using a series

1200 Parr automatic adiabatic oxygen bomb calorimeter and following the ASTM procedure (ASTM #D 2382-65). The char analyses were performed by Galbraith Laboratories, Knoxville, Tennessee.

RESULTS AND DISCUSSION

In previous work with phosphorus-containing flame retardants [1], it was assumed that essentially all of the phosphorus applied to the fabric remained in the char after combustion. However, in the present work, the char was studied more closely in an attempt to find a correspondence between the structure of the char and the flame retardant's efficiency. The first step was to analyze the chars for total phosphorus and nitrogen content. The result of these analyses along with those for samples treated with H_3PO_4 as a reference are shown in Table 1.

There are several observations about the data in Table 1 which seem to be indicative of the mode of char formation. Compounds such as Ia, Ib and II, whose efficiency is thought to depend upon reaction in a phosphoramide form, yield chars with nitrogen contents which are a function of the amount of phosphorus retained. On the other hand, fabrics treated with compounds Ic, Id and III yield chars which have nitrogen contents which are independent of the amount of phosphorus present and are actually almost constant. Finally, by comparing the data in Table 1 with the efficiency plots shown in Figures 1 and 2, it can be seen that the phosphorus retention parallels the efficiency of the retardant. The least efficient retardant tested, compound III, had an average retention of 27%, while the best system, compound II fixed with melamine, was found to yield an average value of almost 100%. The mechanistic implications of these data have not yet been interpreted completely; but it seems to point to a process involving the direct interaction of the retardant with the cellulose rather than with levoglucosan or some other cellulose decomposition product. If the reaction involved these decomposition products, the phosphorus retention would not be expected to be independent of the initial concentration of flame retardant as was observed.

The only exception to this behavior was found with compound II when it was not fixed to the cotton. In this case the phosphorus retention increased from 32% to 84% as the add-on of II increased from 3% to 22%. However, this apparent anomaly can perhaps be explained as the result of the formation of a polyphosphoramide such as IV at the higher add-on levels. The formation of this polyphosphoramide results in an *in situ* fixing of the flame retardant.

$$\mathrm{II} \xrightarrow{\Delta} \underset{\mathrm{IV}}{\left(-\underset{\substack{|\\ }}{\overset{CH_3}{\overset{|}{N}}} - \overset{O}{\overset{\|}{\underset{\substack{|\\ N\\ H\diagup\ \ \diagdown CH_3}}{P}}} - \right)_n} \xrightarrow{\text{Cell-OH}} \underset{\mathrm{V}}{\left(-\overset{CH_3}{\overset{|}{N}} - \overset{O}{\overset{\|}{\underset{\substack{|\\ O\\ \diagdown \text{Cell}}}{P}}} - \right)_n}$$

Table 1.

COMPOUND	%P[1]	$\%P_A$[2]	$\%P_C$[3]	$\%N_A$[4]	$\%P_A/\%P_C$	ΔH^o_{CHAR}[5], CAL/GM
Ia	2.73	5.35	5.67	1.72	0.94	5074
	2.30	4.28	5.05	1.68	0.94	5199
	1.67	3.80	4.20	1.64	0.90	5433
	0.82	2.81	3.16	0.95	0.89	5843
Ia + Melamine	2.20	4.25	4.75	2.13	0.90	5349
	1.66	3.76	4.25	1.91	0.89	5568
	1.30	3.23	3.76	2.01	0.86	5752
	0.75	2.63	3.02	1.39	0.87	5904
Ib	2.96	4.43	6.93	1.56	0.73	5290
	2.84	4.31	5.92	1.57	0.64	5330
	1.98	3.85	5.84	1.39	0.66	5625
	1.40	3.20	4.75	1.07	0.67	5720
Ic	1.98	4.00	6.45	1.25	0.62	5601
	1.40	3.09	5.07	1.29	0.61	5744
	1.01	2.55	4.40	1.30	0.58	5830
	0.80	2.29	4.15	1.17	0.55	6001
Id	3.66	4.40	12.6	0.64	0.35	5263
	2.90	4.08	11.0	0.17	0.37	5484
	1.58	3.04	7.18	0.31	0.42	5657
	1.33	3.02	6.34	0.64	0.48	5698
II	4.04	7.25	9.10	3.12	0.80	5202
	2.10	4.59	6.29	2.71	0.73	5759
	1.63	3.26	5.25	2.40	0.62	5902
	1.21	2.69	4.84	2.31	0.56	6060
	0.70	1.66	5.26	1.44	0.32	6627
II + Melamine	1.76	3.65	3.77	11.09	0.97	5669
	0.87	2.26	2.40	8.66	0.94	6008
	0.53	1.98	1.75	7.76	1.13	6147
	0.49	1.79	1.90	7.25	0.95	6383
III	2.79	4.68	20.2	1.35	0.23	6100
	2.01	4.08	15.6	1.16	0.26	6097
	1.35	3.55	11.2	1.32	0.32	6099
	1.04	2.91	10.2	0.99	0.28	6286

[1]%P applied to sample in the flame retardant. [2]$\%P_a$ in the char is determined by analysis. [3]$\%P_c$ in the char calculated by assuming none of the phosphorus was volatilized on combustion. [4]%N in the char as determined by analysis. [5]Heat of combustion of the char as determined by static oxygen bomb calorimetry.

Polymer IV would then be capable of reaction with cellulose to form V which, in turn, would pyrolyze to form char. At low add-ons of the retardant this process is insignificant and most of the volatile compound II is vaporized off before it can effectively interact with the cellulosic substrate, thus yielding low phosphorus retention. As the add-on of the retardant increases, the formation of the poly-phosphoramide becomes the dominating process and this *in situ* fixing process increases the retention to 80%. This would also explain the observation that the P/N ratio in the char is nearly 1/2 at low add-ons of compound II but becomes approximately 1/1 as the flame-retardant concentration increases.

Evidence supporting this postulate is obtained from thermogravimetric analyses of mixtures of compound II with aluminum oxide as inert filler. These studies show that the residue remaining at 350°C increases from 37 to 44% as the concentration of II is increased from 10 to 20 weight percent in the mixtures. These data then anticipate the large increase in efficiency which was observed when II was fixed to the cellulose substrate with a resin treatment.

A similar effect was found when the heats of combustion of the chars were examined. In Figure 5, these heats as determined by static oxygen bomb calorimetry, $(\Delta H^{O'}_c)_{char,}$ are shown plotted against the log % phosphorus of the chars. All of the compounds of type I react with cellulose to yield chars having the same dependency of heat of combustion on phosphorus content. Compound II, when fixed with a melamine resin, also produces chars which follow this same relationship. This would seem to indicate that all of these compounds attack the cellulose in essentially the same way and thus leave similar chars. This, presumably, involves a series of phosphorylation and pyrolysis steps [3]. Without the melamine resin present, compound II can react in the same way but also has the options of volatilizing or polymerizing. The former has no effect on the nature of the char but the latter forms a polymeric phosphoramide which remains with the char and therefore influences its heat of combustion. Thus, its plot of ΔH_c vs log (%P) is displaced from that of the phosphonates.

Compound III also produces chars which are noticeably different from those produced by the phosphonates. However, in this case, it seems that the difference is simply the result of the fact that there is almost no interaction between the cellulose and the phosphine. This is in agreement with its ineffectiveness as a flame retardant.

These results also seem to have some mechanistic significance. If the important step in the mechanism of action of a phosphorus-containing retardant is the phosphorylation step, then non-reactive phosphorus may be volatilized and lost. This is probably the basis of the correlation we have observed between efficiency and phosphorus retention. It is important to emphasize at this point that these are comparisons of molar efficiencies and not absolute flammability. It is possible to make almost any cotton fabric flame resistant with high enough add-ons of almost any phosphorus compound. With this in mind, it can be pointed out that for

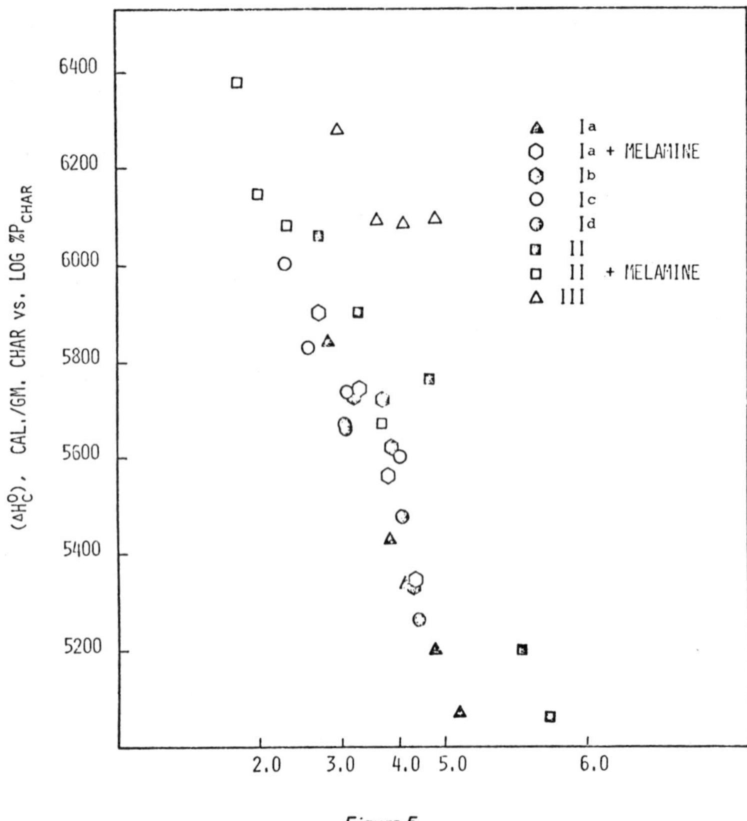

Figure 5.

compounds showing approximately equal phosphorus retention, the more efficient retardant will be the one containing nitrogen, especially at higher add-on levels.

The relationship between the $(\Delta H^{O'}{}_{c})_{char}$ and log %P shown in Figure 5 strongly suggests that all the chars have the same basic chemical structure. These data would be consistent with a mechanism whereby the cellulose pyrolyzes by two separate pathways. This also offers further substantiation for the postulate that these flame retardants interact directly with the cellulose rather than exerting their predominant effect on the levoglucosan or other pyrolysis products. Schematically, this can be shown as follows:

$$\text{Cell—OH} \xrightarrow{\text{heat}} \text{levoglucosan} \xrightarrow{\text{heat}} \text{flammable gases}$$

or

$$\text{Cell—OH} + \text{FR} \xrightarrow[\text{heat}]{\text{phosphorylation}} \text{Cell—O—P(=O)(R)(R')} \xrightarrow{\text{heat}} H_2O + \text{dehydrated char}$$

From the total char analysis it has been possible to calculate an empirical formula of $(C_6H_2O)_x \cdot (PO_{y-n}N_n)_y$ for the char resulting from the phosphoramide treated samples. This formula indicates extensive dehydration of the cellulose substrate.

Chars formed from fabrics where the heat liberated by the burning is very low and the % char very high show less complete dehydration and an empirical formula of $(C_6H_{4.5}O_2)_x \cdot (PO_{y-n}N_n)_y$. This can be interpreted to mean that in these cases flame resistance results principally from elimination of the fuel.

Activation energies from TGA, as determined by the ratio method [5], show that the E_a's decrease from – 45 Kcal/mole for samples treated with III to 8 – 10 Kcal/mole for fabrics containing resin fixed II. This is also consistent with the hypothesis that these phosphorus-containing flame retardants interact directly with the cellulose and not its decomposition products.

An important implication of these results is that there seems to be an inherent limit to the efficiency of a simple organophosphorus flame-retardant system which has only condensed phase activity in cellulosic substrates. Investigations are currently in progress which should more completely elucidate the chemical nature of the phosphorylation and dehydration reactions and establish more clearly the efficiency limits.

ACKNOWLEDGMENT

The authors wish to thank Cotton Incorporated for partial support of this work and Ciba-Geigy, Inc. for the preparation of special samples of Pyrovatex CP.

REFERENCES

1. K. Yeh and R. H. Barker, *Text. Res. J.* 41, 932 (1971).
2. M. J. Drews, Kwan-nan Yeh, and Robert H. Barker, *Textilveredlung, 8*, 180 (1973).
3. A. Zahir, U.S. Patent 3,374,292 (1968).
4. L. Horner and H. Hofmann in W. F. Forest, Ed., "Newer Methods of Preparative Organic Chemistry," p. 207 (1963).
5. R. W. Michelson and I. N. Einhorn, *Thermochim. Acta, 1,* 147 (1970).

A. BASCH AND M. LEWIN

Israel Fiber Institute
(formerly – the Institute for Fibers
and Forest Products Research)
Jerusalem, Israel

THE INFRARED DETERMINATION OF LEVOGLUCOSAN FORMED DURING THE PYROLYSIS OF CELLULOSE

(Received December 8, 1972)

ABSTRACT: Levoglucosan is a primary product of cellulose pyrolysis and is largely responsible for its flammability. An infrared method has been developed to measure the amount of levoglucosan in the tar generated by the combustion of cellulose. The tar is condensed on an ATR crystal and the percent levoglucosan is calculated from the spectrum by a ratio method.

Results show that the amount of levoglucosan produced is dependent on the type of cellulose as well as the purity of the sample.

LEVOGLUCOSAN (1,6-anhydro-g-D-glucopyranose) has long been recognized as a primary product of the pyrolysis of cellulose and has been considered the major fuel which contributes to the flammability of cellulose [1]. Thus the main function of flame-retardant treatments is thought to be either prevention of levoglucosan formation or decomposition of the levoglucosan initially formed to char and water [2]. Levoglucosan has been determined in tars trapped from pyrolysis or obtained by vacuum distillation of cellulose. The analytical methods employed include colorimetric determination as a non-reducing sugar [3], hydrolysis and estimation of the resulting glucose [4], and gas-chromatographic determination [5]. Holmes and Shaw [1] developed an infrared method for levoglucosan. They dissolved pyrolysis tars in dimethyl formamide and determined levoglucosan by its absorption band at 930 cm^{-1}. In a later paper [6] the method was modified to allow for background absorption due to other tar components.

We have developed an infrared method which measures the percentage of levoglucosan present in the tar generated when cellulose is burnt. This method utilizes a reflectance crystal for collection of the tar and its presentation to the spectrophotometer, resulting in experimental simplicity and high sensitivity. The following advantages are realized:

1. The levoglucosan is determined under conditions simulating burning to a considerable degree, and not under distillation conditions.
2. The amount of cellulose required in only 5–20 mg. per determination.

3. The analysis is rapid, with one complete determination, including pyrolysis and measurement, requiring only 20 minutes.

EXPERIMENTAL METHOD

Spectra were run on a Perkin-Elmer Model 257 grating spectrophotometer. Samples coated or deposited on a 45° KRS-5 crystal were mounted in a Wilks Model 12 double beam internal reflection attachment. A clean KRS-5 plate was placed in the reference beam. The spectrometer slit control was set to number 7, which increases the available energy by a factor of two over the usual setting for transmission spectra. This is done to offset the energy losses inherent in the ATR system. The resolution is not significantly reduced by this procedure.

A major difficulty arises from the background due to the presence of materials other than levoglucosan in the pyrolysis tars. An attempt was made to overcome this difficulty by using an actual pyrolysis tar as a "solvent" for the levoglucosan. Therefore, the method was calibrated with a solution of levoglucosan in tar obtained by pyrolyzing cellulose which might be expected to provide calibration samples close in overall composition to those likely to be encountered in actual combustion tars. The tar was obtained by heating two grams of Whatman No. 1 filter paper over a low flame in an open Erlenmeyer flask until charred. The residue was extracted with hot methanol, filtered through anhydrous sodium sulfate and the methanol evaporated under vacuum. A portion of the tar was silylated, and the extracted silyl derivatives [7] analyzed by gas chromatography [5]. The tar was found to contain 14.0 ± 0.3% levoglucosan. Aliquots of this tar in methanol were enriched by the addition of known amounts of pure crystalline levoglucosan (M.P. 185–188°C), up to a levoglucosan content of 30.4%. Spectra of the original and enriched tars were obtained.

Pure levoglucosan has a number of sharp bands which would appear to be suitable for quantitative determination, as shown in Figure 1. The most prominent band, at 1045 cm^{-1}, most likely due to a C–OH stretch, is present in the spectra of many carbohydrates (e.g. glucose, ribose) and is thus not specific enough to be used. The next most prominent band is at 1033 cm^{-1}, probably a C–O–C stretch. In addition, a series of bands of lesser intensities occur between 1010 cm^{-1} and 960 cm^{-1}. Inspection of the spectrum of the initial tar (which contained 14% levoglucosan) revealed that the absorption at 1133 cm^{-1} was reduced in intensity yet still clearly defined even in a weak spectrum. This band was therefore selected to represent levoglucosan in our determination.

Since the quantity and distribution of sample on the ATR plate is variable and cannot be controlled, the usual method of measuring a suitable absorption band cannot be used, and the measurement must be made relative to an internal standard. The C–H stretch at 2900 cm^{-1} representing the total organic matter in the tar was therefore chosen as our internal standard and we measure the 1133 cm^{-1} levoglucosan absorption relative to the 2900 cm^{-1} band. The ratio of these two

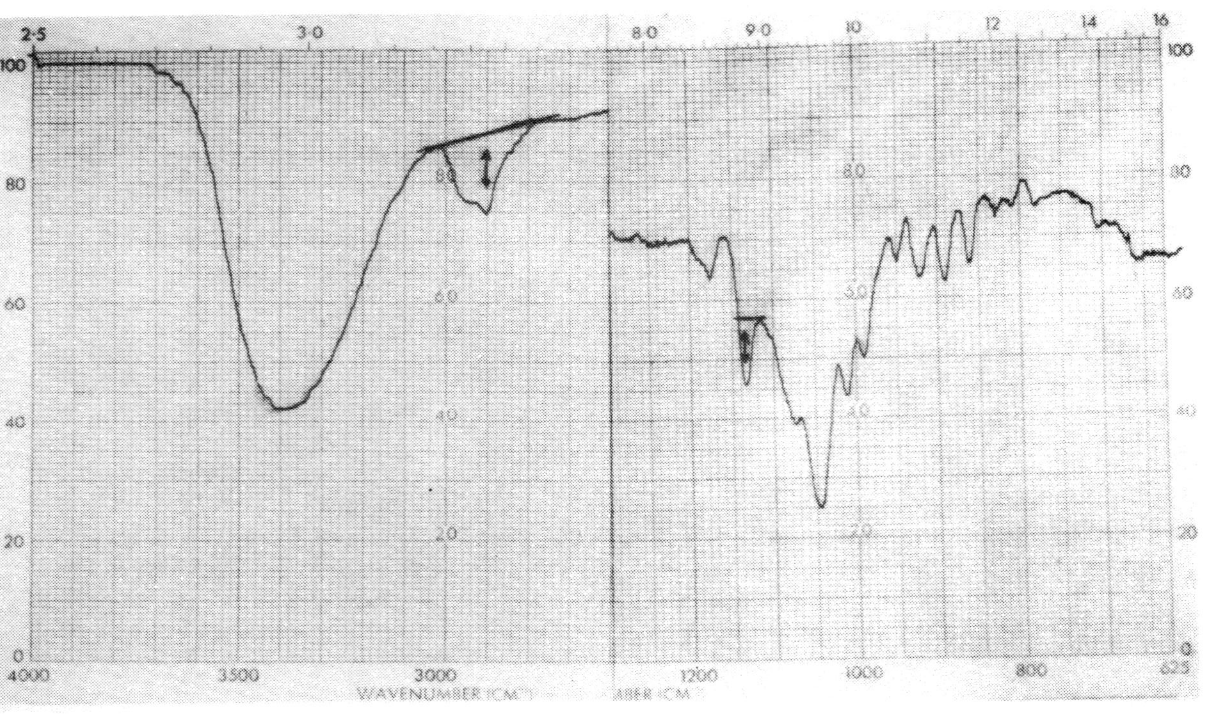

Figure 1. *Measurement of peaks for levoglucosan determination.*

absorptions thus represents a ratio of levoglucosan content to total organic matter in the tar.

The percent transmission of the band at 2900 cm^{-1} was measured using as a baseline a line drawn from the upper edges of the band at about 3000 cm^{-1} and 2760 cm^{-1}. The percent transmission of the band at 1133 cm^{-1} was measured, with the baseline taken as the percent transmission at 1115 cm^{-1}. This method of measuring the baseline for the 1133 cm^{-1} band was chosen because inspection of pyrolysis spectra indicated broad, variable background absorption in this region, upon which the sharp band at 1133 cm^{-1} was superimposed. Measuring the baseline from the band edge at 1115 cm^{-1} largely eliminated this background and yielded more consistent results. No correction was made for the contribution of levoglucosan to the band at 2900 cm^{-1} since our experience has been that such correction does not improve the precision of the results [9]. These measurements are illustrated in Figure 1. The ratio of the relative strengths of these two bands was calculated.

Cellulose samples in the form of fabric or yarn were pyrolyzed in a Wilks Model 40 Pyrolyzer. The sodium chloride windows were removed to allow free air access and to protect them from the moisture evolved during the pyrolysis. Approximately 5–20 mg. of sample were placed directly on the nichrome firing element. An ATR crystal was positioned in its normal place in the pyrolyzer above the sample and the sample pyrolyzed at a nominal temperature of 900–1000°C. A spectrum was obtained of the condensate which deposited on the crystal. If the spectrum was too weak, a second sample was pyrolyzed and the vapors collected on top of the first. The ratio of the band strengths at 1133 cm^{-1} and 2900 cm^{-1} was calculated as described above.

RESULTS AND DISCUSSION

The calibration data are presented in Table 1 and the corresponding calibration curve illustrated in Figure 2. The maximum deviation of the points from the curve is 0.5% which is similar to the uncertainty in the levoglucosan content of the samples (accuracy of the gas-chromatographic analysis).

Table 1.

Percent Levoglucosan in Tar	Ratio 1133/2900 cm^{-1}
14.0	0.24
17.3	0.42
20.3	0.55
25.7	0.75
30.4	0.86

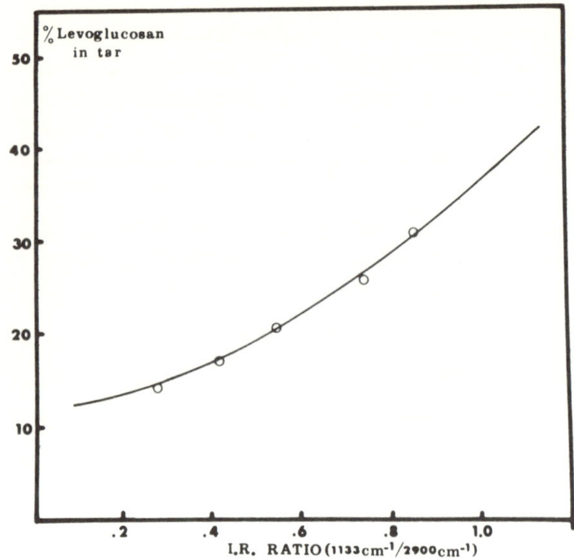

Figure 2. *Calibration curve for levoglucosan determination.*

Application of this method to the pyrolysis of cellulose yielded data with an average precision of ± 3% levoglucosan, with occasionally larger deviations. Typical results obtained for several samples are listed in Table 2.

The reproducibility of the determinations may be inherently limited by the method of tar collection. The collection is not only non-quantitative, but allows for the possibility of fractionation of tar component during migration from the pyrolyzing cotton to the collecting ATR plate. It is therefore necessary to do all experiments in duplicate at least.

The condition of the ATR plate used for the determinations must be rigorously controlled. The plate must be free of scratches, which will cause light scattering in the low wavelength region of the spectrum and affect the band at 2900 cm^{-1} more than the band at 1133 cm^{-1}, thus distorting their ratios. The plate must be completely free of all organic contamination, due to the high sensitivity of the ATR system. The slightest trace of a fingerprint will give rise to a noticable absorption in the C—H stretch region. After insuring the cleanliness of the plate by running a blank spectrum, it should be handled with polyethylene gloves throughout the determination.

The method is sufficiently sensitive to detect differences in levoglucosan contents of pyrolysis tars due to differences in purity or treatment of a cellulosic material. As is seen in Table 2, the levoglucosan content of the tar produced by burning ramie more than doubles in going from native (degummed) ramie which has not been thoroughly cleaned to hydrolyzed ramie.

In our investigations incompletely cleaned celluloses have consistently shown

Table 2. Reproducibility of Levoglucosan Determination

Sample	Ratio 1133/2900 cm^{-1}	% Levoglucosan in Tar	Average % in Tar
Cotton, Purified	0.68	24.0	
	0.56	20.7	
	0.69	24.3	
	0.72	25.3	23.6 ± 2.9
Ramie, delignified	0.45	18.2	
	0.36	16.3	17.2 ± 1.0
Ramie, scoured	0.79	27.4	
	0.64	22.9	25.2 ± 2.2
Ramie, hydrolyzed	1.08	38.9*	
	1.19	44.3*	37.3
	0.83	28.8	
Textile rayon	0.19	13.2*	
	0.29	14.7	14.0 ± 0.8
Textile rayon,	0.57	20.9	
purified	0.40	17.0	19.0 ± 2.0

*Extrapolation

lower levoglucosan contents as compared to the same materials after purification. Similar results were reported by Holmes and Shaw [1] and are mirrored in the DTA scans of raw and purified cotton [8]. This technique affords a new tool with which examination of the effects of impurities and additives upon the pyrolysis of cellulose can be carried out. Furthermore, it may serve for further elucidation of the role of levoglucosan in pyrolysis and combustion studies.

REFERENCES

1. F. A. Holmes and C. J. G. Shaw, J. Appl. Chem. *11*, 210 (1961).
2. W. L. Parker and A. E. Lipska, "A Proposed Model for the Decomposition of Cellulose and the Effect of Flame Retardants." Report submitted to U.S. Naval Radiological Defense Lab. NRDL – TR – 69.
3. M. Dubois, K. Giles, J. K. Hamilton, P. A. Rebers and F. Smith, Nature, *168*, 167 (1951).
4. R. F. Schwenker and E. Pacsu, Industrial and Eng. Chem., Chem. and Eng. Data Ser. *2*, 83 (1957).
5. Y. Halpern, Y. Houminer and S. Patai, Analyst, *92*, 714 (1967).
6. G. A. Byrne, D. Gardiner and F. A. Holmes, J. Appl. Chem. *16*, 81 (1966).
7. R. D. Partridge and A. H. Weiss, J. Chrom. Sci., *8*, 553 (1970).
8. R. L. Hebert, M. Tryon and W. K. Wilson, Tappi, *52*, 1183 (1969).
9. A. Basch and E. Tepper, Appl. Spectros., in press.

Avraham Basch

Avraham Basch is a senior investigator in charge of the analytical section of the Israel Fiber Institute, Jerusalem. He received his B.A. from Yeshiva College in 1953 and his M.A. in physical chemistry from Columbia University in 1955. His research interests include analytical polymer characterization, analytical chemistry and the pyrolysis and combustion of cellulosics.

Menachem Lewin

Professor Menachem Lewin is Director of the Israel Fiber Institute and Head of the Department of Polymer and Textile Chemistry at the School of Applied Science and Technology of the Hebrew University, Jerusalem. He received his M.Sc. in Chemistry in 1945 and Ph.D. in Physical Chemistry in 1947 at the Hebrew University, Jerusalem.

He received in 1959 the Habif prize of the University of Geneva for research in cellulose chemistry. His research interests include: polymer and fiber structure, oxidation, pyrolysis and flame-retardance of polymers.

Worked in 1969 and 1970 as visiting scientist at the R & D division of J. P. Stevens in Garfield, New Jersey.

NARASIMHAN RANGAPRASAD, C. M. SLIEPCEVICH,
AND J. REED WELKER

The University of Oklahoma
Flame Dynamics Laboratory
Norman, Oklahoma 73069

THE PILOTED IGNITION OF COTTON FABRICS

(Received September 8, 1973)

ABSTRACT: Piloted ignition times were measured for several cotton fabrics. The results are generalized into a single curve based on a simple lumped-parameter heat transfer equation and compared with the ignition times of other thin cellulosic materials from the literature.

INTRODUCTION

THE LITERATURE CONTAINS considerable data on the piloted ignition of wood [7, 9, 11, 13, for example] and some data on piloted ignition of polymers [4, 11], but little data on the piloted ignition of fabrics [12]. This work was undertaken to measure and correlate the piloted ignition times of cotton fabrics subjected to thermal radiation. Piloted ignition times are generally considered to be the shortest time in which ignition can occur. They therefore represent the minimum time required for a fabric to begin burning when exposed to a heat source. Piloted ignition occurs faster than spontaneous (non-piloted) ignition because the pyrolysis products do not have to be heated to the ignition point by the energy source. Piloted ignition is more realistic than spontaneous ignition in ordinary fires involving clothing because the clothing must be quite near a flame or other heat source in order to ignite unless the source is quite large.

Two general types of fabrics were used, one obtained especially for these studies and one obtained locally. The special materials were furnished by the Southern Regional Research Center, U.S. Department of Agriculture. The fabrics were unbleached, ecru in color, and woven using the same basic fiber and construction, the only major difference being the fabric weights, which were 0.019, 0.020, 0.022, 0.024, 0.026, 0.029, and 0.035 g/cm^2. Two fabrics were purchased locally as examples of commercially available fabrics. One of these commercial fabrics was bleached white and the other was black; each weighed 0.013 g/cm^2.

PROCEDURE

The piloted ignition times were measured using 10 cm square samples in the ignition cabinet described by Koohyar, *et al*, [5] and used previously to study the ignition of wood [12] and polymers [4]. The cabinet is designed to provide a known level of radiant heat flux incident on the sample surface. The radiant energy source may be either a hydrocarbon flame or high temperature tungsten filament lamps. Flame radiation is used because it simulates the radiant heat from an actual fire much better than any artificial source, including the gas-fired radiant panel. Tungsten lamp radiation is used to provide a comparison with other ignition data from the literature.

The sample is positioned vertically in a sample holder that keeps the exposed surface smooth and facing the radiant energy source. A water-cooled shield is placed between the sample and the energy source until the source has reached steady state. The radiant flux is monitored continuously during this warmup period. When the source has reached steady state the shield is rapidly withdrawn and the sample is exposed to the radiant flux. The pyrolysis products rise from the sample past a three-dimensional gas flame near the top of the sample. This gas pilot flame is designed so that the ignition time is independent of its location. Ignition is recorded by a photosensitive cell that views across the front of the sample.

The incident radiant flux (or irradiance) is varied by changing the distance between the sample and the source or by injecting oxygen at the base of the flame. In the case of the tungsten lamps, the lamp filament temperature is maintained constant because of the effect of radiant energy distribution on ignition time [4, 12]. Incident irradiances up to 3 cal/cm^2-sec can be provided for either flame radiation or tungsten lamp radiation. Complete details of the procedure are given by Rangaprasad [8].

RESULTS AND DISCUSSION

Figure 1 shows the typical result of a series of ignition experiments for the lightest and heaviest of the ecru fabrics. The ignition time is dependent on three parameters, the incident irradiance, the fabric weight and the wavelength of the incident radiation. The dependency on all three can be predicted from a simplified heat balance.

Consider the fabrics to represent a portion of an infinite slab and assume the slab is thin enough that no thermal gradient exists within its boundaries. If the slab is inert and opaque, and if it is assumed that no heat is lost from the slab by convection or reradiation, the heat balance gives

$$c\, w_o\, \Delta T_i = \bar{\alpha} H_i\, t_i \tag{1}$$

c	=	specific heat
w_o	=	weight of material per unit area
ΔT_i	=	temperature rise at ignition
$\bar{\alpha}$	=	average absorptance
H_i	=	incident irradiance
t_i	=	ignition time

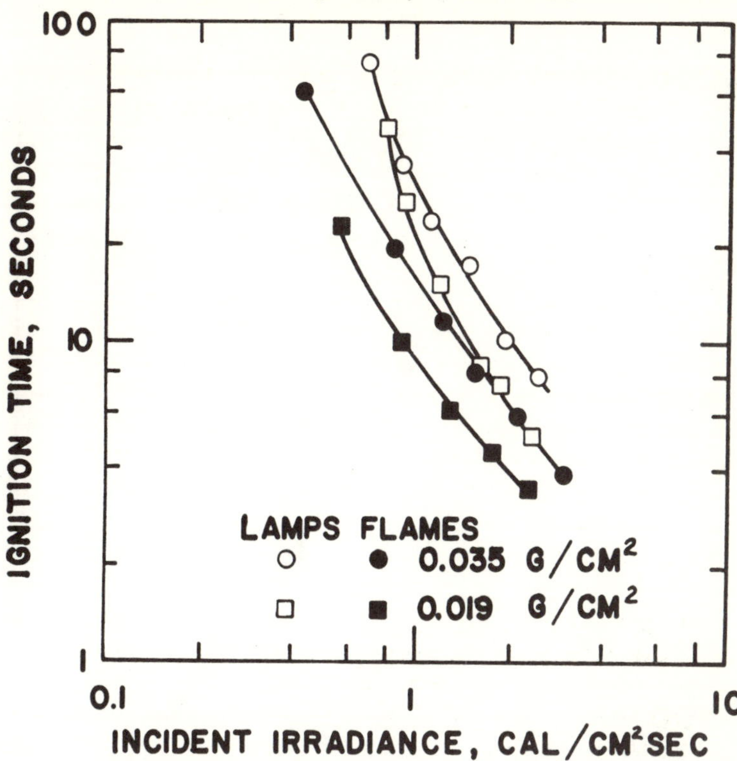

Figure 1. Piloted ignition time for cotton fabrics.

Equation 1 obviously does not account for all the factors that influence ignition because it is so simplified, but it does indicate how the four major factors, incident irradiance, fabric weight, absorptance, and ignition time are related. Figure 1 includes the tacit assumption that the primary variables are the ignition time and the incident irradiance. That assumption is valid because those two factors vary over the widest range during an ignition experiment. However, the fabric weight and average absorptance also affect the ignition time, the fabric weight directly and the average absorptance inversely, as indicated by Equation 1.

Consider first the fabric weight. If the fabric is thin enough to preclude thermal gradients, the heat required to raise the fabric to the ignition temperature is directly related to the fabric weight (or the product of density and thickness). Figure 2 shows the result when the ignition times are corrected for changes in fabric weight. The ratio t_i/w_o is shown to be dependent only on the incident irradiance and the source of radiation. Only the data for fabrics having unit weights of 0.019 and 0.035 g/cm^2 are shown in Figure 2.

The source of radiation causes differences in ignition times because of differences in the spectral absorptance of the fabric. These differences occur not only for fabrics [12], but for wood [13] and synthetic polymers [4] as well. In fact, it is a good generalization that spectral absorptances are important in ignition of

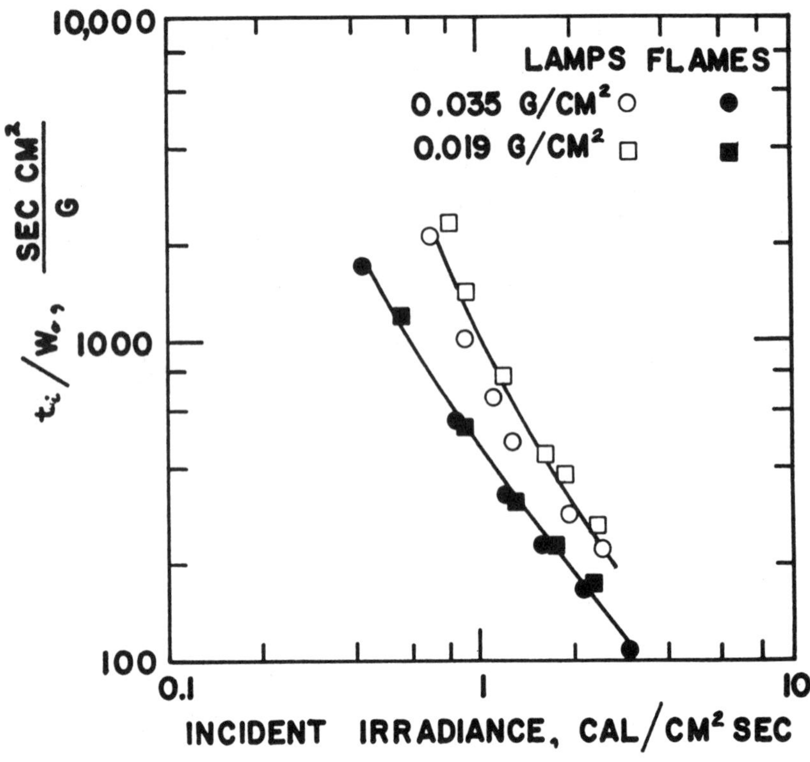

Figure 2. Piloted ignition time corrected for fabric weight.

nearly all light-colored combustible materials. The average absorptance varies with exposure time because of charring of the exposed surface, so for convenience, the average absorptance of the unheated material is used to show the effect. The average absorptance is calculated as described in an earlier paper [12], and because spectral data are not available for the fabrics used in this work, the average absorptances from that paper are used. Their values are given in Table 1 for reference.

It is interesting to note that the average absorptance for flame radiation is about the same regardless of the fabric color and that the average absorptance for tungsten lamp radiation is slightly higher than that for flame radiation for a black fabric while the average absorptance for natural fabrics is significantly lower for lamp radiation than it is for flame radiation. Similar behavior was found for natural and black painted wood and for light and dark synthetic polymers [4, 13].

Figure 3 shows a plot of the fabric ignition data in which the ignition times are considered a function of the initially absorbed irradiance, $\bar{\alpha}H_i$. Data for all fabric weights between 0.013 and 0.035 g/cm^2 are shown. In order to utilize the grouping of parameters suggested by Equation 1, the ignition time is divided by $w_o\ c\ \Delta T$.

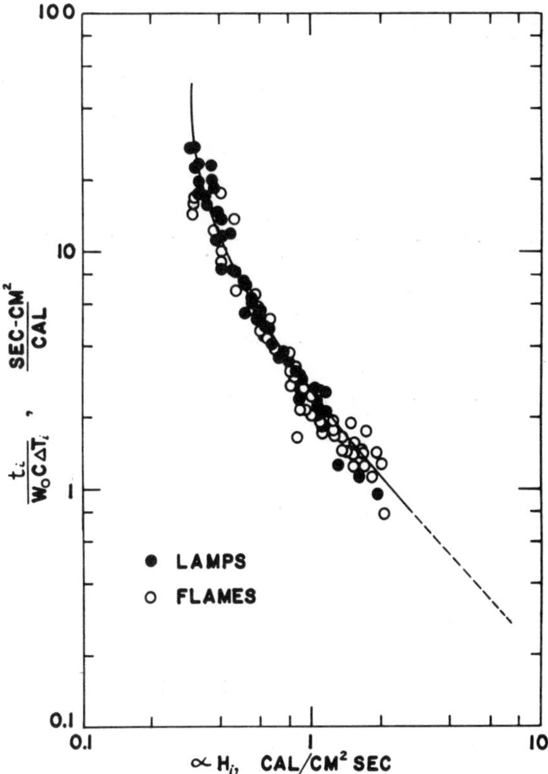

Figure 3. Generalized correlation of piloted ignition data for cotton fabrics.

The temperature rise, ΔT, is taken as the temperature rise at which the maximum weight loss rate is recorded when the material is subjected to thermogravimetric analysis [8]. The result is a fairly tight grouping of the ignition data around a single line, with the degree of scatter of the data approximately within the accuracy that ignition data can be measured. At low irradiances, where the ignition times are longest, the data scatter over a larger range and tend to curve upward. In practice, an asymptote is reached for average absorbed irradiances below about 0.3 cal/cm^2-sec; although the material will char and perhaps glow at lower heating rates, it does not give a flaming ignition. The slope of the curve at higher irradiances is approximately -1, a result that would be expected according to Equation 1, which predicts the inverse relationship between ignition time and heating rate. At higher heating rates the heat losses, which are approximately constant regardless of heating rate, are of lesser importance. At low irradiances, heat losses are important and in fact limit the temperature rise of the samples.

The literature contains several examples where the "critical irradiance" is determined by plotting irradiance versus the reciprocal of ignition time and extrapolating the curve to "infinite" time. That same procedure can be followed with the data on Figure 3, giving the result shown in Figure 4, where $\bar{\alpha}H_i$ is plotted versus $w_o\, c\, \Delta T/t_i$. The line in Figure 4 represents the same data as the line in Figure 3; the individual data points are omitted for clarity. The resulting line in Figure 4 is straight and intercepts $\bar{\alpha}H_i$ at the critical *absorbed* irradiance of about 0.3 cal/cm^2-sec. The total steady state energy loss of 0.3 cal/cm^2-sec required to balance that critical absorbed irradiance would require a fabric temperature of about 260°C if (1) both sides of the fabric lose heat by convention and radiation, (2) the convection coefficient is 2.5 (10^{-4}) cal/cm^2-sec-°C as suggested by Alvares, *et al*, [3], and (3) the average emittance of the fabric is assumed to be 0.8, a value consistent with measurements [12]. The calculated fabric temperature attained at steady state for critical irradiance thus approaches the temperature at which degradation begins to occur for slowly heated cotton fabrics [8].

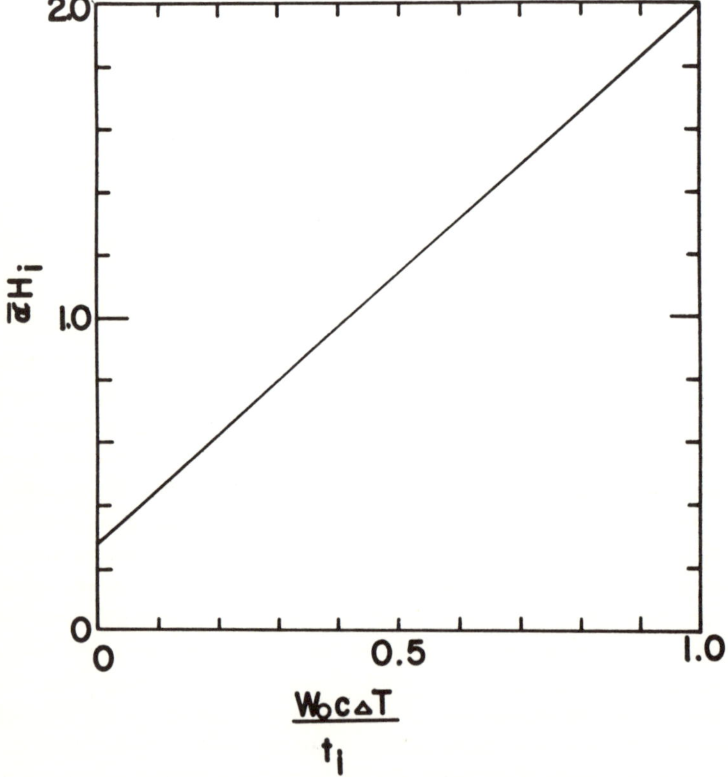

Figure 4. Extrapolation of ignition data to determine critical irradiance.

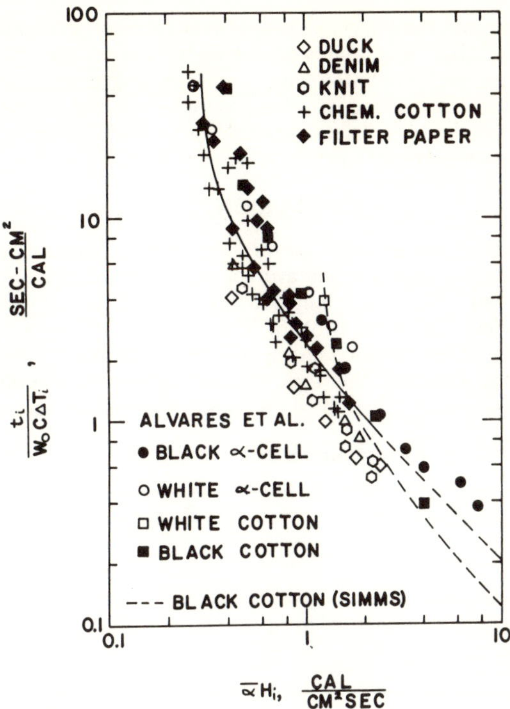

Figure 5. Comparison of ignition data for thin cellulosic materials.

The literature contains data that can be compared to those presented here. Simms [10] presented unpiloted ignition data for fabrics, and his results might be expected to be similar to the data of Figure 3 at high irradiances. Figure 5 compares the curve Simms drew through his data with the curve from Figure 3. The comparison is fair at higher irradiances, but the asymptote is reached at higher irradiances by the unpiloted data, as would be expected. In addition, some data from the work of Alvares [1], Lee and Alvares [6], and Alvares and Anderson [2] are shown; some of these points are for α-cellulose and some are for cotton fabrics. They are all for non-piloted ignition and the ignition times should be higher than for piloted ignition at low irradiances. In fact, it is quite unusual to obtain unpiloted flaming ignition of cellulosic materials at such low irradiances. Most ignition tests, such as those of Simms [10], have found the extrapolated "critical irradiance" to be higher than those at which Lee and Alvares [6] were able to ignite cotton fabrics.

Figure 5 also compares data on piloted ignition of commercially obtained cotton duck, denim, and knit fabrics from earlier work at the University of Oklahoma [11]. Finally, some piloted ignition data for chemical cotton and filter paper taken by Rangaprasad [8] are shown. The commercial fabrics (duck, denim, and knit) ignited a little faster than the fabrics tested in this work, the chemical cotton in about the same time, and the filter paper required longer exposure, especially at low irradiances. Chemicals added to these samples during processing may have influenced their ignition times. Flame-retarded or treated fabrics will not have the same ignition times as the fabrics in these tests.

CONCLUSIONS

Figure 3 shows that the primary factors in the time required to ignite cotton fabrics are the incident irradiance, the average absorptance, and the fabric weight. The "critical absorbed irradiance," below which piloted ignition will not occur, is about 0.3 cal/cm^2-sec. The correlation of Figure 3 can be used to predict piloted ignition times not only for cotton fabrics, but to a first approximation, most thin cellulosic materials as well. For cellulosic materials more than about 0.1-cm thick, the correlation of Reference 12 can be used to estimate piloted ignition time.

Table 1. Average Absorptances for Cotton Fabrics (12)

Fabric Color	Radiation Source	
	Flames	Tungsten Lamps
Black	0.75	0.82
Natural	0.72	0.46

REFERENCES

1. N. J. Alvares, Private communication (May 1967).
2. N. J. Alvares, and T. H. Anderson, "Ignition Hardening of Cellulosic Materials," USNRDL-TR-67-125, U.S. Naval Radiological Defense Laboratory, San Francisco (10 August 1967).
3. N. J. Alvares, P. L. Blackshear, and K. A. Murty, "The Influence of Free Convection on the Ignition of Vertical Cellulosic Panels by Thermal Radiation," Central States Section, Combustion Institute Meeting, University of Minnesota (March 1969).
4. J. Hallman, J. R. Welker, and C. M. Sliepcevich, "Ignition of Polymers," *SPE Journal, 28,* 43 (1972).
5. A. N. Koohyar, J. R. Welker, and C. M. Sliepcevich, "An Experimental Technique for the Ignition of Solids by Flame Irradiation," *Fire Technology, 4,* 221 (1968).
6. B. T. Lee, and N. J. Alvares, "The Effect of Irradiance on the Ignition and the Rate of Fire Spread along Composite Cellulosic Materials," 1967 Spring Meeting, Western States Section, The Combustion Institute, La Jolla, California (April 1967).
7. E. B. Moysey, and W. E. Muir, "Pilot Ignition of Building Materials by Radiation," *Fire Technology, 4,* 46 (1968).
8. Narasimhan Rangaprasad, "The Piloted Ignition of Thin Materials," Ph.D. dissertation, University of Oklahoma, Norman, Oklahoma (1972).
9. D. L. Simms, "On the Pilot Ignition of Wood by Radiation," *Combustion and Flame, 7,* 253 (1963).
10. D. L. Simms, "Ignition of Cellulosic Materials by Radiation," *Combustion and Flame, 4,* 293 (1960).

11. J. R. Welker, and C. M. Sliepcevich, "Susceptibility of Potential Target Components to Defeat by Thermal Action," University of Oklahoma Research Institute Report No. OURI-1578-FR, Norman, Oklahoma (1970).
12. J. R. Welker, H. R. Wesson, and C. M. Sliepcevich, "Ignition of Alpha-Cellulose and Cotton Fabric by Flame Radiation," *Fire Technology, 5*, 59 (1969).
13. H. R. Wesson, J. R. Welker, and C. M. Sliepcevich, "The Piloted Ignition of Wood by Radiation," *Combustion and Flame, 16*, 303 (1971).

Narasimhan Rangaprasad

Narasimhan Rangaprasad received his B.E. at the University of Madras in 1967, his M.E. at the Indian Institute of Science in 1969, and his Ph.D. at the University of Oklahoma in 1972. He is presently employed in India.

C. M. Sliepcevich

C. M. Sliepcevich is a Research Professor of Engineering at the University of Oklahoma at Norman, Oklahoma. He received his B.S. degree in Chemical Engineering from the University of Michigan in 1941, his M.S. degree in Chemical Engineering from the University of Michigan in 1942, and his Ph.D. in Chemical Engineering from the University of Michigan in 1948. Before joining the faculty of the University of Oklahoma in 1955, he was a faculty member at the University of Michigan and a senior chemical engineer at Monsanto Chemical Company. His experience includes thermodynamics, process control, reaction kinetics and catalysis, high pressure design, energy scattering, and cryogenics.

J. Reed Welker

J. Reed Welker is a Research Engineer at the University of Oklahoma Research Institute at Norman, Oklahoma. He received his B.S. degree in Chemical Engineering from the University of Idaho in 1959, his M.S. degree in Chemical Engineering from the University of Idaho in 1961, and his Ph.D. degree in Chemical Engineering from the University of Oklahoma in 1965. His experience includes flame dynamics, mass transfer thermodynamics, and atmospheric dispersion of pollutants.

George L. Drake, Jr., Leon H. Chance and
W. A. Reeves

*Southern Regional Research Center**
New Orleans, Louisiana 70179

A LOOK AT FLAME RETARDANTS BASED ON PHOSPHORUS COMPOUNDS**

(Received May 22, 1974)

ABSTRACT: There are several flame retardants, although not ideal, which can be used to impart flame resistance to cellulosic textiles. Several factors effect the flame retardancy of cellulose: (1) the physical state of the fiber; (2) compounds of nitrogen and phosphorus selected; (3) deposited alkali and alkaline earth metal salts; and (4) the oxidation state of the phosphorus in the finish. Durable flame retardants for textiles discussed include those based on THPC, N-methylol dialkyl phosphonopropionamide, trimethylol phosphoric triamide, and those based on a reactive vinyl phosphorus ester.

INTRODUCTION

Current trends in textile finishing are to impart functional qualities that provide consumers with materials having improved performance characteristics. In many cases finishes are applied for aesthetic or convenience effects, but in others the application is for the protection of the consumer. Finishes that have no protective or safety aspects need to meet vague performance standards that satisfy only the individual consumer. A finish that is designed to protect the user must meet rigid standards to provide a margin of safety under extreme conditions of use. The protective finish that is currently receiving the most widespread attention of the textile industry is the one for imparting flame resistance. This attention is due in part to the education of the public; however, most of it is due to a projection of legislative trends which are requiring higher performance standards for an ever-widening variety of industrial, household, and wearing apparel products. Higher performance standards are being sought because of facts such as those published by the Office of Information and Hazard Analysis-Fire Technology Division, NBS, concerning accidents in the United States due to fires. A few of these pertinent facts are listed below:

---

*One of the facilities of the Southern Region, Agricultural Research Service, U.S. Department of Agriculture.

**Presented at the Seminar on Flammability of Consumer Products, February 21–22, 1974, in Atlanta, Georgia.

Two million burn accidents occur annually. (This estimate includes only burns serious enough to require medical attention or at least one day's restriction of activity).

From 12,000 to 13,000 deaths occur annually as a result of burns. (This number represents approximately 10% of all deaths from accidents.)

Approximately 10,000 hospital beds are occupied daily by patients with severe burn injuries. (This occupancy represents 2% of all available hospital beds and accounts for over 3½ million bed/days annually).

Seventy-five to one-hundred thousand burn victims require hospital care averaging 70 days. (Average hospital costs for burn patients are $160–$180 per day.

Hospital costs for burn victims are close to a billion dollars a year. These costs do not include loss of time and productive capacity. Ninety percent of burn patients must be cared for in general and community hospitals because of the shortage of specialized burn care.

The most common causes of burns are: flames – approximately 40%; scalds from hot liquids – approximately 30%; hot objects – approximately 10%; electrical arcs – approximately 6%; chemicals – approximately 5%; steam – approximately 5%; semi-liquids/semi-solids – approximately 4%.

The greatest burn hazards for children are heaters and stoves – approximately 33%; hot liquids or vapors – approximately 24%; electrical appliances – approximately 7%; chemical action – approximately 4%; explosion or fire of play equipment – approximately 3%; combustible fluid – approximately 2%; matches – approximately 2%; other or unknown – approximately 25%.

About 75% of burns occur at home (inside or outside); about 20% occur at work; about 1.5% occur on streets or highways; and 3.5% results from other or unknown causes. Most burn deaths are in the winter months of November through March.

Flammable clothing is one of the major contributing problems. Many articles of clothing on the market today explode into flames within 10 seconds after exposure to fire.

Crippling injuries from burns in Texas in 1969 were more than double the number of paralytic cases of poliomyelitis during the 1954 epidemic.

Burn deaths in Texas in 1969 were more than four times the number from polio during the 1954 epidemic. (540 burn deaths/134 polio deaths)

Since cotton is one of the major fibers used in production of civilian, military, and household items, there is concern over its flammability.

Cotton cellulose supports combustion, burns rapidly, and has some afterglow. Flaming takes place in the gaseous phase and consists of the rapid oxidation of volatile byproducts from combustion. Smoldering combustion is the direct oxidation in the solid state of chars resulting from earlier stages of pyrolysis. This type of combustion requires as little as 1% O_2 to be self-sustaining. Therefore, for some end-use items there is a need to impart to it a degree of both flame and smolder resistance.

NATURE OF COTTON CELLULOSE AND APPROACHES FOR ACHIEVING FLAME RETARDANCY

Cotton in the form of fabric is subjected to flame-retardant chemicals for reaction to impart flame resistance. The physical state of the fiber has a very definite bearing upon the degree of flame resistance one can obtain. Rowland has published a good description of how cotton undergoes chemical modification [1].

The cotton fiber is a complex, well-ordered unit that is generated during the growth cycle of the cotton plant in the form of a long, hollow tube, approximately 20 microns in width by 25,000 microns (1 in) in length. Each fiber has a twisted convoluted ribbon-like structure, with a somewhat furrowed surface, a section of which is shown in A of Figure 1. It is composed of a multitude of microstructural units which are packed in close proximity and are microfibrils; these units become evident when an expanded portion of a cross section of a cotton fiber is examined under high magnification, as in electron micrographs B and C of Figure 1. The surface of a slab torn out of the cotton fiber (D of Figure 1) shows the microfibrils, at high magnification, in an arrangement which they have in the fiber. These microfibrils are composed from cellulose molecules, which in turn consist of more than 3,000 D-glucopyranosyl units, such as shown in E of Figure 1, joined in a linear chain.

The hydroxyl groups in cotton cellulose, upon which we depend for the crosslinking reactions and for the development of performance properties, are buried in a catacomb-like labyrinth in the cotton fiber. Only a fraction of these potential sites for chemical reaction are actually accessible to the reagent in conventional finishing operations. Moreover, there are two different types of hydroxyl groups in the fibers of cotton cellulose; i.e., secondary hydroxyls at carbon atoms 2 and 3, and primary hydroxyls at carbon atom 6 of each D-glucopyranosyl unit (E of Figure 1). These three hydroxyl groups react at different rates and the linkages developed exhibit different stabilities.

How does the chemical reagent reach the hydroxyl groups in the bulk of the fiber? While the cross section of the native fiber, or a mercerized fiber, shows no pores or channels even at the high magnification of an electron micrograph, the cross section of a fiber, swollen as a result of wet embedment in methacrylate polymer [2], exhibits a selective concentric expansion with the development of a pattern such as shown in B of Figure 1. The expansion and pore development that the fiber undergoes in aqueous solutions of crosslinking agents are probably of this same type but considerably lesser in degree. The pores of decrystallized cellulose have been estimated by gel permeation chromatography to accommodate molecules of sugars having molecular weight ranging downwards from approximately 1800 [3]. Sugars having molecular weights slightly below 1800 find very few pores of adequate size; however, with decreasing molecular weight of the sugar, an increasing number, or volume, of pores becomes available to accommodate the sugar molecules. Preliminary assessment of pores in fibrous cotton by the same method has

Figure 1. Features of the cotton cellulose fiber which are involved in the crosslinking reaction: A, a section of a fiber; B, an electron micrograph of an expanded cross section of a cotton fiber; C, a portion of the expanded cross section under highest magnification; D, an electron micrographic view of the longitudinal structure and microfibrils in a slab torn from a cotton fiber; and E, two D-glucopyranosyl units which constitute a portion of the molecular chain of cellulose.

indicated a permeability limit in the range of 2800 [4]. It is quite evident, then, that the normal type of crosslinking flame-retardant agent, which has a molecular weight in the range of 100–200, finds many pores or channels through which it may reach the hydroxyl groups of the cellulose molecules in the microstructural units.

Several techniques have proved useful in the production of flame-retardant cellulosic fabrics:

A. Simple substitution of the cellulose
B. Coating fibers with thermoplastic polymer
C. Deposition of thermosetting polymer within the fiber
D. Complex substitution of cellulose combined with internal polymer deposition
E. Combinations of any of the above

The task of developing an effective, durable flame-retardant finish for cotton is difficult because of the many requirements the finish must meet. To be satisfactory for most uses, a flame retardant must: (1) be easy to apply, principally with existing finishing equipment from a water solution; (2) be effective with few additives to avoid extensive increases in weight; (3) produce a finish durable to laundering and drycleaning; (4) leave the fabric air permeable; (5) be physiologically inactive; (6) render the fabric resistant to afterglow; (7) not stiffen the fabric appreciably; (8) cause little or no loss in strength or change in dye shade; and (9) be reasonable in cost.

Phosphorus-containing compounds are by far the most important class of compounds used to impart durable flame resistance to cellulose. Used in conjunction with phosphorus are the elements nitrogen or bromine, or both. Reeves, Perkins, Piccolo and Drake [5] recently reported some of the chemical and physical factors influencing flame retardancy.

Proper selection of the nitrogenous polymer or compound to be used with phosphorus-containing flame retardants is very important. The combination can aid flame resistance, have a neutral effect, or reduce flame resistance contributed by the phosphorus. Amide and amine nitrogen generally aid flame resistance, whereas nitrile nitrogen can detract from it. Amide nitrogen from nylon fiber is unusual in that the nylon detracts from flame resistance. Several factors may contribute to this effect. Perhaps the most important reason for the nylon effect is that the amide is part of a thermoplastic polymer having a high percentage of combustible hydrocarbon. Cyanamide, which contains an amine and nitrile nitrogen, contribute flame resistance much the same as do amides. This effect is due to the conversion of cyanamide to urea and dicyandiamide during treatment of the fabric. This study indicates that flame retardants such as those containing cyanoethyl phosphonate groups would contribute much less flame resistance than the corresponding amino carbamoylethyl phosphonate.

Nitrogen contributes most efficiently to phosphorus-containing flame retardants when it is present in low concentrations relative to the phosphorus. N/P atomic ratios of about one in the fabric contribute maximum flame retardancy per unit weight of nitrogen. Much of the nitrogen lost during charring must contribute flame resistance by acting in the vapor phase; otherwise, the efficiency of nitrogen would drop more rapidly as the N/P atomic ratio in the fabric is increased. Many factors should be taken into consideration in deciding the amount of nitrogen to use with phosphorus. Some factors are (a) cost – nitrogenous agents may be much less expensive than the phosphorus being replaced, (b) tensile properties of the fabric, and (c) hand.

Small amounts of phosphorus contribute more flame resistance per unit weight of phosphorus than large amounts. If complete flame resistance is desired, it is usually necessary to use large, less efficient quantities of phosphorus. All flame retardants examined decrease in effectiveness per unit weight of retardant as the add-on of retardant is increased. Thus, the first 5% of flame retardant added to a fabric is more efficient than the second or third 5%. These and other data indicate that the most efficient flame-retardant system will contain two retardants – one acting in the solid and the other in the vapor phase.

In the absence of nitrogenous materials, essentially all of the phosphorus in a flame-resistant fabric can be accounted for in the char produced by pyrolyzing the fabric in air. The phosphorus also can be accounted for in the char when nitrogen is present as an amide, pseudo amide, or amine. With nitrile nitrogen in the fabric, there is a significant loss of phosphorus from the char and reduced efficiency of the phosphorus. The amount of nitrogen in the char is dependent on the N/P atomic ratio in the flame-resistant fabric. High ratios in the fabric are decreased substantially during the charring, whereas ratios slightly above 1 are reduced very little. The most stable ratio is 1. Based on atomic ratio and insolubility in water, acids, and alkali, it is assumed that the thermally stable residual compound may contain phosphorus oxynitride.

Any bromine present in a flame-resistant fabric escapes to the tar or vapor phase during pyrolysis in air and appears to have little or no effect on the amount of phosphorus remaining in the char. Bromine contributes to the flame resistance of phosphorus-containing flame retardants, although it acts independently of the phosphorus. To do this, it must act almost completely in the vapor phase.

FACTORS AFFECTING FLAME RESISTANCE OF FIBERS

While phosphorus-containing flame retardants have some unique features such as glow resistance, these retardants require proper application and care to assure maximum efficiency and durability. Alkali and alkaline earth metal ions reduce or inhibit the flame resistance contributed by phosphorus. These ions need not be in direct union with the retardant to exhibit their effect; mere presence in the fiber or fabric is adequate. Calcium acetate is more detrimental to flame retardants than calcium phosphate. Calcium and magnesium can be picked up by flame-resistant fabric from wash water during laundering. This pick-up can occur through ion exchange properties of the retardant and through precipitation in the fiber by combining with phosphate from detergents or fatty acids from soap. The amount of phosphate salts picked up by the fabric depends on the nature of the flame retardant and the amount of phosphate and calcium (or magnesium) in wash water. Pickup of these phosphates can lead to erroneous results when the phosphorus content of fabric is used as a measure of durability of the flame retardant. Calcium and magnesium soaps in a fabric similarly can offset flame tests that are often used to measure durability. The adverse effects of the foreign materials can be eliminated by rinsing the laundered fabric occasionally in dilute acid (Figure 2). Pickup of copper ions from solution has little effect on flame resistance.

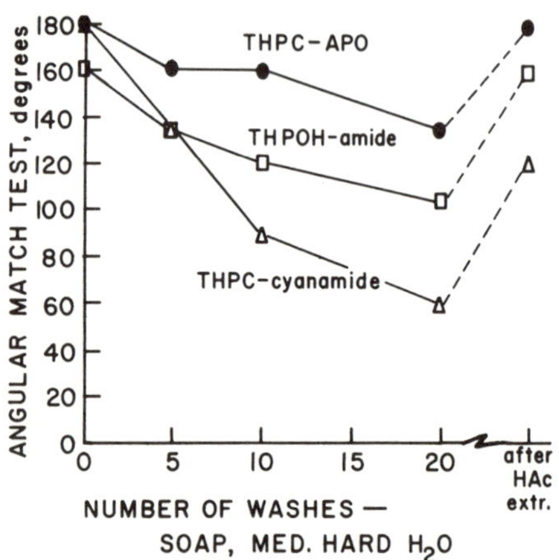

Figure 2. Flame resistance of three cotton fabrics after repeated laundering and after acetic acid extraction.

The oxidation state of phosphorus in the finish affects the durability of flame-retardant fabrics. THPC retardants with trivalent phosphorus are less durable to laundering than are the retardants that contain the corresponding phosphine oxide structure. Thus, the application of THPC-type retardants should include an oxidation step.

Retardants are generally degraded, at least to a small extent, when exposed to sunlight. This degradation becomes a significant problem for some retardants when the treated fabrics are laundered and line dried. The unoxidized retardant based on the reaction of "THPOH" with ammonia is particularly sensitive to sunlight. A small amount of pigment can protect some flame retardants from the destructive action of sunlight. Pressing, which is sometimes part of the laundry cycle for apparel and household goods, also decreases the effectiveness of some flame retardants. The ester groups of alkyl phosphonates are hydrolyzed at the pressing temperature to produce sites for ion exchange, which is detrimental when the fabrics are laundered. Another factor affecting the durability of flame-retardant finishes is bleaching. None of the known commercial flame-retardant finished fabrics can pass the DOC FF-3-71 test after being bleached 50 times with hypochlorite bleaches.

DURABLE FLAME-RETARDANT FINISHES WHICH ARE COMMERCIAL OR WHICH ARE BEING EVALUATED ON A COMMERCIAL SCALE

The most widely used durable flame retardants for apparel, household goods, and certain military items, such as tent liner fabrics, are based upon tetrakis (hydroxymethyl) phosphonium chloride (THPC) prepared in high yield from formaldehyde, phosphine, and hydrochloric. THPC is a water-soluble, crystalline compound produced in acid. In the U.S., it is sold by Hooker Chemical Co.[1] and Aceto Chemical Co., and in Europe by Albright and Wilson Ltd., England, and in Canada by American Cyanamid Co. The methylol groups of THPC react readily with amines and amides [6]. During a study of the physical and chemical properties of aminized cotton the value of THPC as a flame retardant was first recognized [7]. A flame retardant based upon THPC and an amide was then developed. The formulation consists of THPC, urea, and trimethylolmelamine (2:4:1 mole ratio) plus various auxiliaries [8]. Normally about a 35% solution is applied (70% wet pickup) to an 8 oz cotton fabric, which is then dried and heat cured. The finish passes the Standard Vertical Flame Test and is durable to both laundering and drycleaning. Its primary limitations are the stiffness imparted to some types of fabrics, and a reduction in fabric tearing and tensile strength. Durability of the finish is inversely related to the concentration of urea in the treating solution. Urea is needed to improve tensile and tear strength. Thiourea can be substituted for urea in about equal weights to give a finish that is actually more durable to laundering and alkaline boiling. Trimethylolmelamine is preferred over N-methylol amides.

When methylated methylolmelamines are substituted for trimethylolmelamine, treated fabrics have a softer hand, but the amount of phosphorus fixed is lower. Generally, the same is true when other N-methylolamides such as dimethylolethyleneurea, dimethylolethyltriazone, dimethylol-4, 5-dihydroxy cyclic ethyleneurea, and dimethylol carbamates are substituted for trimethylolmelamine in the formulation.

Sodium hydroxide is used in the formulation to neutralize the free hydrochloric acid that is usually present in the THPC. Any excess sodium hydroxide neutralizes or reacts with THPC and converts it to other P-methylol compounds. Other bases, including alkanolamines, can be used instead of sodium hydroxide.

---

[1] [1] Use of a company or product name by the Department does not imply approval or recommendation of the product to the exclusion of others which may also be suitable.

Wetting and softening agents are the main auxiliaries used in the formulation. Generally, about 0.1% of a wetting agent is used. The amount of softener used can vary considerably. The softener can have a significant effect on tearing strength; 0.5% cetylamine or stearamide causes an increase of about 25% in tear strength, which is retained through as many as 15 laundry cycles. Usually laundering reduces the tearing strength.

Addition of sodium sulfite to react with free formaldehyde reduces premature polymerization in the treating formulation and produces fabrics with a better hand, less stiffness, and higher tearing strength. The amount of sulfite can vary from about 0.9 up to 2.0 moles per mole of THPC [9]. A polymerization catalyst, such as magnesium chloride or an amine hydrochloride, is also necessary in this formulation.

The THPC-amide flame retardant has been modified in several ways in an effort to improve the properties of the treated fabric and/or to reduce the cost. The bromoform adduct of the allyl ester of phosphonitrilic chlorides can be produced in an aqueous emulsion using a peroxide catalyst and then added to the THPC-amide formulations [10]. The adduct is a very efficient flame retardant, and less of the combined formulation is required than of the THPC-amide alone. The major objection to this modification is that the overall cost is increased. Another adduct that has been added to the THPC-amide formulation is the bromoform adduct of triallyl phosphate. Again cost was the deterrent factor.

A recent modification for use on polyester/cotton sheets was made by the Piedmont Section of the AATCC [11]. Tris (2,3-dibromopropyl-phosphate), which is water-insoluble, can be emulsified and added directly to the THPC-amide formulation. The brominated phosphate penetrates both the polyester and the cotton and produces a very durable finish for the blend fabric. The presence of the phosphate with the THPC-amide prevents loss in lightfastness of certain dyes [12].

Antimony oxide-polyvinyl chloride mixed with the THPC-amide formulation and applied to cotton fabrics of over 5 oz per sq yd imparts excellent flame resistance durable to more than 50 laundry cycles. This finish is used primarily on military fabrics, work clothing and hospital cubicle curtains.

Recently Donaldson *et al*, [13] have discovered that sodium phosphate salts catalyze the reaction of THPC and urea. They found that cotton textiles treated with a solution containing 26.6% THPC, 8.4% urea, 4% dibasic sodium phosphate, and sufficient NaOH to adjust the pH to 6, dried at 85°C, and cured at 160°C for 1 to 2 minutes could pass the DOC FF 3-71 flame test after 50 laundry cycles. Samples treated with this finish without softeners retained 90% and 65–70% of their breaking and tearing strength, respectively. Industry is evaluating this finish at present.

A finish based on N-methylol dialkyl phosphonopropionamide [14] was introduced in the United States in 1968 by the Ciba Co. under the trademark Pyrovatex C.P. A typical treating solution consists of 40% Pyrovatex C.P., 5% melamine formaldehyde resin (such as Cyanamid's Aerotex 23 Special) 2.0% polyethylene softener (25% solids), 0.5% phosphoric acid, 0.1% wetting agent, and

52.9% water. Fabric is wet out with the solution, padded to about 75% pickup, dried, cured for about 4 minutes at 330°F, and given an alkaline afterwash. This finish has low toxicity, an excellent hand, and is durable to washing and tumble drying. Tensile strength is decreased about 20–30% and tearing strength about 30%. Effective treatments require an add-on of 25–35%. The finish is sensitive to acid sours, and its heat stability could be improved. Cost is relatively high.

The Monsanto Chemical Company has recently introduced a new flame-retardant finish for cellulosic textiles called their MCC 100/200/300 finish. It consists of a three-component system. One component is trimethylphosphoric triamide, $(CH_3NH)_3\ P = O$; the second component is a methylolmelamine aminoplast; and the third component is an amine hydrochloride catalyst.

A water solution of the chemicals is applied to fabric using a pad-dry-cure technique. With an add-on of about 20–24%, the flame retardancy is retained after 50 laundry or dry cleaning cycles and is reported to pass the DOC FF-3-71 test. Due to the stiffness imparted to fabric by this finish, its use is presently limited to cellulosic floor coverings and drapery fabrics. It is applied to carpets by a kiss roll or spray technique during the backcoating operation. An add-on of approximately 4–8% is required to impart adequate flame retardancy to the carpets. This flame retardant has the same drawbacks as other flame retardants with respect to chlorine bleaching and laundering in hard water with soap and carbonate detergents.

Stauffer Chemical Company has recently developed a new type of durable flame-retardant finish for cellulose. It is called Fyrol 76.

Fyrol 76 is a reactive vinyl phosphorus ester that contains 22.5% phosphorus. It is a clear water-soluble liquid, which is supplied as 100% solids. It is designed for use in conjunction with methylolacrylamide (NMA) using a free radical catalyst such as potassium persulfate. A typical formulation is shown below:

Formulation	Cotton Flannel	Knit
Fyrol 76	20.0%	17.0%
NMA	22.0%	19.0%
$K_2S_2O_8$	0.5%	0.5%
Chelon 100	0.1%	0.1%
Triton X-100	0.1%	0.1%
Softener	4.0%	4.0%
Water	53.3%	59.3%

Fabric is wet out with the formulation using conventional padding followed by drying and curing, or the wetted fabric can be simultaneously dried and cured for 0.2 to 10 minutes at from 250°F to 500°F, the temperature for cure being based on the time. Conventional curing is usually done at 300°–350°F. The wetted fabric is also suitable for rapid steam curing or radiation curing techniques. Little or no odor is evolved during curing. This finish differs from other durable flame-retardant finishes based on phosphorus in that the curing is accomplished by a free radical mechanism.

Sleepwear fabric requires approximately 25–30% add-on to pass the DOC FF 3-71 test. The fabric has an excellent hand, good strength, and some permanent-press properties. Approximately 80–100% tensile strength and about 65% tearing strength are retained. Chlorine bleach and some carbonate detergents reduce the durability of this finish, whereas perborate bleaches do not affect the durability of the finish. As with the THPC type finishes, an oxidation step is generally used.

The treatments described above are for the most part cellulose reactive and therefore also impart some wrinkle- and rot-resistance to the fabric. A method of polymerizing the phosphorus compounds inside the fiber, with little or no reaction with the fiber, produces flame-resistant fabrics without causing strength losses has met with success. Since there is essentially no crosslinking with the cellulose this finish does not impart wrinkle- or rot-resistance to the cellulose. The most successful technique is to chemically "fix" the polymer with anhydrous ammonia.

The original process utilizing ammonia to insolubilize methylol phosphorus polymers inside cotton to impart flame resistance was developed in 1953 and patented in 1956 [15]. THPC, trimethylolmelamine, and urea were padded onto fabric and dried. The fabric was then exposed to ammonia and/or ammonium hydroxide which immediately insolubilized a polymer containing phosphorus and nitrogen inside the fiber. Strength of the treated fabric was essentially unchanged. Slight modifications of this process have been made to increase the stability of the pad bath and to make it more attractive for application using commercial laundry equipment.

A water-soluble precondensate of THPC and urea [16, 17] is suitable for the process. After impregnation with the precondensate the fabric is dried, exposed to ammonia vapor, and then passed through ammonium hydroxide. The ammonia causes polymerization within the fiber; subsequent treatment with aqueous ammonium hydroxide insures the production of a highly insoluble polymer. The finish is flame resistant even after repeated laundering and drycleaning. Tearing strength is slightly reduced but breaking strength is essentially unchanged. This process has been used in Europe for about 10 years. The ammonia gas step can be eliminated by incorporating diammonium sulfite in the treating solution. Ammonia is released during the drying step and initiates the polymerization reaction.

A new, simple, chemical fixation process uses the THPOH-NH_3 system [18]. THPOH is made by adjusting the pH of THPC to about 7.2 with concentrated NaOH. Fabrics are impregnated with the THPOH solution, dried to about 3.5–15% moisture, and exposed to anhydrous NH_3 gas for from 15 seconds to two minutes. An 8.5 oz twill or sateen so treated to about a 13% add-on will pass the standard vertical flame test. This process also can be used on very light-weight fabrics – a 2 oz fabric requires about 20% add-on. Stiffness is essentially unaffected; breaking strength remains unchanged or sometimes increases; and, with a softener, reduction in tearing strength is about 10%. The basic steps in the process are illustrated in Figure 3.

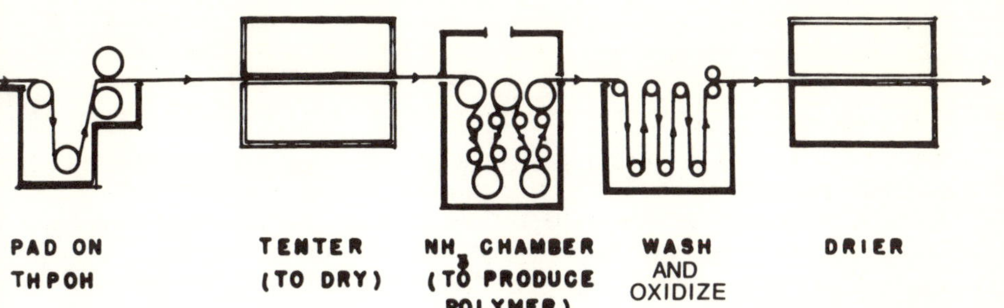

Figure 3. Schematic diagram showing the basic steps in the THPOH-NH_3 process.

The THPOH-NH_3 finish requires peroxide oxidation after the ammonia cure in order to impart outstanding durability to the finish and to remove any odor that might be attributed to the trivalent phosphorus in the polymer structure. This finish is being applied commercially to cotton sleepwear by the United Merchants under the trade name "Firestop" and also by Avondale Mills.

SUMMARY

As discussed above, there are several flame retardants which, although not ideal, can be used to impart excellent flame resistance to cellulosic textiles. These finishes are based mainly on formulations containing phosphorus. Some flame retardants can impart other properties such as wrinkle-, rot-, and mildew-resistance to fabric. At present no one finish is suitable for fabrics of all weights and constructions. Flame retardants suitable for use on polyester-cotton blends are needed. It is probable that no such durable flame retardant exists at the present time. Progress is being made in designing a flame retardant for manmade fibers. With the passage of the amendment to the Flammable Fabrics Act, research efforts to develop suitable flame-retardant processes have increased markedly, and new advances can be expected.

REFERENCES

1. S. P. Rowland, Amer. Dyest. Rep. *58* (22) 15–18 and 41 (1969).
2. M. L. Rollins, and V. W. Tripp, in "Methods in Carbohydrate Chemistry," Vol. III, R. L. Whistler, Ed. Academic Press, New York, 1963, pp. 356, 364.
3. L. F. Martin, N. R. Bertoniere, F. A. Blouin, M. A. Brannan, and S. P. Rowland, Textile Res. J. 39, (to be published).
4. L. F. Martin, private communication.
5. W. A. Reeves, R. M. Perkins, B. Piccolo, and G. L. Drake, Jr., Text. Res. J. *40* (3), 223–231 (1970).
6. W. A. Reeves, and J. D. Guthrie, Ind. and Eng. Chem. *48* (1), 64–67 (1965).
7. W. A. Reeves, O. J. McMillan, and J. D. Guthrie, Text. Res. J. *23* (8), 527–532 (1953).
8. J. D. Guthrie, G. L. Drake, Jr., and W. A. Reeves, Amer. Dyest. Rep. *44* (10) 328–32 (1955).

9. R. F. Zimmerman, and G. M. Wagner, (to Hooker Chemical Co.), U.S. Pat. 3,247,015 (1966).
10. C. Hamalainen, W. A. Reeves, J. D. Guthrie, Text. Res. J. *26* 145–149 (1956).
11. Piedmont Sect. Amer. Assoc. Textile Chem. And Colorists, Amer. Dyest. Rep. *57* (10) 373–377 (1968).
12. Piedmont Sect. Amer. Assoc. Textile Chemists and Colorists, Text. Chem. Color. *1* (5), 124–129 (1969).
13. D. J. Donaldson, F. L. Normand, G. L. Drake, Jr., and W. A. Reeves, "A Durable Flame Retardant Finish for Cotton Based on THPC and Urea." (In Press)
14. R. Aenishanslin, C. Guth, P. Hoffman, A. Maeder, and H. Nachbur, Text. Res. J., *39* (4), 375–381 (1969).
15. W. A. Reeves, and J. D. Guthrie, (to U.S. Secretary of Agriculture), U.S. Pat. 2,772,188 (1956).
16. H. Coates, (to Albright and Wilson Ltd.), U.S. Pat. 3,096,201 (1963).
17. G. L. Drake, Jr., W. A. Reeves, and R. M. Perkins, Amer. Dyest. Rep. *52* (16), 41–44 (1963).
18. J. V. Beninate, E. K. Bolyston, G. L. Drake, Jr., and W. A. Reeves, Amer. Dyest. Rep. *57* (25), 47–51 (1968).

Elwood J. Gonzales, and Sidney L. Vail

Southern Regional Research Center *
New Orleans, Louisiana 70179

DIMETHYL AND DIETHYL PHOSPHORAMIDATES AS FLAME RETARDANTS FOR COTTON

(Received July 13, 1974)

ABSTRACT: Cotton fabrics were made flame retardant with diethyl (DEPA) or dimethyl (DMPA) phosphoramidates and various cellulosic crosslinking agents such as trimethylolmelamine (TMM), dimethyloldihydroxyethyleneurea (DMDHEU), dimethylolpropyleneurea (DMPU), and tetramethylolglycouril (TMGU) in the presence of an acidic catalyst. DEPA-finished fabrics were characterized by very good flame retardant properties and high oxygen index (O.I.) values when TMM was the crosslinking agent but durability of the flame retardancy to home-type laundering was only moderate. DEPA and DMPA finishes had minimum release of formaldehyde during curing operations.

INTRODUCTION

The formation of P—N linkages during pyrolysis of cotton treated with flame retardants containing phosphorus and nitrogen has been suggested [7] as a possible basis for the synergistic effects of phosphorus and nitrogen observed with these retardants. Presumably the P—N linkage facilitates the phosphorylation of the cellulose so that dehydration of the cellulose is favored during pyrolysis rather than production of combustible gases. Similarly, in other studies [3], flame retardant finishes obtained from reagents containing the P—N linkage appeared to be inherently more efficient than finishes (believed to contain the $P{-}CH_2{-}N$ link) prepared from conventional THPC type treatments. Therefore, we have selected two compounds, diethyl phosphoramidate (DEPA) and dimethyl phosphoramidate (DMPA),

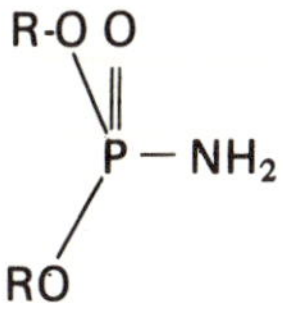

$$(RO)(RO)P(=O){-}NH_2 \qquad R = CH_3\text{-DMPA} \qquad R = C_2H_5\text{-DEPA}$$

containing P—N linkages for further study.

---

*One of the facilities of the Southern Region, Agricultural Research Service, U.S. Department of Agriculture.

These compounds have been used by researchers to flameproof cellulosic textiles since the late fifties. Hechenbleikner and Glade [4, 5, 6] first used the compounds in formulations with derivatives of melamine, notably methylolated melamines and methylated methylolmelamines. In many instances, they did not isolate DMPA and DEPA as pure compounds.

In some instances, they obtained a fair degree of flame retardancy (3–5 inches char length) as measured by the standard vertical flame test. In our study, we have prepared and used highly pure products and have attached DMPA and DEPA to cotton cellulose, not only with trimethylolmelamine (TMM), but with other cellulosic crosslinking agents such as dimethyloldihydroxyethyleneurea (DMDHEU), dimethylolpropyleneurea (DMPU), and tetramethylolglycouril (TMGU). We have studied the efficiency of the application of DEPA and DMPA to cotton printcloth and flannelette in various reactive systems, and evaluated the flame retardant properties of the resultant finishes.

MATERIALS AND METHODS

Chemicals

Diethyl phosphoramidate (DEPA) was prepared according to the method of Malowan [8]. *Anal*: Calcd: P, 20.26; N, 9.15; *lit.* [8], mp 46–47°. Found: P, 20.43; N, 9.09; mp 47–48°.

Dimethyl phosphoramidate (DMPA) was prepared according to the method of Malowan [8] by reacting 220.1 grams (2 moles) dimethyl phosphite in 307.6 grams (2 moles) CCl_4 and 800 grams benzene with ammonia gas for 5 to 8 hours, filtering off the NH_4Cl, and stripping off the solvents under aspirator vacuum at 50–60°C until crystallization occurred. *Anal*: Calcd: P, 24.80; N, 11.20. Found: P, 25.00; N, 10.54.

Dimethyloldihydroxyethyleneurea (DMDHEU) was prepared according to a method described in the patent literature [1]. Calcd.: N, 15.73. Found: N, 15.67; mp 76.5°.

Dimethylolpropyleneneurea (DMPU) was prepared according to the patent literature [13]. Calcd: N, 17.50; lit: mp 104–106°. Found: N, 17.17; mp 102–104°.

Other chemical products used were as follows[1]: Resloom HP (trimethylolmelamine (TMM)) from Monsanto Chemical Co.; Permafresh 113B [45%(wt) DMDHEU solution] from Sun Chemical Co.; tetramethylolglycouril (TMGU) from Fine Organics, Inc.; hexamethylenetetramine (HEX) from Eastman Kodak; Triton X-100 from Rohm and Hass; and catalyst H7 [35% solution of buffered $Zn(NO_3)_2$] from Rohm and Haas.

---

[1]Use of a company or product name by the Department does not imply approval or recommendation of the product to the exclusion of others which may also be suitable.

Methods

Nitrogen was determined by the Kjeldahl method. Formaldehyde was determined by the chromotropic acid method employing spectrophotometric techniques [10]. Phosphorus, chlorine, magnesium, and calcium analyses were carried out by X-ray fluorescence using a General Electric XRD-6 spectrometer [11].

Durability to laundering of the treated fabrics was determined by repeated washing and drying in an agitator-type washer and gas dryer under normal conditions for cotton using the AATCC detergent.

Fabric properties which were measured by ASTM procedures are: breaking strength and elongation at break, strip method (ASTM D39-49); tear strength, Elmendorf method (ASTM D1424-63); crease recovery, Monsanto method (ASTM D1295-67); oxygen index (O.I.) values (ASTM D2863-70). The O.I. instrument is manufactured by MKM Machine Tool Co., Inc., Jeffersonville, Ind. Other fabric properties determined were: stiffness, Tinius-Olsen method [12]; flame resistance [2] (DOC-FF-3-71); and strip test method (match angle) of flame resistance [9].

Fabric Treatments

We used fabrics that had been desized, scoured, and bleached; cotton printcloth (80×80) weighed 3.2 oz/yd^2 and cotton flannelette weighed 4.1 oz/yd^2. Fabric samples (11×13 in.) were padded, two dips and two nips, to 87% wet pickup for printcloth and to 100% wet pickup for flannelette with aqueous solutions containing the phosphorus compound, cellulosic crosslinking agent and catalyst in definite weight percentages. Padded samples were then dried 7 min at 60°C and cured 3 min at 160°C in a forced-air oven. Triton X-100 was added as a wetting agent in runs 5, 6, 8, and 9 of Table 1. After curing, samples described in Table 1 were washed 15 min in hot tap water and ironed dry on the cotton setting. Samples described in Table 3 were washed in a home-type automatic washer using 5 ml Triton X-100, hot wash, warm rinse, low water level, 4 min delicate wash cycle and finally tumbled dry in an automatic gas dryer for 30 min.

RESULTS AND DISCUSSION

Diethyl Phosphoramidate (DEPA) and Crosslinking Agents

Data in Table 1 represent the efficiency of the reaction of DEPA with cotton cellulose in the presence of cellulosic crosslinking agents. With printcloth (runs 1 to 4), an increase in the concentration of DEPA in the pad bath from 5 to 30%, when the TMM concentration is held constant at 10%, decreased efficiency of reaction as determined by bound phosphorus in the finished fabrics. The mole ratios (Table 1) of DEPA to TMM (runs 1–4) indicate that more DEPA was available to react with TMM, thereby decreasing the amount of N-methylol groups available to react with cotton. Comparison of runs 5 and 2 which were similar except for base fabrics indicated no advantage in efficiency or other properties. Comparison of runs 7 and

TABLE 1. EFFICIENCY OF THE REACTION OF DIETHYL PHOSPHORAMIDATE (DEPA) IN THE PRESENCE OF VARIOUS CROSSLINKING AGENTS WITH COTTON CELLULOSE[1]

Run No.	DEPA		Crosslinking Agent[2]			Fabric Type[3]	Add-On, %	N, %	HCHO, %	P %	Efficiency % (based on P)
	Wt. %	Moles	Wt. %	Moles	Name						
1	5	0.032	10	0.046	TMM	P	11.2	3.64	1.84	0.50	63
2	15	0.098	10	0.046		P	10.5	3.51	1.63	0.76	32
3	20	0.131	10	0.046		P	11.1	3.63	1.79	0.65	21
4	30	0.196	10	0.046		P	9.9	3.21	1.84	0.88	18
5	15	0.098	10	0.046		F	9.6	3.54	1.75	0.70	33
6	15	0.098	20	0.093		F	33.7	7.93	4.97	1.90	85
7	20	0.131	10	0.046		F	15.0	4.34	2.67	1.10	30
8	15	0.098	20	0.112	DMDHEU	F	20.8	3.00	4.21	1.35	57
9	20	0.131	20	0.112		F	23.1	3.13	3.58	1.62	50
10	15	0.098	10	0.038	TMGU	F	12.3	2.21	3.23	1.05	39
11	20	0.131	10	0.038		F	12.5	2.03	2.81	1.02	29
12	15	0.098	5	0.019	TMGU	F	13.6	3.48	2.93	0.96	36
			5	0.023	TMM						
13	15	0.098	10	0.038	TMGU	F	20.4	4.26	4.47	1.26	51
			5	0.023	TMM						
14	20	0.131	5	0.019	TMGU	F	13.7	3.34	2.61	0.97	28
			5	0.023	TMM						
15	15	0.098	10	0.063	DMPU	F	8.4	1.24	1.26	0.55	20

1/ All formulations also contained 2.8%(wt) catalyst H7. Fabrics were padded two dips-two nips to 87% (printcloth) or 100% (flannelette) wet pickup, dried at 60°C for 7 min , cured at 160°C for 3 min , and given a 15 min wash in hot tap water and ironed dry. Runs 5,6,8,9 contained 1% Triton X-100 in the pad bath. Roll pressure was 60 lbs. in all runs except 6 and 15 in which it was 40 lb.

2/ TMM=trimethylolmelamine; DMDHEU=dimethyloldihydroxyethylene urea; TMGU=tetramethylolglycouril; DMPU=dimethylolpropylene urea

3/ P=printcloth; F=flannelette

3, however, showed that efficiency was higher with flannelette than with print cloth. When the concentration of TMM was increased to 20% (run 6), all the fabric properties increased markedly.

Efficiency of the reaction was reduced with crosslinking agents dimethyloldihydroxyethyleneurea (DMDHEU) and tetramethylolglycouril (TMGU) (compare runs 8 and 9, and 10 and 11); other properties such as add-on, bound nitrogen, formaldehyde and phosphorus, however, were not changed greatly. Replacement of part of the TMGU in run 10 with TMM (run 12) increased only the amount of nitrogen bound in the finished fabric. When the TMGU in the pad bath in run 10 was supplemented with TMM (run 13), add-on, nitrogen, formaldehyde, and phosphorus increased appreciably. The efficiency of the reaction decreased from 36 to 28% in the TMGU-TMM mixed crosslinking agent system when the concentration of DEPA in the pad bath was 20% (compare runs 12 and 14).

Dimethylolpropyleneurea (DMPU) was less efficient than TMM when allowed to react with DEPA and bind the DEPA into cotton (Table 1). Comparison of runs 15 and 5 shows this conclusion in the add-on, nitrogen, formaldehyde, and phosphorus determinations.

Table 2 reports physical properties of fabric samples from treatments specified in Table 1. Examination of match angles and O.I. values indicate that fabrics from runs 2, 3, 6, 7, 8, 9 and 13 had the best flame retardant properties. Moderate degrees of dry crease recovery were imparted to fabrics treated with DEPA-TMM (runs 5, 6); DEPA-TMGU (runs 10, 11); and DEPA (TMM plus TMGU) runs 12, 13, 14). Degrees of dry and wet crease recovery were highest with DEPA-DMDHEU (runs 8, 9). The best combination of flame retardant and stiffness properties was run 8. Run 6, in which 20% (wt) TMM was used, had the largest stiffness value. Generally strength retention was good even on the treated fabrics having moderate degrees of dry crease recovery and especially for runs 8 and 9 which had good dry and wet crease angles. A characteristic of the treatments listed in Table 1 was the minimal amount of formaldehyde odor detected during drying and curing operations.

Dimethyl Phosphoramidate (DMPA) and Crosslinking Agents

Data in Table 3 represent the efficiency of the reaction of DMPA with cotton cellulose in the presence of TMM or DMDHEU. The initial efficiency of 27% decreased to 19% when the DMPA concentration in the pad bath was increased to 30% (wt). The nitrogen, formaldehyde, and phosphorus values in run 2 which are similar to those in run 1 demonstrate this decreased efficiency excellently. Efficiency of reaction with cotton cellulose was decreased when the concentration of DEPA was increased in the pad bath (see Table 1). Since molecules of TMM and DMPA have six atoms and one atom of nitrogen, respectively, the N/HCHO/P ratios for runs 1 and 2 indicate that a small amount of polymerization occurred and that a molecule of HCHO was lost in the reaction with cotton. Identical treatment with DEPA, using the same concentrations and fabric substrate (run 7, Table 1), gave a similar N/HCHO/P ratio of *8.1/2.5/1.* If the crosslinking agent was changed from TMM to DMDHEU (run 3) the efficiency of the fixation of DMPA increased from 27 to 39% and again, a small amount of polymerization and loss of formaldehyde were indicated by the N/HCHO/P ratios.

Table 4 reports the physical properties of fabrics from treatments given in Table 3. Fabric from run 1 had very good flame retardance as determined by the match

TABLE 2. PHYSICAL PROPERTIES OF COTTON CELLULOSE FINISHED WITH DIETHYL PHOSPHORAMIDATE (DEPA) IN THE PRESENCE OF VARIOUS CROSSLINKING AGENTS FROM TABLE 1.

Run No.	Match Test Angle deg	O.I.	Breaking Strength (W), lb	Elongation (W),%	Wrinkle Recovery, (W+F), degrees Dry	Wrinkle Recovery, (W+F), degrees Wet	Tinius-Olsen, Stiffness, X10^4 in.-lb
1	90	0.225	40	6.4	255	228	4.0
2	135	0.250	50.7	7.8	193	190	15.5
3	135	0.247	52.8	8.4	172	226	52.0
4	105	0.262	59.4	9.1	144	226	69.3
5	115	0.241	26.2	5.1	246	204	6.1
6	180	0.289	37.3	2.6	236	244	262.0
7	135	0.261	30.4	5.3	216	231	31.0
8	135	0.240	23.3	2.7	276	271	13.5
9	135	0.252	28.6	6.2	280	274	18.5
10	105	0.241	20.6	4.2	250	246	6.5
11	105	0.238	19.8	3.7	258	235	4.5
12	105	0.253	19.7	5.1	248	227	9.8
13	135	0.270	20.8	3.7	268	241	29.0
14	105	0.259	22.0	4.2	243	242	24.5
15	45	0.207	21.0	5.1	222	198	3.2
UPC	0	0.182	45.5	7.9	168	151	3.0
UFC	0	0.182	33.1	6.7	169	176	3.2

UPC = untreated printcloth control

UFC = untreated flannelette control

test and O.I. Although the match angle, 135°, was similar with DEPA (run 7, Table 1), the O.I. value with DMPA (0.270) was greater than that of DEPA by .009. Their wrinkle recoveries were nearly equivalent but the stiffness value with DMPA (17.2 $\times$ 10^{-4} in.–lbs) was slightly better than that with DEPA (31.0 $\times$ 10^{-4} in.–lbs). The breaking strength of the DEPA-treated fabric at 30.4 lb was 9 lb better than that of DMPA-treated fabrics. An increase in the concentration of DMPA to 30% (wt) gave a fabric with better breaking strength but slightly more stiffness. With DMDHEU

Table 3. Efficiency of the Reaction of Dimethyl Phosphoramidate (DMPA) in the Presence of Trimethylolmelamine (TMM)1/ or Dimethyloldihydroxyethyleneurea (DMDHEU)1/ and an Acid Catalyst2/ with Cotton Cellulose3/

Run No.	DMPA, % (wt.)	Crosslinking Agent % (wt.)	Crosslinking Agent Name	Add-on %	N, %	HCHO, %	P, %	N/HCHO/P Ratio	Efficiency % (based on P)
1	20	10	TMM	13.9	4.51	2.42	1.20	8.3/2.1/1	27
2	30	10	TMM	14.8	4.45	2.51	1.25	7.9/2.1/1	19
3	20	10	DMDHEU	12.5	1.78	1.61	1.72	2.3/1/1	39

1/ Permafresh 183 K (45% (wt.) aqueous solution of DMDHEU); Resloom HP (TMM).

2/ Catalyst H7 (a buffered 35% (wt.) zinc nitrate solution) 2.8% (wt.) in treating bath.

3/ Cotton flannelette swatches were padded two dips, two nips through squeeze rolls adjusted to 60 lb. roll pressure to give approximately 100% wet pickup, dried 7 min. at 60°C, cured 3 min. at 160°C, washed in Triton X-100 in washing machine, tumble-dried 30 min. in gas dryer.

Table 4. Physical Properties of Cotton Cellulose Finished with Dimethyl Phosphoramidate (DMPA) in the Presence of TMM or DMDHEU from Table 1.

Run No.1/	Match Test Angle deg.	O.I.	Breaking Strength, (W), lb.	Elongation, (W), %	Tear Strength, Grams	Wrinkle Recovery, (W+F), degrees Dry	Wrinkle Recovery, (W+F), degrees Wet	Tinius Olsen, Stiffness, x 10^4 in.-lb.
1	135	0.270	21.1	5.7	1000	217	231	17.2
2	135	0.271	30.7	6.4	950	208	220	32.5
3	120	0.255	18.0	5.0	667	244	231	3.6
UFC2/	0	0.182	33.1	6.7	1983	169	176	3.2

1/ Run numbers same as those in Table 1.

2/ UFC = untreated cotton flannelette control.

(run 3) as the crosslinking agent antiflame properties were not quite as good as with TMM but wrinkle recoveries, both wet and dry, were enhanced and the fabric had a very supple hand.

Durability to Laundering Studies

Cotton flannelette was treated with a solution consisting of 20% (wt) DEPA, 20% (wt) TMM, and 2.8% (wt) catalyst H7 by padding through rolls at 60 lbs pressure to 100% wet pickup, dried 7 min at 60°C, cured 3 min at 160°C, washed in hot tap water 15 minutes and air dried overnight. Properties listed in Table 5

TABLE 5. LAUNDERING STUDY OF COTTON FLANNELETTE FINISHED WITH 20% (WT) DIETHYL PHOSPHORAMIDATE (DEPA), 20% TRIMETHYLOL MELAMINE (TMM) IN THE PRESENCE OF 2.8%(WT) CATALYST H7 [1]

Properties of Finished Fabric	Number of Launderings					
	0	10	20	30	40	50
N,%	7.31	6.79	6.34	6.35	6.33	6.46
P, %	1.62	1.33	1.32	1.24	1.23	1.18
HCHO,%	4.64	4.14	4.03	3.88	3.78	3.76
Breaking strength(w), lbs	40.7	33.6	34.5	31.9	32.1	31.3
Tinius-Olsen, Stiffness X 10^{-4} in.-lb	197.5	8.7	6.3	6.8	5.5	5.5
Wrinkle recovery (W+F), degrees						
Dry	184	219	224	229	221	230
Wet	222	223	212	218	212	206
Match Test, degrees	180	180	180	180	180	180
O. I.	0.320	0.293	0.302	0.303	0.304	0.300
Char length, in.	2.2	7.3	4.3	9.7, BEL [2]	7.8, BEL	5.5, BEL

1/ Fabrics were padded two dips-two nips through rolls at 60 lb pressure to 100% wet pickup, dried 7 min at 60°C, cured 3 min at 160°C, washed 15 min in hot tap water and air-dried. Subsequent laundering was done as described in the experimental section. Initial add-on was 28.2%.

2/ BEL = Burned entire length.

were determined before laundering and after successive intervals of 10 laundering cycles up to and including 50 cycles. Although nitrogen, phosphorus, and formaldehyde contents decreased with increase in the number of washings and dryings, the high match angle and O.I. values still indicated good flame retardancy. Char lengths were very good through 20 laundering cycles. Even after 50 launderings, 95% of the strength of untreated cotton flannelette, 33.1 lbs, was retained. After water washing but before any laundering stiffness values of the cured fabric were high (197.5×10^{-4} in.–lbs); one washing and drying lowered this value to 33.5. Stiffness values determined after extended launderings were all very good. About 0.1% calcium was found in the laundered fabrics and remained nearly constant through 50 launderings. Magnesium, initially determined at 0.99% before laundering, decreased to 0.3–0.7% very erratically. Addition of sesquimethylolurea (urea reacted with 1.5 moles of formaldehyde) to the pad bath in 5% (wt) concentration decreased the initial stiffness value from 197.5 to 31.3. However, match angles were decreased from 180 to 135 degrees and O.I. values from 0.320 to 0.273.

After 50 home-type launderings, the flame retardancy of DMPA finished cotton flannelette of run 1, Table 3, was not as durable as that of DEPA finished fabric; its match angle decreased from 135 to 90 degrees.

SUMMARY

Diethyl phosphoramidate (DEPA) was attached to cotton flannelette with various cellulosic crosslinking reagents such as trimethylolmelamine (TMM), dimethyloldihydroxyethyleneurea (DMDHEU), dimethylolpropyleneurea (DMPU), and tetramethylolglycouril (TMGU) in the presence of an acidic catalyst. Very good flame retardant properties were imparted to the cotton when TMM was the crosslinking agent. The best wet and dry wrinkle recovery properties were obtained with DMDHEU as the crosslinking agent. Laundering studies indicated taht the DEPA-TMM finish was moderately durable to home-type washing and tumble drying even though match test angles and O.I. values remained high through 50 washing and drying cycles.

Dimethyl phosphoramidate (DMPA) was also attached to cotton flannelette with TMM and DMDHEU in the presence of an acid catalyst. Although good flame retardance was obtained with TMM, the flame retardancy of the finished fabric was not as durable as DEPA to home-type launderings. Finishes with DEPA and DMPA had minimal amount of formaldehyde release during the curing operation.

ACKNOWLEDGMENT

We thank Mr. John Mason and Mrs. Ann Black for nitrogen and formaldehyde analyses; Mr. Biaggio Piccolo for determinations of phosphorus, chlorine, magnesium, and calcium; Mr. Matthew F. Margavio for determination of oxygen index values; and members of Textile Testing Research Group for determination of the physical properties of treated fabrics.

REFERENCES

1. Badische Anilin und Soda Fabrik, "Improvements in the Production of Nitrogenous Condensation Products," British Patent 720, 386 (Dec. 15, 1954).
2. Federal Register 36, No. 146 (1971).

3. A. W. Frank, E. J. Gonzales, M. F. Margavio, S. L. Vail and W. A. Reeves, "New or Unusual Approaches to Flameproofing," presented at the 13th Cotton Utilization Research Conference, New Orleans, La., May 14–16, 1973.
4. N. Glade and I. Hechenbleikner, "Textile Fire Retardant Treatment," U.S. Patent 2,828,228 (Mar. 25, 1958).
5. N. Glade, "Textile Treatment with Novel Aqueous Dispersions to Achieve Flame-Resistant and Water-Repellant Finishes," U.S. Patent 2,971,929 (Feb. 14, 1961).
6. I. Hechenbleikner, "Aqueous Flameproofing Compoisitions and Cellulosic Materials Treated Herewith," U.S. Patent 2,832,745 (April 29, 1958).
7. J. E. Hendrix, G. L. Drake, Jr., and R. H. Barker, "Pyrolysis and Combustion of Cellulose. III. Mechanistic Basis for the Synergism Involving Organic Phosphates and Nitrogenous Bases," J. Appl. Polymer Sci., *16*, 257–274 (1972).
8. J. E. Malowan, *Inorganic Synthesis* Vol. 4, p. 77, J. C. Bailar, Jr., editor-in-chief, McGraw-Hill (1953).
9. W. A. Reeves, O. J. McMillan, Jr., and J. D. Guthrie, "Chemical and Physical Properties of Aminized Cotton," Textile Res. J., *23*, 527-532 (1953).
10. W. J. Roff, "The Determination of the Formaldehyde Yield of Cellulose Textiles Treated with Formaldehyde, Urea-Formaldehyde, or Melamine-Formaldehyde," J. Textile Inst. Transactions *47*, T309-T318 (1956).
11. V. W. Tripp, B. Piccolo, D. Mitcham, and R. O'Connor, "X-ray Fluorescence Analysis of Modified Cottons," Textile Res. J. *34*, 773-777 (1964).
12. U.S. Federal Supply Service "Textile Test Methods," Federal Test Method Standard No. 191, Method 5202.

(3)

LUDWIG REBENFELD AND BERNARD MILLER

Textile Research Institute
Princeton, New Jersey

THE SPECIAL CASE OF TEXTILES IN THE FLAMMABILITY OF POLYMERIC MATERIALS*

(Received June 24, 1974)

ABSTRACT: A wide variety of synthetic polymer man-made fibers have come to play a dominant role in the textile industry. The advent of these fibers has changed the character of the textile industry from a two-fiber (cotton and wool) to a multifiber industry which in turn has made possible a virtually unlimited number of multicomponent blends. With increasingly broad and stringent flammability standards being imposed on textile materials, it is necessary that we understand the chemical and physical factors that govern flammability behavior. This paper outlines some of the special problems posed by textile materials with regard to flammability. Included among these special problems are the multifiber character of most textile products, the high surface-to-volume ratio of textiles, the variability of chemical composition, and the unpredictability of shape and configuration of the final textile product. Specific experimental data are presented, obtained with the TRI Flammability Analyzer, showing the dependence of flame temperature on fiber content in several mixed fiber systems. Burning rate studies on multilayered systems are also presented. These data reveal clearly that the flammability behavior of many multicomponent fiber systems cannot be reliably predicted from the behavior of the parent species.

INTRODUCTION

EVERY THOUGHTFUL person is aware that a worldwide revolution has taken place in the materials industry within the past three decades. In that period, a whole new class of materials, namely the synthetic or man-made organic polymers, has emerged and evolved from its status as a relatively obscure laboratory curiosity to an important class of materials. The increasing use of these synthetic polymeric materials as general engineering elements of construction and as specific functional materials in terms of design and performance can be documented in many ways. A particularly striking example has been given by Gratch [1] in terms of plastics consumption in automobiles. The average plastics consumption per car has grown from essentially zero in 1950 to approximately 20 lbs in 1960, and to well in excess of 100 lbs in 1971. Predictions indicate that the 1975 plastics consumption will be

*Based on a presentation to the NAS/NRC/NMAB ad hoc Committee on Fire Safety Aspects of Polymeric Materials in Princeton, NJ, on March 25, 1974.

approximately 200 lbs per car, an order of magnitude increase in a 15-year period. Polymers are beginning to take a dominant position in the world of materials. Indeed, in many specific domains polymers have already completely displaced the more traditional organic, metallic, and ceramic substances.

This paper will examine the place of textile fibers within the polymeric materials industry, and indicate the special problems posed by textile materials with regard to their flammability behavior.

FIBER CONSUMPTION

The growing use of man-made polymers is nowhere better illustrated than in the case of textile fibers. In Table 1 is shown the U.S. polymer production for 1971 and 1972, indicating that in both those years approximately 20% of this production was in the form of fibers and filaments. The extent to which man-made fibers have come to dominate the U.S. fiber scene is shown in Table 2. In 1960 cotton accounted for 65% of the fiber used by the U.S. textile industry. In 1973 the cotton share of U.S. fiber consumption decreased to approximately 29% and man-made fibers accounted for nearly 70% of the market share. As can also be seen in Table 2, there is no apparent levelling off in the increasing use of man-made fibers, although it appears that the cotton and wool consumption is not decreasing. In other words, the growth in fiber consumption is in the area of man-made fibers. It should be emphasized that man-made fibers refers to those fibers that are manufactured from synthetic polymers as well as those fibers that are manufactured from naturally occurring polymers, principally cellulose, either in the form of regenerated material (rayon) or in the form of a chemical derivative (acetates). The principal fiber types are listed in Table 3, showing clearly that the major growth has taken place in the man-made fibers based on synthetic polymers. Within this category, the polyester fibers have exhibited spectacular growth. While the data shown in Tables 1 to 3 are restricted to U.S. consumption, essentially the same trends would be revealed by data on a worldwide basis.

*Table 1. U.S. Polymer Production**
(Billion Pounds)

	1971	1972
Rubber	4.95	5.45
Fibers	6.15	7.32
Plastics	21.10	24.20
Total	32.20	36.97

* Source: Chemical & Engineering News, May 7, 1973 (p. 10).

Table 2. Fiber Consumption by U.S. Textile Mills (Billion Pounds)*

	Cotton	Wool	Man-Made	Total
1960	4.19	0.41	1.87	6.48
1970	3.81	0.24	5.50	9.56
1971	3.95	0.19	6.53	10.67
1972	3.84	0.22	7.57	11.65
1973	3.65	0.17	8.80	12.62

* Source: U.S.D.A. Economic and Statistical Analysis Division via Textile Highlights, American Textile Manufacturers Institute, Inc., Washington, D. C., December 1973.

Table 3. U.S. Man-Made Fiber Production (Billion Pounds)*

	1960	1970	1971	1972	1973
Rayon	0.74	0.88	0.92	0.96	0.89
Acetate	0.29	0.50	0.48	0.43	0.46
Nylon	0.41	1.36	1.60	1.97	2.18
Polyester	0.11	1.43	1.76	2.24	2.77
Olefin	0.01	0.20	0.26	0.35	0.42
Acrylic	0.14	0.49	0.55	0.63	0.74
Glass	0.17	0.47	0.47	0.57	0.69
Other	0.01	0.05	0.07	0.10	0.13

* Source: Textile Organon, January-February 1974, Textile Economics Bureau, Inc., New York.

There are two important conclusions to be drawn from these fiber consumption statistics. The first is that synthetic polymer fibers have come to be the dominant fibers in the U.S. textile industry. The second even more important point is not that one fiber type has increased in consumption more than another, but that the character of the U.S. textile industry has changed dramatically from a two-fiber industry (cotton and wool) to a multifiber industry. The textile industry has participated fully in the previously described worldwide materials revolution.

A particularly important consequence of the multifiber character of the textile industry is that most textile products that are being made available to the consumer today are composed of more than one fiber type. With the advent of a multifiber textile industry, it became clear that certain advantages, in terms of both aesthetics and performance, could be achieved by mixing or blending two different fiber types into a yarn or fabric from which the final product is produced. Two-component blends have become commonplace, and there are certain articles of apparel, such as men's shirts, where it is nearly impossible to obtain anything else. Approximately 50% of the fibers consumed by the U.S. textile industry are in the form of fiber blends. Furthermore, an increasing number of three- and four-component blends can be expected in the future. These multifiber products will be carefully designed and engineered on the basis of material properties to satisfy certain key aesthetic and end-use performance demands. From a materials science point of view, we must consider multicomponent blends as new materials with new and unique combinations of chemical and physical properties. Thus, the advent of a multifiber textile industry has created in the world of textiles a virtually unlimited number of different materials from which the common articles of apparel, household, and industrial textiles are manufactured.

FLAMMABILITY OF TEXTILES

In considering the flammability of textiles, it seems reasonable to divide the overall problem into two broad classifications, namely, the central problem, and the special problems which are peculiarly associated with textiles. Clearly the central problem is the fact that the common textile fibers are composed of organic polymers and that these polymers contain extremely high amounts of carbon and hydrogen which, in the presence of air oxygen, are susceptible to ignition. Once ignition has taken place, the variety of endo- and exothermic thermal transformations and chemical reactions are adequate to propagate the flame. Quite simply, nearly all of the common organic polymers are flammable in a 21% oxygen environment. This is the central problem in the flammability of textiles, which of course is also the central problem in the flammability of all forms of polymeric materials.

Regarding the special problems associated with the flammability of textiles, it has already been pointed out that textile materials are composed, more frequently than not, of several different polymer types. The blended textile materials pose unique problems, and as will be shown, the flammability behavior of a blended or multicomponent fiber structure cannot be directly predicted from the behavior of the components [2, 3]. In many mixed fiber systems there are complex chemical and physical interactions between the components and between the thermal degradation products of the components which make the flammability behavior of the mixed system quite unpredictable. These interactions pose a major problem in the development of suitable flame retardants for each one of the fiber types. A fiber that has been modified in some way to achieve flame resistance or retardance may not be equally retardant when it is a component in a blended or mixed system.

Another one of the special problems associated with textile flammability arises from the fact that there are always numerous additives or finishing agents that are added to fibers by the time the final textile product is manufactured. Anywhere from 2 to 10% of the weight of a textile material is nonfibrous in character. Thus, the chemical composition of a product which may be labelled 50% cotton and 50% polyester is in reality much more complicated. Such a product would normally contain two or three structurally complex dyestuffs, usually with a high degree of aromaticity, amounting to about 2% of the weight of the textile product, and about another 5% of the weight of the product in the form of nitrogenous high molecular weight resins and additives to provide light stability, antistatic properties, soil resistance, durable-press properties, and a host of other aesthetic and performance characteristics. As a result, the exact chemical composition and structure of a textile product is not really known either by the producer or by the ultimate consumer. Yet, the influence of chemical composition and structure on thermal stability and flammability behavior is general knowledge.

Another factor that causes special problems for textile polymeric materials is the extremely high surface-to-volume ratio that characterizes all textile-like substances. Clearly, this has a profound effect on flammability behavior, and even more so on our ability to predict and eventually to control this material characteristic. It is no exaggeration to say that the physical structure of textile materials is ideally designed to make them highly flammable. All textile fabrics are porous with only a fraction of the nominal volume being fiber. At least 50%, and in many cases much more, of a textile fabric is air. One simply cannot imagine any other form of polymeric material being as susceptible to flamming as a textile fabric.

There is another factor that is rather peculiar to textiles. The final polymeric textile product that is manufactured by the textile industry is a two-dimensional planar sheet-like material. Yet, invariably, that is not the final polymeric product that causes the potential hazard to the ultimate consumer. The two-dimensional fabric is transformed to a complex three-dimensional product, which may be an article of clothing, a curtain, draperies, or a piece of furniture. Everyone is aware that the flammability behavior of any material is radically dependent on its shape and on its attitude and position with respect to the ignition source and the direction of the propagating flame front. In textiles, the unpredictable three-dimensionality of the material in the hands of the ultimate consumer poses special and unique problems in terms of performance criteria.

Finally, there is another factor that is really quite general and influences the flammability behavior of all materials, but that is in fact of particular importance with textiles. This factor hinges on the inability to control the environment when a textile material is exposed to an ignition source. In this case, "environment" is used in the broadest possible sense to include such factors as temperature, relative humidity, fluid flow in the gaseous phase, proximity to other solid substances, and of course the chemical and physical properties of these solids. There is a large body of research data being accumulated by various research groups indicating that the flammability properties of a textile material are dramatically influenced by these environmental characteristics.

APPROACHES TO THE TEXTILE FLAMMABILITY PROBLEM

The 1967 amendments to the United States Federal Flammable Fabrics Act, and the subsequent promulgation of flammability performance standards for many textile products, have spurred an intense technical effort on the part of the chemical, fiber, and textile industries. This effort is motivated both by the need to produce textile products that will satisfy the stringent performance standards, and by the desire of industry to provide the consumer with the best possible products. In the development of textile materials that would satisfy flammability performance criteria, a number of approaches have been taken, some of which have already achieved considerable success. These approaches may be summarized as follows:

1. Chemical modification; applicable to cotton and regenerated cellulosic fibers such as the various rayons.
2. Additive finishing of fibers; principally applicable to cotton, rayon, and wool. The chemical finishing agent may be covalently bound to the fiber, which would make this approach the same as No. 1 above. In principle this approach is appropriate to all fibers.
3. Development of polymer variants of existing types for extrusion as man-made fibers. This approach is intrinsically quite difficult since even slight variations in the chemical structure of the base polymers, such as the polyesters, polyamides, acrylics, polyolefins and others, tend to influence adversely the development of crystallinity, orientation, and other morphological features characteristic of fibers.
4. Incorporation of a flame-retardant chemical into the spinning dope prior to fiber extrusion by either melt, dry, or wet spinning processes. This is an approach that has been taken successfully in the case of rayon, cellulose acetate, and polyester fibers, and may possibly become quite generally accepted for other man-made fibers. The chemical additives that will impart the desirable flame retardancy must be compatible with the spinning dope, must be stable to the high temperatures during extrusion (melt spinning), and must not interfere with the development of fiber structure after extrusion.
5. Development of entirely new fiber types with thermal and flammability properties that are their principal points of interest. These fibers may be referred to as specialty fibers.
6. Special geometric design of yarn, fabric, and garment structure to minimize flammability.

COMPONENTS OF TEXTILE FLAMMABILITY

In order that significant progress be achieved along any one or all of the approaches outlined above, it is necessary to understand quantitatively and rigorously the factors that control the flammability of textile materials. Unfortunately, flammability is not an intrinsic thermodynamically precise physical property. In many ways it is more of a legal term that has many different meanings, and it has been found useful to consider the flammability of textiles in terms of several behavioral components which are listed in Table 4. Unlike flammability, these behavioral components can be quantitatively and reproducibly measured. This means that it is possible to determine systematically how various chemical and physical factors influence these components. It is only by means of such systematic studies that real progress can be expected in understanding the controlling chemical and physical mechanisms that govern textile flammability; and it is only through such understanding that we can expect the development of effective flame-retardant textile materials.

Table 4. Behavioral Components of Textile Flammability

1. Ease of ignition
2. Rate of burning
   - flame propagation rate
   - mass burning rate
3. Flame temperature
4. Heat emission
5. Smoke generation
6. Toxic by-products
7. Extinguishability

TRI FLAMMABILITY ANALYZER

When the modern era of research on textile flammability began several years ago, it became immediately apparent that the lack of suitable measurement techniques posed a significant problem in itself. Many people attempted to adapt some of the varied testing procedures that had been developed in the past for control purposes, but it became obvious that these test methods could not be used for research purposes, since their primary purpose was to give pass-fail results and not reproducible quantitative data that would provide a sense of direction to the research scientist. It therefore became a primary task to develop laboratory techniques that would allow the generation of such data.

Among the various components of textile flammability, the rate of burning is obviously a most important one. While the ideal situation would be one in which textile materials did not ignite under any set of conditions, the more realistic situation reflects the fact that textile materials do ignite, and thus their rate of burning becomes of critical importance. In the normal burning process, after ignition has taken place, the flame front moves along the specimen creating a continuously changing situation with respect to the gaseous environment, boundary conditions, and the heat transfer processes in the vicinity of the flame front. As a result, burning rate studies in the conventional mode of a moving flame front generate data of poor reproducibility. A far better set of experimental conditions would be achieved in an arrangement wherein the flame front was stationary in space. To achieve such an arrangement, the specimen would have to be fed into the flame front, with the rate of fabric feed required to maintain the flame stationary being equal to the rate of burning. Furthermore, in examining again the behavioral components of textile flammability (Table 4), it is apparent that many of these items such as flame temperature, heat emission, and smoke generation, could be measured more precisely if the flame were stationary in space.

In the final design of the TRI Flammability Analyzer [4], which is shown in Figure 1, feeding fabric into a flame at a known and variable rate is accomplished by mounting the sample on the rims of a pair of linked wheels rotating on a horizontal axis, and arranging for the continuous adjustment of rotational speed. The operator ignites the fabric by allowing a pilot flame to impinge momentarily on the sample. The ignition source is removed, and the speed of the wheels adjusted until the flame is observed to remain stationary at some predetermined point of the circular path. The rate of fabric movement is then equivalent to the rate of flame propagation. An important feature of the Flammability Analyzer is that any direction of flame progress can be measured with this system. With the flame kept at the top of the fabric path (i.e., at 12 o'clock), burning is in effect horizontal. However, the same type of experiment can be performed with the flame remaining at 3 o'clock (vertical upward burning), at 9 o'clock (candle-like downward burning) or at any predetermined angle of inclination. The instrument has been fully described by Miller and Meiser [5], and the burning rates of various textile systems have been determined.

Figure 1. *TRI Flammability Analyzer.*

FLAMMABILITY BEHAVIOR OF BLENDS

The TRI Flammability Analyzer has made possible many important studies of the flammability characteristics of textile materials. Notable among these are those relating to the effects of chemical composition on flammability behavior, such as is achieved by the systematic variation of fiber content in blend fabrics. Using flame temperature as the response criterion, examples of radically different behavior patterns have been observed for some of the common mixed textile systems. In Figure 2 are shown the flame temperatures of polyester/cotton mixtures as a function of the oxygen content of the immediate environment during burning. In this case, as in all other systems studied, flame temperature is linearly dependent on the oxygen concentration. This dependence (in terms of the slope of the generated lines) can be a sensitive indicator of any changes in flammability behavior. In this particular case it is obvious that at any given oxygen concentration the flame temperatures of all mixtures of polyester and cotton are essentially the same as the flame temperature of pure cotton. One can see that during normal burning at 21% oxygen the polyester material in the mixed systems will experience a temperature nearly 100 degrees higher than it would if it were burning by itself. Even if this increased temperature did not produce new chemical reactions, the effect on the kinetics of the combustion processes would obviously be significant. Thus, we would predict from these data that polyester/cotton blends should show somewhat enhanced flammability over what might be expected from them on the basis of the two component materials. This has been confirmed by a variety of evaluation techniques, including oxygen index measurements and the determination of the heat emitted from such burning systems.

Another system studied in this manner has been the combination of polyester and wool fibers, resulting in flame temperature data as shown in Figure 3. Here we have a clear indication that there is some chemical interaction when these two species are burned together. The oxygen sensitivity of the flame temperatures for the blends is considerably different from that of the two parent species, which coincidentally have just about the same oxygen sensitivities. On the other hand, mixing polyacrylonitrile and wool does not seem to change the flame temperature-oxygen concentration relationships of the two components. Figure 4 shows that mixtures of these two fibers produced flame temperatures that are essentially weighted averages of the temperatures found for the pure components. Here no interaction of any kind seems to be occurring.

In contrast to the above three examples, mixtures of nylon and cotton exhibit a different type of behavior. Figure 5 shows that specific mixtures of these two polymers can result in flame temperatures either higher or lower than that of either cotton or nylon alone.

FLAMMABILITY OF LAYERED SYSTEMS

In nearly all apparel and household uses of textile materials, fabrics are found in

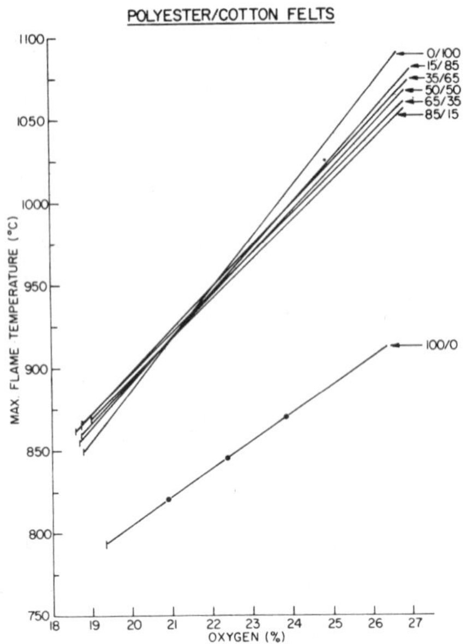

Figure 2. *Dependence of flame temperature on oxygen concentration for a series of polyester/ cotton felts.*

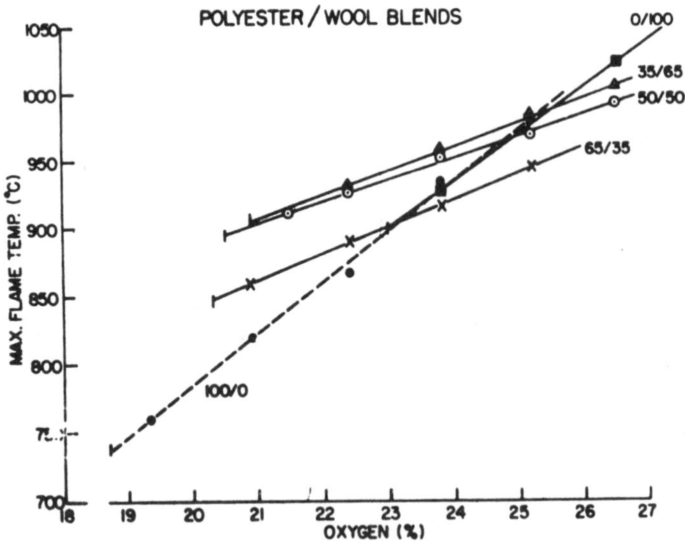

Figure 3. *Dependence of flame temperature on oxygen concentration for a series of polyester/wool blends.*

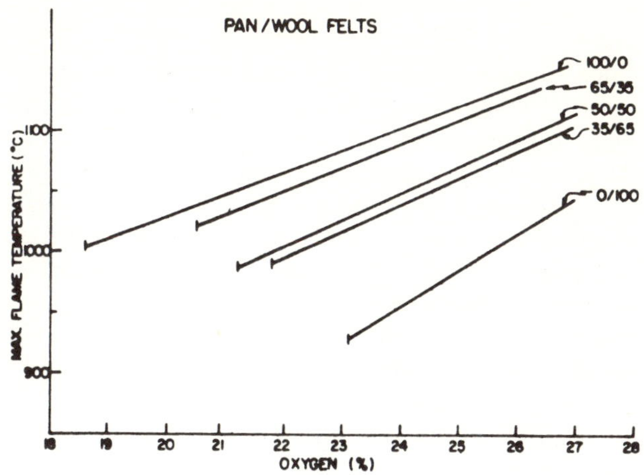

Figure 4. *Dependence of flame temperature on oxygen concentration for a series of polyacrylonitrile/wool felts.*

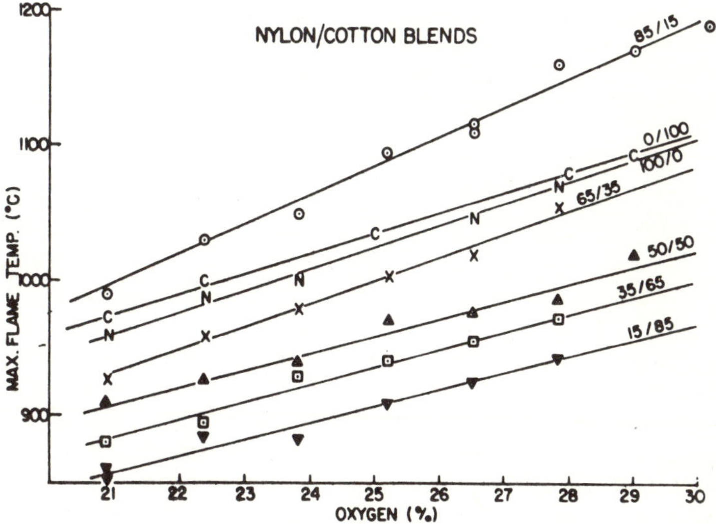

Figure 5. *Dependence of flame temperature on oxygen concentration for a series of nylon/cotton felts.*

close proximity to each other in various geometric arrangements. One such arrangement that can be visualized readily is that of a double layer. Obviously there are many possible double layer systems, such as two fabrics of different fiber content, two identical fabrics, an untreated fabric and a flame-retardant treated fabric, and several others. A particularly interesting case, recently studied by Miller and Turner [6], is that of cotton fabric double layers, where one of the fabrics has been treated with diammonium phosphate (DAP) as a model flame retardant for cellulosic fabrics. The effects of add-on of DAP on flammability has been described by Tesoro and Meiser [2]. As expected, flammability is suppressed by DAP, and, for example, the oxygen index increases linearly with DAP add-on. At intermediate and high add-on levels the flame propagation rate is also decreased, as expected.

Some rather interesting observations are made, however, when the effects of a low add-on level are considered. In Table 5 are shown the effects of a 1% add-on of DAP on various flammability parameters. The major point of interest here is that while flame temperature and heat emission decrease and oxygen index increases, the flame propagation rate is significantly greater than it is for the untreated material. These data confirm an original observation by Coppick [7] in 1947 who reported that inorganic phosphate and sulfamate flame retardants increase the burning rates of cotton fabrics when present at low add-on levels. The mechanism of this flammability enhancement is not known, but it is likely connected with the effects of the flame retardant during the initial thermal decomposition of the cellulose.

Table 5. Effect of 1% Diammonium Phosphate Add-On on the Flammability Parameters of Cotton Fabric

	Untreated Cotton	Cotton + 1% DAP
Propagation Rate, in/min	4.6	6.1
Flame temperature, °C	890	800
Heat emission, cal/oz	47,000	28,000
Oxygen index	0.178	0.186

From the point of view of flammability hazard, the enhancement in burning rate at low add-on levels of flame retardant becomes of great importance in double layer systems. The question arises whether a double layer composed of an untreated and a treated fabric behaves in accordance with the net add-on of the double layer systems. Depending upon the add-on level of the treated fabric, the net add-on of the double layer may be low enough to bring the system into the domain of enhanced flammability. The pertinent data obtained by Miller and Turner [6] confirm that the effect of a flame retardant can be transferred to a neighbor fabric, and that this transfer can produce a more rapidly burning system. When a treated fabric is placed underneath a layer of untreated material, the burning rate of the double layer almost exactly equals the rate for a combination of two identical layers with the same net average add-on. Thus, a combination of untreated cotton over a layer of 2.2% DAP-treated cotton burns at the same rate as two layers of

1.1% DAP cotton. For the untreated over 1.1% treated combination, there is an almost 100% increase in burning rate over that of the untreated material. On the other hand, if the treated material is placed above the untreated, only a slight transfer of the effect of the DAP is observed, except at the highest add-on level, where presumably the excess of retardant is sufficient to accomplish some transfer.

Two-layer burning rate studies have also been carried out with a nylon 66 fabric, using thiourea as a flame-retardant treatment, and with a polyester fabric treated with tris(2,3-dibromopropyl) phosphate. Each material was treated with its respective flame retardant to achieve two levels of add-on, one twice as much as the other. It was therefore possible to have the same net add-on with the retardant either in one or in both of the fabrics comprising the double layer. The results for nylon indicate that a combination of untreated and treated fabrics will burn much slower than two layers of untreated material. For the same net add-on of thiourea a somewhat slower burning rate is produced by having all of the retardant in one layer rather than distributing it equally in both. Treating the second (upper) layer so as to double the net add-on caused only a small additional reduction in burning rate. Polyester fabrics exhibited a similar behavior in that a mixture of untreated and 10% treated fabrics burned much more slowly than did two layers of the untreated. However, in comparing the relative effectiveness of the alternative ways of achieving a net 5% add-on, it seems that putting the additive in both layers produces more retardancy than when only one layer is treated.

Obviously, the transfer effects in the DAP-cotton system are particularly interesting because of the unique enhancement of flammability at low add-on levels. Weight loss studies during programmed heating in air of model DAP-cotton systems has indicated that the transfer effect takes place through the gas phase. Although additional studies are needed to identify the conditions required for transfer and to explain the transfer phenomenon in fundamental chemical and physical terms, it is obvious that both fabric manufacturers and users must be alert to the possibility of enhanced flame propagation in certain multilayer structures.

CONCLUSION

There are numerous general problems associated with the flammability of polymeric materials and a number of rather special problems associated with the flammability of textile materials. The studies of textile flammability behavior that have been summarized in this paper just barely scratch the surface of this technologically complex problem. While we are beginning to generate a body of knowledge concerning the chemical mechanisms of flammability and flame retardancy, there are many other factors that need to be understood. For example, the physical form of the textile material greatly influences flammability behavior, particularly burning rates. Such factors as yarn twist, yarn ply, fiber and yarn linear density, weave or knit construction, bulk density and others play an important role. A number of research workers [8, 9] have reported the effects of these nonchemical constructional form factors on flammability. It is becoming quite clear that a great deal can be done in minimizing flammability by combining the chemical and physical approaches to the problem. Other important items regarding the potential hazard to the ultimate consumer include those dealing with the three-dimensional character of a textile material when it is fabricated into a garment, a curtain, a piece of furniture and other end-products. Many new complex technical problems arise from the unique shape of a textile material when it is placed into its final configuration. We must also consider the dynamics of the burning process. Rarely, after accidental ignition, is the textile material stationary. The panic that follows ignition creates a complex situation which is at best only vaguely approximated by the standard testing procedures [10] that are being promulgated.

With the near certainty that textile materials will be subjected to expanding standards, it is necessary that continuing effort be exerted in understanding the details of the total flammability picture. Only on the basis of a sound understanding of the problem can we hope to realize the ultimate goal of providing the consumer with products that satisfy both his demand for functional performance and the government's demand for safety.

REFERENCES

1. S. Gratch, "Polymer Engineering and its Relevance to National Materials Development," F. R. Eirich and M. L. Williams, editors, Institute of Materials Research, University of Utah, Salt Lake City, June 1973.
2. G. C. Tesoro, and C. H. Meiser, Jr., "Some Effects of Chemical Composition on the Flammability Behavior of Textiles," Textile Res. J. *40*, 430–436 (1970).
3. G. C. Tesoro, and J. Rivlin, "Flammability Behavior of Experimental Blends," Textile Chem. Colorist *3*, 156–160 (1971).
4. U.S. Patent No. 3,667,277, June 6, 1972.
5. B. Miller, and C. H. Meiser, Jr., "A Method for Measuring the Burning Rates of Fabrics," Textile Chem. Colorist *3*, 118–122 (1971).
6. B. Miller and R. Turner, "Transfer of Flame Retardant Effects," Textile Res. J. *42*, 629–633 (1972).
7. S. Coppick, in "Flameproofing Textile Fabrics," Ed. R. W. Little, ACS Monograph Series, Reinhold Pub. Corp., 1947, pp. 50–54.
8. B. Miller, and B. C. Goswami, "Effects of Constructional Factors on the Burning Rates of Textile Structures. Part I: Woven Thermoplastic Fabrics," Textile Res. J. *41*, 949–955 (1971).
9. A. S. Tweedie, M. T. Mittow, and D. M. Wiles, "Rate of Burning Measurements and the Evaluation of Fabric Flammability," Canadian Textile Journal *88*, No. 11, p. 73 and *88*, No. 12, p. 79, (1971).
10. Standard for the Flammability of Children's Sleepwear, DOC FF 3-71 (as amended). Textile Chem. Colorist *4*, 71–76 (1972), and *6*, 130–131 (1974).

Ludwig Rebenfeld

Dr. Rebenfeld received a B.S. in textile chemistry from Lowell Technological Institute in 1951, and a Ph.D. in organic chemistry from Princeton University in 1955. He joined the staff of Textile Research Institute in Princeton, New Jersey in

1954 and was named President of TRI at the beginning of 1971. He is also a Visiting Lecturer with rank of Professor in the Department of Chemical Engineering at Princeton University.

Dr. Rebenfeld is Secretary-Treasurer of The Fiber Society, Inc., a Fellow of the British Textile Institute, a Fellow of the American Institute of Chemists, a Fellow of the Canadian Institute of Textile Science, and a member of the Board of Trustees of the Philadelphia College of Textiles and Science. He has served as President of the National Council for Textile Education and as Chairman of the 1966 Gordon Research Conference on Textiles. In 1968 Dr. Rebenfeld was presented with the Fiber Society Award for Distinguished Achievement in Fiber Science, and he has been named as the recipient of the 1974 Harold DeWitt Smith Memorial Award given by Committee D-13 on Textile Materials of the American Society for Testing and Materials.

Bernard Miller

Bernard Miller has been Associate Director of Research at Textile Research Institute since 1970. He joined the full-time staff at the Institute in 1966 and also was a Postdoctoral Fellow there during 1958–59. He holds Bachelor and Master's degrees from the Virginia Polytechnic Institute and a Ph.D. degree from McGill University, all in chemistry. In addition to several years of industrial research for the duPont and Celanese organizations, he taught for a number of years at American University and Lowell Technological Institute. He was a Fiber Society National Lecturer for 1973–74, and is active in the Information Council of Fabric Flammability. His major fields of interest are the thermal and combustion behavior of polymers, fabric flammability and the surface properties of fibrous materials, on all of which he has lectured and published extensively.

ILDO E. PENSA, STEPHEN B. SELLO

J. P. Stevens & Co., Inc.
Technical Center
Garfield, NJ 07026

WALTER BRENNER

Chemical Engineering Department
Polytechnic Institute of New York
Brooklyn, NY, 11201

FLAMMABILITY BEHAVIOR OF POLYESTER/CELLULOSIC FIBER BLENDS

(Received March 7, 1974)

ABSTRACT: The flammability characteristics of polyester/cotton fiber blends treated with selected model flame-retardant systems have been studied. Thermogravimetric analysis was employed to ascertain the predominant pyrolysis mechanism.

Condensed and vapor phase pyrolysis mechanisms can be differentiated by a newly developed empirical parameter derived from the thermogravimetric data. Major determinants of the mode of pyrolysis were found to be the type of flame-retardant systems employed and the composition of the fiber blends.

INTRODUCTION

BLENDS OF SYNTHETIC fibers with the more common natural fibers have achieved commercial prominence on account of the superior aesthetic and functional properties offered by such blends. Thus a polyester/cotton fiber blend combines the comfort and moisture absorption of cellulose with the shape retention and strength of polyester to give a desirable balance of overall performance properties. However, blending also alters the burning characteristics of the components and may cause difficult flammability problems. During combustion, the cotton fiber in the blend forms a carbonaceous grid that prevents the thermoplastic polyester from melting away from the heat source. As a result of this "scaffolding effect", many blends of thermoplastic and non-melting fibers are more flammable than the synthetic component. Only limited information is available in the literature on the combustion characteristics of polyester/cellulosic blends [1, 2] and on the interaction between the pyrolysis products of the individual fiber types.

It is an objective of this report to establish a correlation between the thermal properties and flammability characteristics of polyester/cotton fiber blends by investigating the effect of various model flame retarding agents on these blends. It is also attempted to determine the predominant operating mode of selected flame retardant models when applied to polyester/cotton blends.

EXPERIMENTAL

Needle punched webs were used as textile substrates for the application of the flame retarding chemicals. These webs were prepared by mixing polyester and cotton fibers at specified weight ratios and passing the fiber mixture through a Rando Feeder/Webber®, followed by needle punching. The fibers used for the manufacture of webs had the following characterists:

Cotton fiber: Mixture of Low Middling Plus 1-3/32″–1-1/8″ staple length, scoured and bleached stock.

Polyester fiber: Blue C Polyester (Monsanto Co.) semi-dull, 2¼ denier, 1½″, staple length.

The fiber content of the blends (determined by ASTM Method D-629) is summarized in Table 1.

Table 1. Fiber Composition of the Experimental Fiber Blends

% Polyester Component	% Cotton Component
–	100
22.7	77.3
45.6	54.4
73.2	26.8
100	–

Some of the average physical characteristics of these fiber webs are tabulated in Table 2.

Table 2. Physical Characteristics of Needle Punched Fiber Webs.

Properties	Test Method	Average Test Data	Range of Data
Weight	ASTM D-1910	4.35 oz/sq yd	4.2–4.6
Thickness	ASTM D-1777	0.068 inches	0.064–0.081
Air Permeability	ASTM D-73	261 (ft^3) (min^{-1}) (ft^{-2})	240–300

The flame retardant (FR) chemicals were applied from aqueous media containing 10% of an acrylic latex such as Hycar 2679, Hycar 2600X171 (B. F. Goodrich) or Rhoplex K-3 (Rohm and Haas). The fiber webs were impregnated with the FR-containing latex compositions, squeezed through rubber rolls to predetermined wet pick-ups and dried in a forced draft oven at 100–110°C.

Fiber webs were treated with increasing add-ons (0–40% on the weight of the treated web; % OWF) of the following FR chemicals: diammonium phosphate (DAP), diammonium phosphate + urea, ammonium bromide (NH_4Br), ammonium bromide + antimony trioxide (Sb_2O_3) and tris-(2,3-dibromopropyl)-phosphate (TDBP).

Flammability data were generated from tests carried out with the GE Flammability Oxygen Index Tester (3–5) using unsupported web specimens conditioned at 65% RH and 21–24°C. Only the untreated 100% polyester test specimens required stitching with fiber glass thread to hold the melting polymer in the flame front.

Thermal analyses were run on the Du Pont 950 Thermogravimetric Analyzer. An 8–10 mg specimen of the fiber web was placed in the pan, then the temperature was raised to 95–100°C (15°C/min) and held there until no change in weight occurred. The weight scale was reset at 100% and the thermal analysis of the bone dry specimen was continued at a heating rate of 15°C/min. To simulate the conditions that exist during the actual combustion of polymeric materials, air was selected as the environment for pyrolysis. A positive air flow of 30 mls/min was maintained throughout the thermogravimetric measurements to remove, from the immediate surroundings of the specimen, any volatile degradation products formed during pyrolysis. A temperature override usually occurs when untreated 100% cotton is analyzed in air. This condition was avoided by using the selected weight for the test specimen and maintaining its original fibrous character.

DISCUSSION

Distribution of Chemicals Deposited Onto Fiber Surfaces

Chemicals applied from aqueous media to polyester/cotton fiber blends deposit preferentially on the hydrophilic surface of the cotton [6, 7]. Non-polar solvents readily wet hydrophobic surfaces [8] and therefore chemicals applied from such solvents deposit preferentially on the polyester component of the fiber blends. Such selective deposition for FR chemicals was demonstrated on a cotton warp and polyester filling fabric [9]. The hypothesis that each fiber component in a blend should be flame retarded with its specific FR system to render the composite flame resistant has been proposed by several researchers [9–13].

Further studies have also shown that flame retardants effective on both fiber types may impart flame resistant properties to the total composite even if such chemicals are deposited on one of the fiber types at the virtual exclusion of the other [14, 15]. As a result, it was stipulated that as long as the flame retardants are in the "vicinity" of the fibers, flame resistance can be achieved. This was substantiated by electron microscopic studies that showed a non-uniform distribution of the flame retardants between the two fiber types of the blend [14].

Investigations were also carried out on discrete applications of flame retarding chemicals to textile substrates [16, 17]. It was found that flame resistance could be achieved even if the chemicals were applied to the textile substrate in a network pattern [16]. This indicated that uniformity of distribution was not a prerequisite for good flame resistance [14, 16] and that transfer of retarding effects could be obtained [18].

Since most of the chemicals used in this study were water soluble, their application to textile substrates was carried out from aqueous media. In order to reduce preferential deposition of the chemicals from such systems, acrylic latices, showing good adhesion to the polyester fiber, were incorporated into the treatment. These "flame retardant + latex" aqueous systems were applied to the needle punched fiber webs and dried at relatively low temperatures to reduce excessive migration of the deposited chemical.

The adhesion of chemicals to polyester fiber was studied by electron microscopy. When DAP is deposited from an aqueous solution, most of the DAP crystals are held mechanically onto the surface of the polyester fiber (Figure 1). The types of deposition obtained on polyester with the acrylic latex alone and with added DAP are shown in Figure 2, and Figure 3, respectively. It is apparent (Figure 3) that the DAP crystals are embedded in the latex surface and are adhering to the fiber. Microscopic views of DAP-treated polyester/cotton blends (Figure 4 and Figure 5) indicate that DAP is distributed more uniformly between the polyester and the cotton fiber components when it is applied in the presence of the latex.

The flammability properties of polyester/cotton blend substrates, with and without an acrylic latex treatment, were compared. Oxygen index (OI) data (Table 3) show that the effect of the latex on flammability properties of the blends is minimal. It was also found that the presence of latex eliminated the need for fiber glass stitching of the 100% polyester test specimens. The similarity of the thermogravimetric traces obtained on latex-treated and untreated fiber blends also showed the negligible effect of the latex.

Thermal Properties of Unmodified Fiber Blends

A thermogravimetric study [19] has suggested the following sequence for the thermal decomposition of polyester/cotton blends: the cotton component of the blend starts to decompose at about 260°C; most of the heat generated during the exothermic degradation is absorbed by the polyester component; the polyester begins to pyrolyze around 325°C; above 380°C, the cotton component has been completely decomposed and further degradation is essentially due to the polyester. It has also been shown that the residue of the blend at 480°C is less than the sum of the residues produced by each component if pyrolyzed separately. It has been implied [19] that chemical interaction occurred between the fiber components of the blend during thermal decomposition. However, there is no information available on the mechanisms involved in this interaction or on the intermediate products formed.

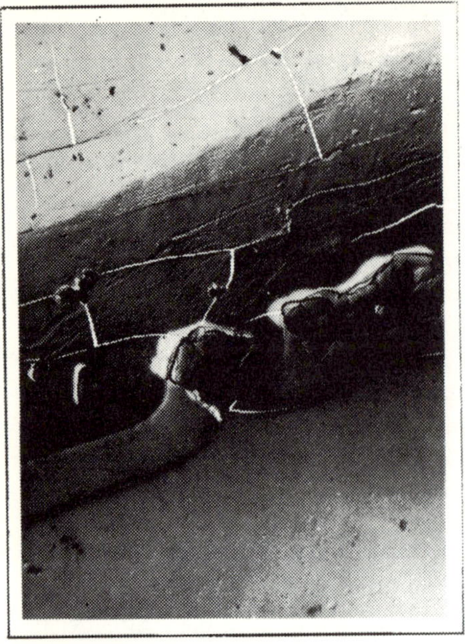

Figure 1. Electron micrograph of polyester fiber treated with diammonium phosphate (3.0% OWF) magnification: 6700X.

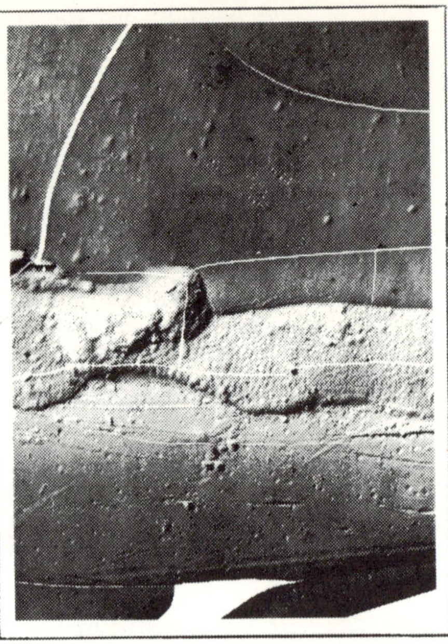

Figure 2. Electron micrograph of polyester fiber treated with acrylic Hycar latex 2600 × 171 (1.9% OWF) magnification: 6700X.

Figure 3. Electron micrograph of polyester fiber treated with diammonium phosphate (4.0% OWF) and acrylic Hycar latex 2600 × 171 (2.0% OWF) magnification: 6700X.

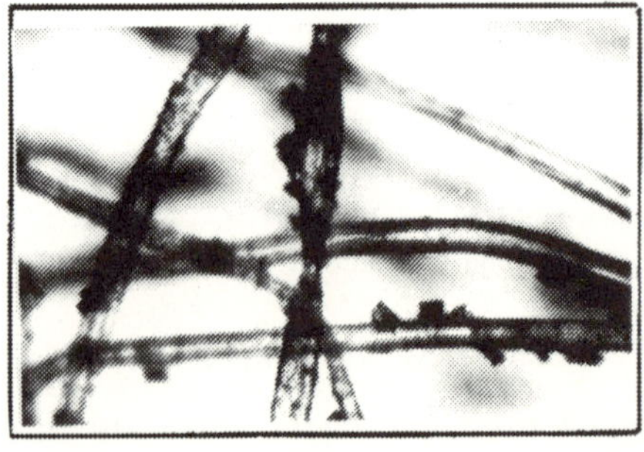

Figure 4. Micrographs of 46% polyester: 54% cotton fiber webs treated with diammonium phosphate (30.6% OWF) and acrylic Hycar lates 2600 × 171 (4.7% OWF) magnification: 220X.

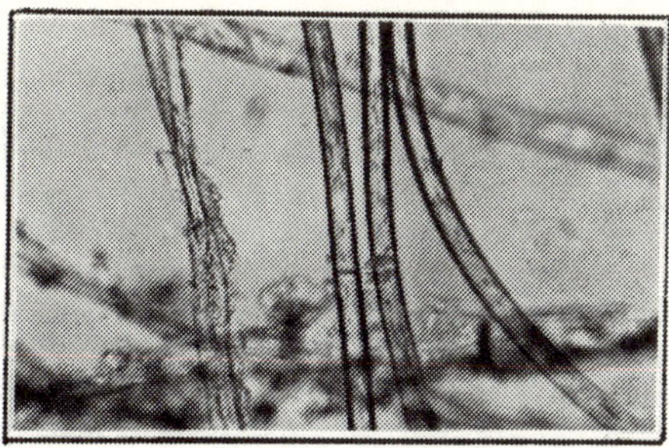

Figure 5. Micrographs of 46% polyester: 54% cotton fiber webs treated with diammonium phosphate (28.8% OWF) magnification: 220X.

Table 3. Effect of Latex Binder on the Oxygen Index (OI) of Polyester/Cotton Fiber Blends

Fiber Blend Composition		Oxygen Index Values	
% Polyester	% Cotton	Webs WITH Latex 4.0-4.7% OWF*	Webs WITHOUT Latex
100	-	.205	.204**
73	27	.203	.202
46	54	.197	.199
23	77	.197	.196
-	100	.191	.192

* % OWF = % solids on the weight of the treated fiber web, i.e. % latex solids + % fiber = 100%

** Specimen tested with fiberglass thread stitch support.

Thermograms of untreated polyester/cotton fiber blends as well as 100% cotton and 100% polyester webs (Figure 6) were developed in this investigation. The bicomponent character of the fiber blends is apparent from the shape of these thermograms. The final char weight of the samples (w_r for 100% cotton webs and w_r^I for 100% polyester and polyester/cotton webs) was computed by extrapolating the main decomposition curve to the abscissa and determining the weight at that temperature (Figure 7).

The data developed from the TGA plots (Table 4) show that the inflection points, indicative of minimal rates of weight loss in the pyrolysis of blends, occur at 375–380°C. This temperature range corresponds to the completion of the main decomposition reaction of 100% cotton and to the beginning of the decomposition of 100% polyester. The latter exhibits a 4.0–4.5% weight loss at 375–380°C.

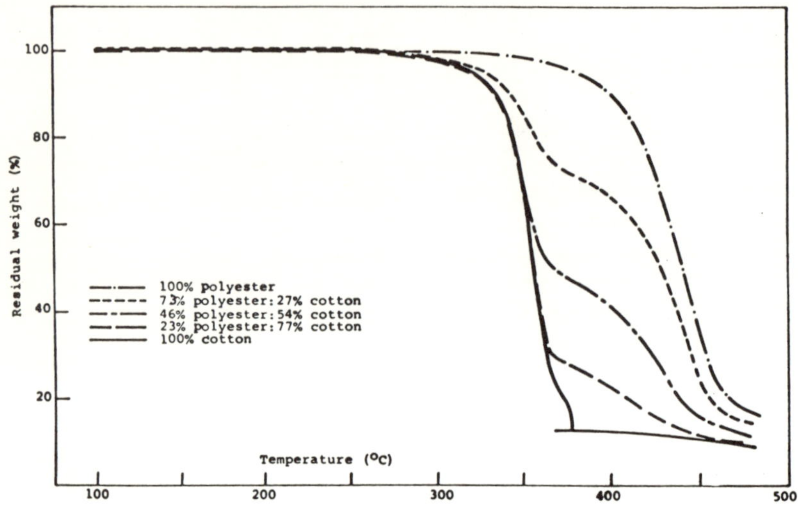

Figure 6. TGA plots for untreated webs of 100% cotton, 100% polyester and polyester/cotton blends.

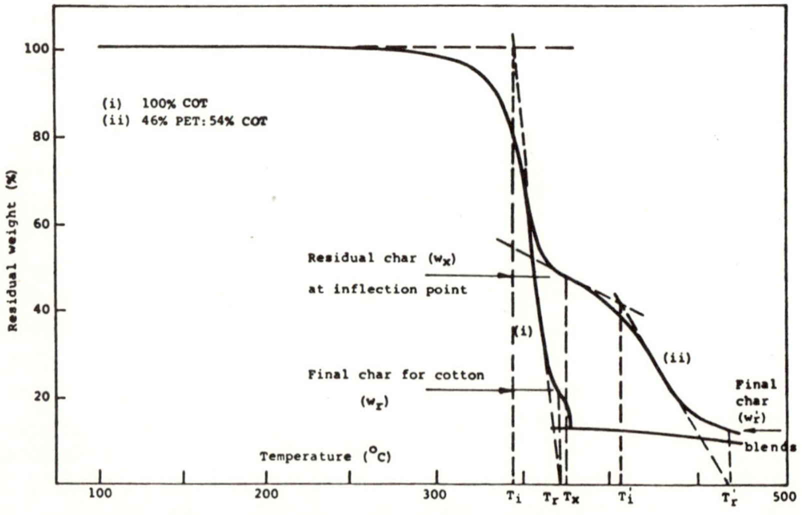

Figure 7. Graphical determination of residual (w_x) and final (w_r, w_r') char weights for untreated 100% cotton and 46% pet: 54% cot fiber webs.

Table 4. Thermogravimetric Data for Untreated Webs of 100% Cotton, 100% Polyester and Polyester/Cotton Blends

Fiber Blend Composition	% Polyester / % Cotton	100 / -	73 / 27	46 / 54	23 / 77	- / 100
COTTON COMPONENT						
Initial Decomposition Temperature	$[T_i]$ (°C)	-	334	333	337	343
Final Char Temperature	$[T_r]$ (°C)	-	-	-	-	371
Final Char Weight	$[w_r]$ (%)	-	-	-	-	20.0
Inflection Point: Temperature	$[T_x]$ (°C)	-	378	375	380	-
Residual Weight	$[w_x]$ (%)	-	71.7	48.0	27.0	-
POLYESTER COMPONENT						
Initial Decomposition Temperature	$[T_i']$ (°C)	406	419	405	393	-
Final Char Temperature	$[T_r']$ (°C)	474	466	467	467	-
Final Char Weight	$[w_r']$ (%)	18.5	16.5	13.0	10.0	-

The OI values for untreated webs of polyester, cotton and their blends are plotted as a function of fiber content (Figure 8). The addition of cotton to polyester results in a progressive decrease of the OI of the fiber webs. The variations in the shapes of the OI plots reported in the literature [9, 20, 21] can be attributed to differences in the textile and FR test parameters. It is, however, important to note that, whatever the shape of the curve, the deviation in OI values is quite small (0.010–0.015).

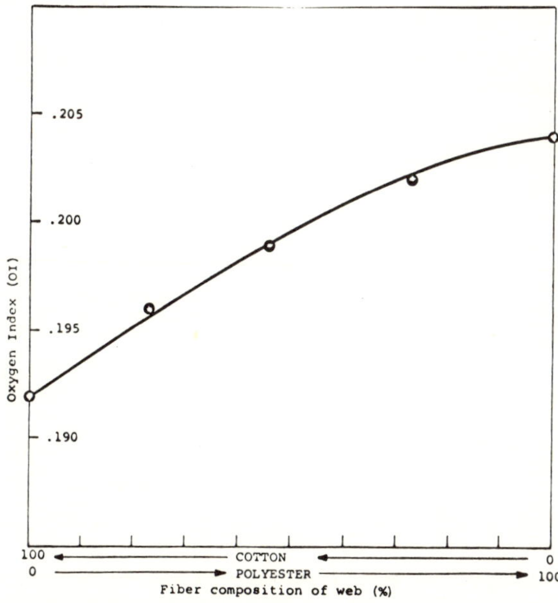

Figure 8. Oxygen index as a function of fiber composition of the webs.

The Residue Number (N_r)

A correlation of experimentally determined final char weights with the flammability properties and flame-retardant add-ons of FR-treated webs could generate useful information as to the main operating mechanism of the FR agents, provided a common and standard base can be used for computation. Therefore the concept of residue number (N_r) was empirically developed to fill this need. The N_r values for FR-treated fiber systems were computed from the following expression:

$$N_r = (w_r/F^*)/w_{ro}$$

w_r = percent final char residue for the FR-treated fiber webs obtained from the TGA data

F^* = fiber fraction in the FR-treated fiber system: (Fiber fraction (F^*)) + (FR fraction (F^*_a)) = 1.0

w_{ro} = percent final char weight of the untreated fiber webs obtained from the TGA data

The experimentally determined final char residues of the FR-treated webs were converted to percent residues with respect to the total fiber portion of the FR/fiber system. Thus a common basis for all final char residues (w_r) has been established. N_r is the ratio of this percent residue and the final char residue (w_{ro}) of the untreated fiber blend.

N_r values were computed from thermogravimetric traces of the polyester/cotton fiber webs treated with increasing add-ons of the various model flame retardants. Plots of N_r as a function of OI values obtained on these treated fiber webs are shown in Figures 9–13.

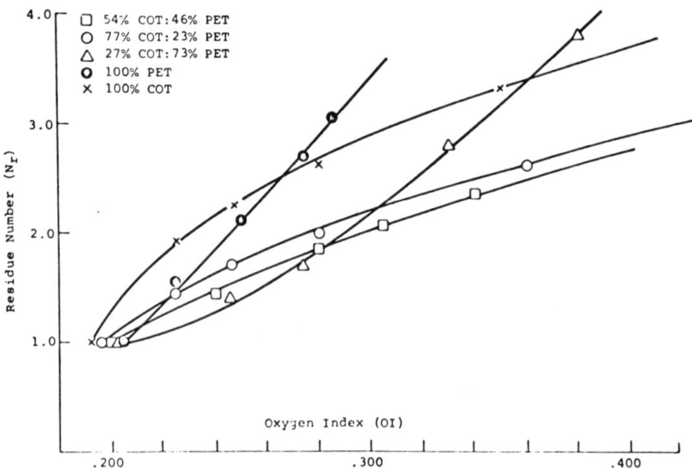

Figure 9. Residue number as a function of oxygen index for webs of 100% polyester, 100% cotton and polyester/cotton blends treated with diammonium phosphate (0–36% OWF).

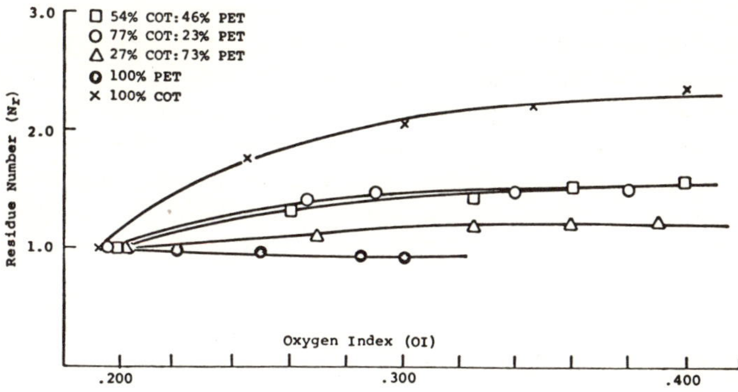

Figure 10. Residue number as a function of oxygen index for webs of 100% polyester, 100% cotton and polyester/cotton blends treated with ammonium bromide (0–40% OWF).

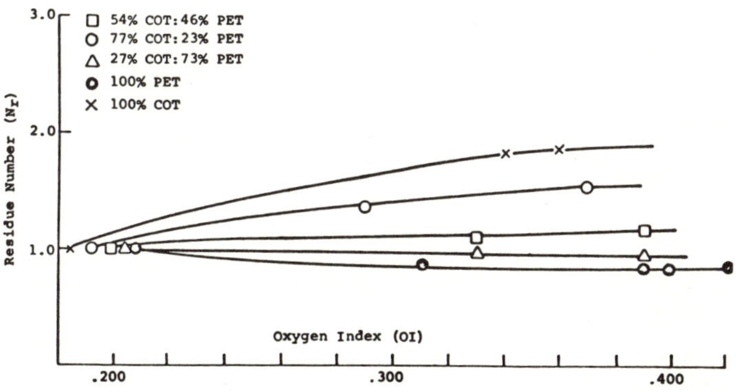

Figure 11. Residue number as a function of oxygen index for webs of 100% polyester, 100% cotton and polyester/cotton blends treated with tris-(2,3-dibromopropyl)phosphate (0–23% OWF).

Evaluation of Experimental Data

As reported in the literature, FR agents may operate predominantly in either the condensed or vapor phase [22–24]. One method for distinguishing between the two modes of flame retardancy is to vary the type of oxidizing environment in which pyrolysis is performed [1, 2]. Condensed phase FR agents are not equally effective on all types of substrates since these agents operate by altering the pyrolysis reactions of the treated substrates. However, their effectiveness is independent of the nature of the oxidizing environment in which thermal decomposition occurs. Vapor phase FR agents operate by "poisoning" the flame reactions, and in most cases their performance is similar on various types of substrates. As a result, their

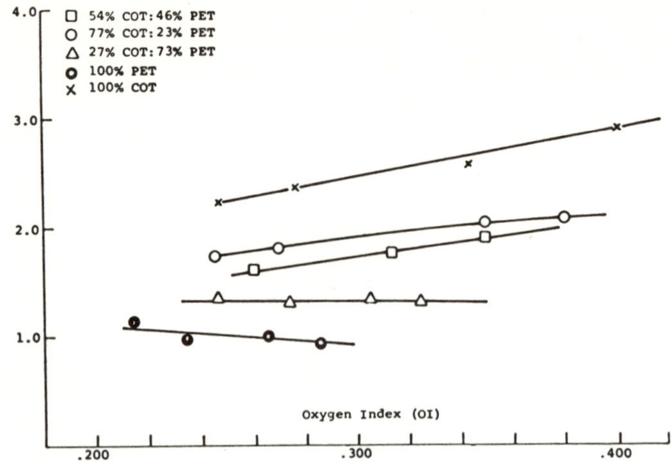

Figure 12. Residue number as a function of oxygen index for webs of 100% polyester, 100% cotton and polyester/cotton blends treated with diammonium phosphate (5.2% OWF) + urea (0–33% OWF). The DAP is kept constant and the urea add-on is varied.

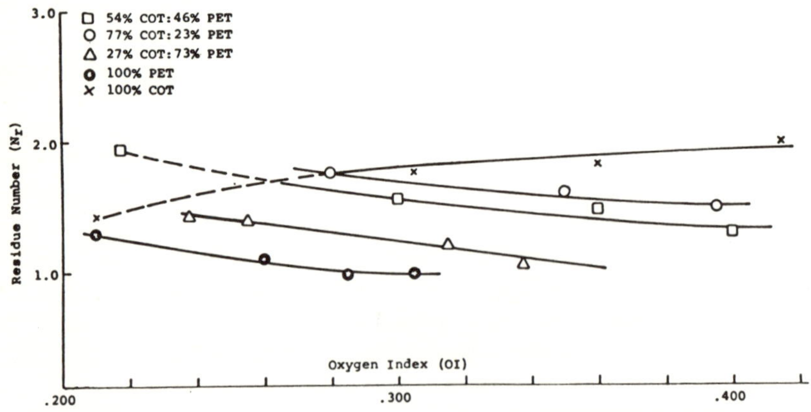

Figure 13. Residue number as a function on oxygen index for webs of 100% polyester, 100% cotton and polyester/cotton blends treated with antimony trioxide (4.2% OWF) + ammonium bromide (0–18% OWF). The Sb_2O_3 is kept constant and the NH_4Br add-on is varied.

effectiveness is dependent on the nature of the oxidizing environment. On the basis of the above, the operating mode of flame retardants can be predicted by plotting the oxygen index (OI) and nitrous oxide index (NOI) as a function of FR add-on and determining whether the two curves are parallel (condensed phase) or divergent (vapor phase) [1].

It has been found that the N_r plots provide a useful technique for determining the operating mode of FR agents. Thus, the pyrolysis mechanisms can be predicted directly from thermogravimetric data.

DAP, as well as most other phosphorus-containing FR agents, functions principally in the condensed phase [22, 23]. Increasing FR add-ons result in the formation of large char residues. In Figure 9 it is seen that the N_r for each blend level increases with increasing OI. Thus N_r values that increase over the whole add-on range (Figure 9), are indicative of condensed phase FR agents.

NH_4Br is primarily a vapor phase FR agent [22, 24]. In this case the N_r reaches a limiting value and remains constant with further increases in OI (Figure 10). For a vapor phase mechanism to be dominant, N_r must be close to 1.0 over the OI range as in the case of 100% polyester treated with NH_4Br (Figure 10). Values greater than 1.0 indicate that the FR agents are effective also in the condensed phase. NH_4Br-treated 100% cotton gives an N_r curve of this latter type (Figure 10). The N_r plots for the various polyester/cotton blends fall between those of the two fiber components. It would appear that as the polyester content of a fiber blend is increased, the condensed phase mechanism decreases until at 100% polyester the vapor phase mechanism becomes dominant.

Similar mechanisms can be predicted for polyester/cotton fiber webs treated with TDBP. The N_r curves (Figure 11) for the various TDBP-treated webs reach a constant value. It is also apparent that the vapor phase mechanism becomes dominant as the polyester content of the blend is raised.

A synergistic FR effect between phosphorus and nitrogen has been reported in the literature [25–28]. DAP + urea was used in this investigation as the model for this type of flame retardant system. As discussed above, it would be expected that the DAP will operate in the condensed phase. However, the urea, when added to the system in increasing levels, behaves as a vapor phase additive. This is evident in the plots of N_r as a function of OI (Figure 12). Here again the significant condensed phase mode of the FR-treated 100% cotton web and the decrease in N_r with the addition of polyester to the fiber structure are apparent. It is surprising to note that the synergistic effect of the urea (nitrogen) on the DAP is basically a vapor phase mechanism. This suggests that synergism may be in certain cases the result of FR systems operating in both phases during pyrolysis.

The $NH_4Br + Sb_2O_3$ combination is representative of a synergistic FR model with in situ formation of the effective FR agent. The suggested mechanism is as follows [22, 29]:

$$R{\cdot}HBr \xrightarrow{\text{heat}} HBr + R$$

$$Sb_2O_3 + 2HBr \longrightarrow 2SbOBr + H_2O$$

$$Sb_2O_3 + 6HBr \longrightarrow 2SbBr_3 + 3H_2O$$

It is reported that SbOBr and/or $SbBr_3$ is the effective flame retardant formed during the burning process [22, 29]. The predominantly vapor phase mechanism of this FR system is evident in the plot of N_r values as a function of OI for fiber webs treated with $NH_4Br + Sb_2O_3$ (Figure 13).

N_r values can also be plotted as functions of FR chemical add-on. Such plots are shown in Figure 14 for DAP (condensed phase FR) and Figure 15 for NH_4Br (vapor phase FR). The similarity between the N_r–OI and N_r-FR add-on curves is evident. Thus the operating mode of FR agents on polyester, cotton and their blends can be predicted from the N_r–FR add-on plots without necessarily measuring the flammability properties of the treated webs.

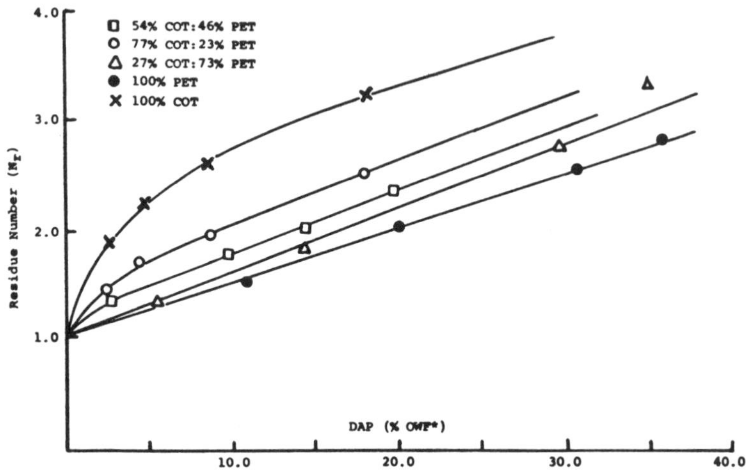

Figure 14. Residue number as a function of FR add-on for webs of 100% polyester, 100% cotton and polyester/cotton blends treated with diammonium phosphate.
**% OWF = % on weight of treated fiber web*
i.e., % DAP + % fiber = 100%

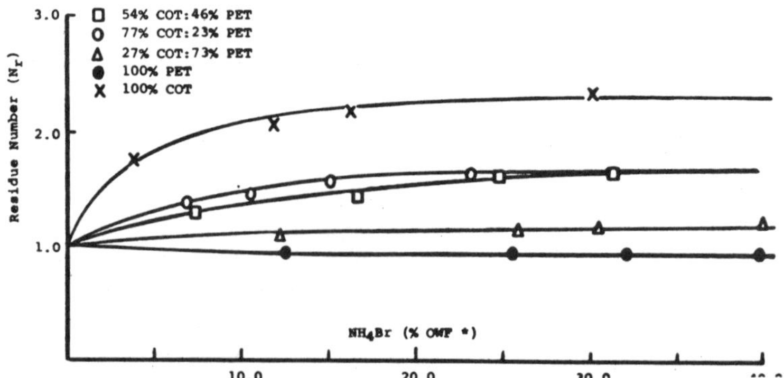

Figure 15. Residue number as a function of FR add-on for webs of 100% polyester, 100% cotton and polyester/cotton blends treated with ammonium bromide.
**% OWF = % on weight of treated fiber web*
i.e., % NH_4Br + % fiber = 100%

The OI values reported in the literature as the threshold for achieving flame resistance fall between .260 and .280 [30]. An OI value of .275 was selected as the threshold for this study. N_r values at OI = .275 were plotted as a function of the fiber composition of the web for the various FR systems (Figure 14).

The curves show graphically the FR modes of operation discussed above. In addition, several other characteristic effects become apparent. The N_r curve for DAP, a condensed phase FR, goes through a minimum as the polyester content of the web is increased. This suggests that for blends, effective flame retardancy is achieved with lower N_r values, and therefore lower amounts of char residues are formed than for the fiber components of the blends. Furthermore, the dominance of condensed phase mechanism for DAP is evident even for the 100% polyester.

N_r curves for predominantly vapor phase FR agents do not show this minimum (Figure 16); instead N_r values decrease with increasing polyester content of the fiber web and the resulting plots approach N_r = 1.0. For 100% cotton and high cotton content blends, the condensed phase effect of the vapor phase FR agents is apparent.

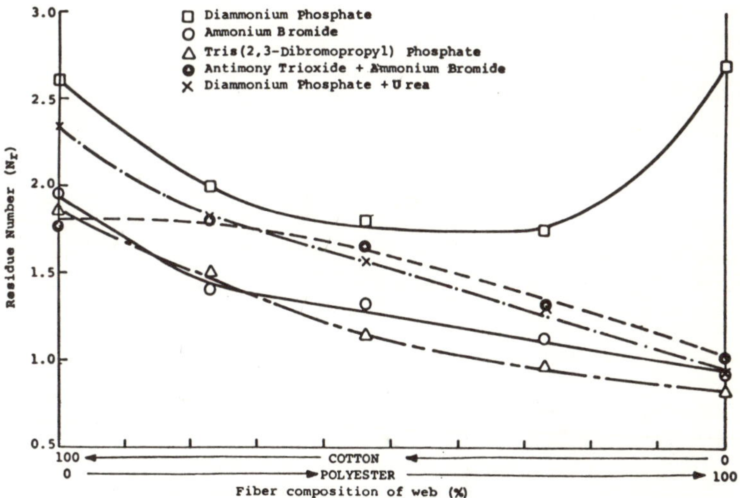

Figure 16. Residue number at OI = .275 as a function of fiber composition for webs treated with model Fr systems.

CONCLUSION

This investigation describes the development of the residue number (N_r) concept as a useful tool for studying the operating mechanism of FR systems directly from thermogravimetric data. N_r represents the ratio of the final char weights obtained from the pyrolysis of FR-modified and unmodified bicomponent fiber blends.

When the FR agent operates mainly in the condensed phase, the N_r values tend to increase with higher chemical add-ons. Predominant vapor phase pyrolysis is associated with relative constancy of the N_r values with increasing chemical add-on and OI. Since the N_r values could provide pertinent information for determining the dominant operative mechanism of FR systems, they were computed for the various model flame retardants studied in this investigation.

It was found that certain FR systems operating in the vapor phase are also capable of effecting the condensed phase, especially when applied to 100% cotton webs. However, as the amount of polyester in the web structure was increased, the FR agents functioned increasingly in the vapor phase.

REFERENCES

1. C. P. Fenimore, and F. J. Martin, Proceedings of the 4th Materials Research Conference, Natl. Bur. Stand. (1970).
2. J. E. Bostic, Jr. and R. H. Barker, "Pyrolysis and Combustion of Polyester. Part II. Effect of Triphenyl Phosphine Oxide as a Flame Retardant," Short course sponsored by PIA at Clemson University, Clemson, S.C. (1974).
3. C. P. Fenimore and F. J. Martin, Modern Plastics, *44* (3), 141 (1966).
4. General Electric, Type FL-101 Flammability Index Tester, Bul. No. 4541K25-001.
5. J. L. Isaacs, *J. Fire Flam., 1* 36 (1970).
6. M. R. Fox, Text. Chemist & Colorist, *1*, 566 (1969).
7. C. J. Bent, T. D. Flynn, and H. H. Sumner, J. Soc. Dyers & Colourists, *85*, 606 (1969).
8. B. Milicevic, Text. Chemist & Colorist, *2*, 87 (1970).
9. G. C. Tesoro and C. H. Meiser, Text. Res. J. *40*, 430 (1970).
10. G. C. Tesoro and J. Rivlin, Text. Chemist & Colorist, *3*, 156 (1971).
11. W. Kruse, Melliand Textilber, *50*, 460 (1969).
12. W. Kruse, Publication No. 5, Gottlieb Duttweiler, Inst. Econ. Social Studies, Ruschlikan-Zurick, Switzerland, 137–161 (1969).
13. R. Barber, F. Moussalli, F. Marascia, J. Bridgeman, J. Firpo, J. Jerome, G. Moses and R. Richardson, Amer. Dyestuff Reptr., *57*, 373 (1968).
14. P. Linden, S. B. Sello and H. S. Skovronek, Textilveredlung, *6*, 651 (1971).
15. P. Hoffmann, "Concept of Flame Retardant Finishes for Polyester/Cotton Blend Fabrics," IFVTCC Congress, Münich, W. Germany (1972).
16. B. Miller, Textile Research Institute, Princeton, N.J., Private Communication (1972).
17. B. Miller and R. Turner, Text. Res. J. *42*, 629 (1972).
18. B. Miller, Sources & Resources, *5* (4), 11 (1972).
19. S. T. Tarim and D. M. Cates, Appl. Polym. Symp., *2*, 1 (1966).
20. J. DiPietro, H. Stepniczka and R. C. Nametz, Text. Res. J., *41*, 593 (1971).
21. H. Stepniczka and J. DiPietro, J. Appl. Polym. Sci., *15*, 2149 (1971).
22. J. W. Lyons, "The Chemistry and Uses of Fire Retardants," Wiley-Interscience, New York, N.Y., 1970.
23. K. C. Frisch, "Flammability Characteristics of Polymers," Polymer Conference Series, Polymer Inst., University of Detroit (1972).
24. F. C. Clarke, III, "Aspects of Fire-Retardance Mechanisms in Polymers: Criteria for Stationary-Phase Retardance," Polymer Conference Series, Polymer Inst., University of Detroit (1972).
25. G. C. Tesoro, S. B. Sello and J. J. Willard, Text. Res. J., *39*, 180 (1969).
26. S. B. Sello, G. C. Tesoro and R. Wurster, Textilveredlung, *5*, 391 (1970).

27. G. C. Tesoro, Textilveredlung, *2*, 435 (1967).
28. G. C. Tesoro, S. B. Sello and J. J. Willard, Text. Res. J. *38*, 245 (1968).
29. J. J. Pitts, *J. Fire Flamm., 3*, 51 (1972).
30. J. J. Willard and R. E. Wondra, Text. Res. J., *40*, 203 (1970).

Ildo E. Pensa

Ildo E. Pensa is Manager of Finishing Development and Technical Service at the J. P. Stevens & Co., Inc., Technical Center, Garfield, NJ. He received a B.S. in Chemical Engineering at the University of Alexandria (Egypt), a Masters degree from Columbia University (1963) and a Ph.D. from New York University (1973). Prior to joining J. P. Stevens in 1959, he was Technical Director at the Flixecourt Plant of Saint Frères S.A. (France). His twenty years experience in the textile field include piece and yarn dyeing, coating systems, durable press, soil release, antistatic and flame resistant finishing. He is also the author of a number of technical publications and patents.

Walter Brenner

Dr. Brenner is Professor of Chemical Engineering at the Polytechnic Institute of New York. He has been active in various phases of textile and polymer technology for the past twenty years. Special topics of interest include durable press, fire safety for textiles and other polymeric materials, effects of high energy ionizing radiation on polymeric materials and fiber reinforced composites. He is also the author of over 60 technical publications, 2 books and a number of patents. He won, in 1967, the S.P.I. Award for the most original work. He has also lectured at the National Bureau of Standards.

Stephen B. Sello

Stephen B. Sello received his Ph.D. in Chemistry from the University of Sciences (Budapest, Hungary) in 1947. He has 25 years industrial and research experience in the field of textile chemistry. He was Associate Professor at the Polytechnical University of Budapest, and in 1959 joined J. P. Stevens & Co., Inc., Technical Center, Garfield, NJ where he is now Manager of Textile Chemistry Research. He is the author of 37 publications and 20 patents in the field of textile chemistry. He is a member of the ACS, the AATCC, the Fiber Society and the New York Academy of Science. He is a Fellow of the American Institute of Chemists and in 1972 was Chairman of the Research Committee of the Information Council on Fabric Flammability.

H. Ishibashi, C. Horiuchi

Textile Research Institute of
Osaka Prefecture, Japan

S. Suga

Suga Test Instruments Co., Ltd., Tokyo, Japan

FLAMMABILITY ON BLENDS OF FLAME RETARDANT FIBER

(Received July 26, 1974)

ABSTRACT: To obtain the basic data necessary for the development of compound products of two-component fiber systems, the flammability and the smoking of experimental blends of two-component fiber systems (spherical assembly and union fabric) were examined.

The results of flammability test on two-component fiber systems in Fiber ball test (spherical assembly), measurement of Oxygen index 45° angle test and vertical test (Doc. FF3-71, AATCC 34-1969) showed that the flammability on two-component fiber system deviated from additive property of each component fiber in almost all cases. The difference of blend effects of flame-retardant fiber between spherical assembly and union fiber were found considerably great.

Flame-retardant fiber generally generates dense smoke, but when nylon or polyester was blended to it the generation of smoke extremely increased. Additive property was not also found in smoke measurement. Burning behavior of these blends were also discussed.

INTRODUCTION

There are two ways to make the textiles flame retardant: one is to postreat with flame retardants and the other is to produce them with flame-retardant materials built-in. The postreatment is convenient because it can be done anytime at demand, but is apt to bring about change of fabric hand and loss of strength, and moreover, it has questions in durability to laundering or drycleaning and in toxicological effects. And it is very difficult to postreat synthetic fibers, its blends or high-piled fabrics.

On the other hand, textile goods made of 100% flame-retardant fiber are superior in the fire-proof and its permanency, but they are unsatisfactory in practical performance, aesthetics (pleasant hand, bright coloration) and cost etc. That is to say, it is difficult to make only one kind of flame-retardant fiber satisfy all the terms of goods now.

From this point of view, it will be of great importance to develop flame-retardant goods by blending fire-retardant fiber with other fibers, to make up each drawbacks and make the most of its merit.

Flammability of two-component fiber systems has been already studied by Kruse [1] and Tesoro [2]. In our study, in order to gain the basic data necessary for developing fire-proof goods of two-component fiber systems by flame-retardant fiber blend, flammability of two-component fiber systems of flame-retardant and combustible fiber or two flame-retardant fibers was systematically examined.

As a method of study, it is ideal to manufacture the goods like drapery and carpet for trial for testing and to obtain practical data and information, but it is almost impossible because there are too many combinations and it takes an enormous time and labor.

Therefore, two experimental blends whose specimens were relatively easy to be prepared and also alike in construction of each goods were prepared. One is spherical assembly as a model of carpets or blanket etc., and the other is union fabric as a model of drapery, sleepwear and apparel.

Fire performance requirement for the textile goods is to be difficult to ignite, and once ignited, to burn with difficulty.

Moreover, lowering of smoke generation has been required to extricate human life in fire in these days. Therefore, smoke generation of the experimental blends was also examined.

EXPERIMENTAL

Test Specimen

Fiber for spherical assembly and yarn for union fabric are shown in Table 1. The construction of fabrics was a plain weave of 36 ends and 36 picks in the grey, using mainly 20/2 cotton counts yarns in both warp and weft. These specimens were scoured in 0.2% soap solution to remove all residual spinning oil. Prior to testing specimens were left in the oven maintaining at 50 ± 2°C for 24 hours. Then they were left in the dessicator with silca gel for more than 2 hours to be cooled.

Test Method of Flammability

1. **Fiber ball test.**—The test was carried out upon the Fiber Ball Test Method [3] of the Japanese Fire Protection Association.
   Two kinds of fibers were taken to be 10.0g in total weight and mixed with hand card. A fiber ball was made of them and set in the ball ring with four wire rings in the circles of longitude and one ring in the equator, like a globe.
   This was hung with wire and exposed to the flame of a meker burner for 30 sec, and the loss of weight (%) due to combustion, residual flame time and after glowing time were determined.
2. **Oxygen index (OI) measurement.**—It has been reported by Mr. Suga [4] that

oxygen index method is very effective for assessing the flame retardancy of plastics and textiles, and this method is adopted in two JIS [5].
In one of them, JIS D 1201, flame retardancy is classified by using oxygen index.
Then, Oxygen index was measured by using ONI Meter Model ON-2D, manufactured by Suga Test Instruments Co., Ltd. This model has Smoke Density Meter, but it was not used in OI measurement. The specimen was set in the burning column using U-typed holder (inner size is 5 cm X 12 cm). Thermoplastic specimens were stitched by three cotton threads in vertical direction so as to prevent from dripping during burning.
Total flow rate of mixture of oxygen and nitrogen gas is 11.4ℓ/min and temperature was maintained at 20 ± 2°C. The ignition flame was put on the top end of specimen and the minimum concentration of oxygen necessary for holding the burning 5 cm long from the top was measured. This minimum concentration was considered as oxygen index.

Table 1. Fiber and Yarn for Experimental Blends

Generic Name	Fiber for Spherical Assemblies		Yarn for Fabrics	
	Denier (d)	Fiber Length (mm)	Count (cotton count)	Twist (time/inch)
Vinylidene(Vinylidene chloride-vinyl chloride copolymer)	10	51	-	-
Polyvinyl chloride(PVC)	15	76	20/2	5
Modacrylic(Vinyl chloride type)	3	76	20/2	10
Modacrylic(Vinylidene chloride type)	5	76	20/2	10
Polychlal(PVA-PVC matrix fiber)	15	76	19/2	10
FR-rayon	15	76	20/2	13
FR-acrylic	15	76	-	-
Wool	about 6	38-76	21/2	12
Nylon(Nylon 6)	15	152	20/2	10
Polyester(Polyethylene terephthalate)	10	51	20/2	12
Vinylon	1.4	44	20/2	7
Acetate	3.6	76	-	-
Acrylic	15	76	18/2	11
Cotton	about 3	12-20	20/2	13
Rayon	7	76	20/2	11
Polypropylene	15	76	10/1	7
Aramid(Aromatic polyamide)	2	50	20/2	9

3. **45° angle test.**—This text was carried on JIS L1091 Method A-1 (size of specimen-35 cm X 25 cm, ignition source-microburner (butane gas), exposure time — 1 min.) [6].

Char surface area, residual flame time and after glowing time were measured.

4. **Vertical test.**–This test was conducted on Doc. FF3-71 of Standard for the flammability of children's sleepwear (exposure time 3 sec.) [7] and AATCC 34-1969 (exposure time 12 sec, three fiber glass seams in specimens of thermoplastic specimens) [8].
   Flammability Tester CS-1 manufactured by Suga Test Instruments Co., Ltd. was used.

5. **Smoke measurement.**–APPARATUS: ONI Meter Model ON-2D was employed. The smoke column of Smoke Density Meter on Uehara-and-Suga system is attached on the top end of the burning column. This system has been already adopted in JIS D 1201 referred to in the previous item. The block diagram of the apparatus is shown in Figure 1. Incandescent lamp (12 V, 50 W) as a light source, silicone photoelectric cell with temperature compensating circuit and visual compensating filter are employed in the light acceptor. Spectral sensitivity of the optical system in smoke measurement must correspond with visual sensitivity of human eyes. Total spectral sensitivity curve of Smoke Density Meter corresponds with standard relative luminosity curve as shown in Figure 2. The air was given to the light source and light acceptor by air-curtain method in order to prevent from depositing on the surface of them (4ℓ/min).

   PROCEDURE: The same method as oxygen index measurement was used except that interior dimensions of U-typed holder are 2.5 cm X 12 cm and Oxygen concentration in the flow-gas was set to be the minimum concentration necessary for burning specimens entirely. The transmittance was recorded from putting the flame to the top of specimen until no smoke could be observed in the chimney of the Meter. The warp direction of specimen was set vertically.

   CALCULATION: Attenuation coefficient (CS), total amount of smoke (C) and smoke generation coefficient (K) were calculated from the following formula.

$$CS = VL \quad \ln Io/I = I/L \quad \ln 100/T \tag{1}$$

where
L = Light path length (m)
Io = Initial Light intensity
I = Emergent Light intensity
T = Transmittance (I/Io X 100)

CS max = Maximum value of CS

$$C = F \int_{o}^{+} CS \cdot dt \; (CS \cdot m^3) \tag{2}$$

where
F = Gas flow rate (m^3/min)
t = Measuring time (min)

Strictly, it is essential that gas flow rate should be determined from the speed of air current in burning, but the flow rate, 0.0114 m^3/min, of mixture of oxygen and nitrogen gas was used here.

$$K = C/W \quad (CS \cdot m^3/g) \tag{3}$$

where C = Total amount of smoke ($CS \cdot m^3$)

W = Weight of burned specimen (g)

Total amount of smoke (C) means the volume of generated smoke converted into the smoke density when Cs=1. Smoke generation coefficient (K) indicates total amount of smoke (C) per 1 g specimen. But the distance which can be seen through smoke when Cs=1 is about 2.7 m.

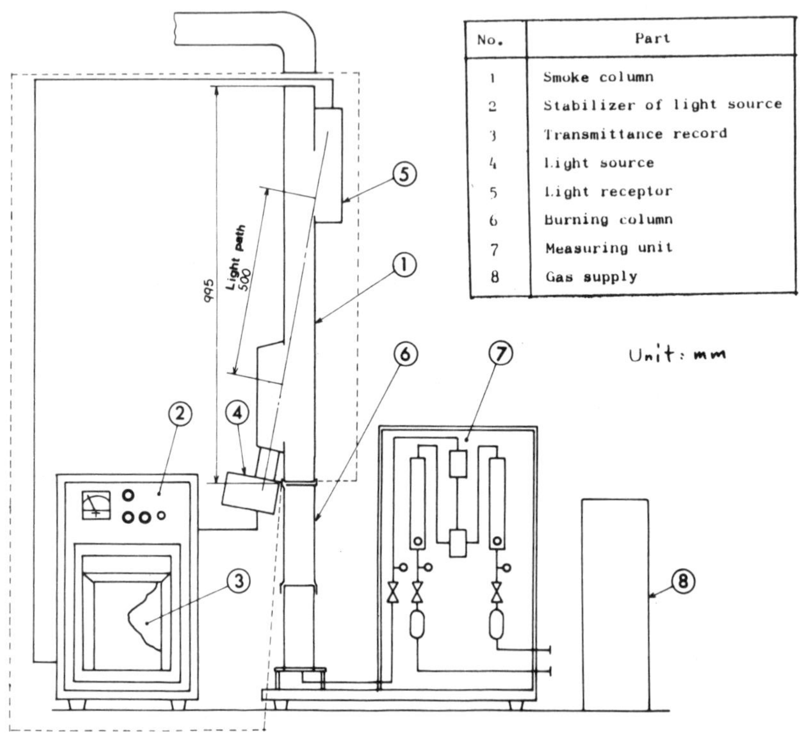

Smoke Density Meter unit is enclosed by dotted lines

Figure 1. *ONI Meter, Model ON-2D.*
(Apparatus for Smoke Measurement by Oxygen Index Tester)

RESULTS AND DISCUSSION

Flammability of Spherical Assemblies

Flammability of textile goods is considerably influenced by their shape, especially the ratio of their surface area to their volume and porosity. Therefore, the constructional factors of spherical assemblies for experiment were compared with that of various goods. As shown in Table 2, the shape of the assemblies is similar to that of carpet or blanket.

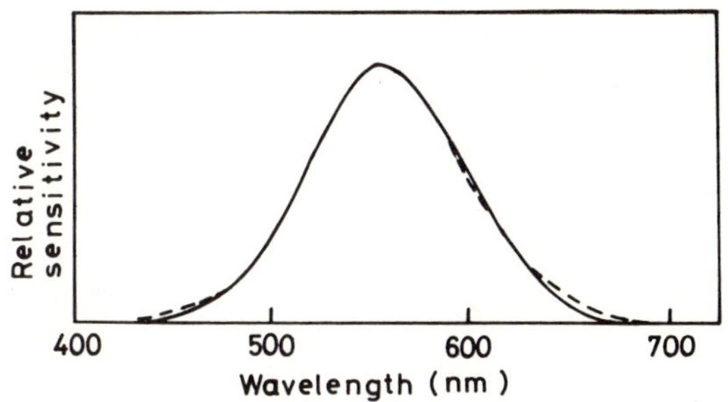

Figure 2. *Total Spectral Sensitivity Curve.*
(------ Standard Relative Luminosity Curve)

Table 2. Construction of Spherical Assembly, Carpet and Blanket

	Apparent Density (g/cm^3)	Porosity* (%)	Ratio of Surface to Volume
Spherical Assembly	0.15	85 - 91	1.2
Carpet	0.09-0.14**	87 - 94**	1.0 - 1.4
Blanket	0.06-0.09	94-- 95	1.1 - 1.3

* **Porosity(%)=(V - V_0) / V × 100**

V: Volume of assembly

V_0: Volume of fiber in assembly

V_0 = W / ρ

(**W: Weight of assembly,** ρ: Specific gravity of fiber)

** **Part of pile**

So, it is possible to say that burning behaviors, such as ease of ignition and burning continuance (self-extinguishing), can be estimated in the fiber ball test.

Polypropelene and polyester are easy to be ignited and almost completely burn out in the fiber ball test of the assemblies of single combustible fibers, though partly melt and fall. But no residue is observed. And rayon is also easy to be ignited and presents flaming combustion but, on the way, turns to non-flaming combustion and burns completely. On the other hand, flame-retardant fiber burns under the exposure to flame, but extinguishes as soon as the flame is removed and presents the self-extinguishing. The results of two-component fiber systems are shown in Figure 3.

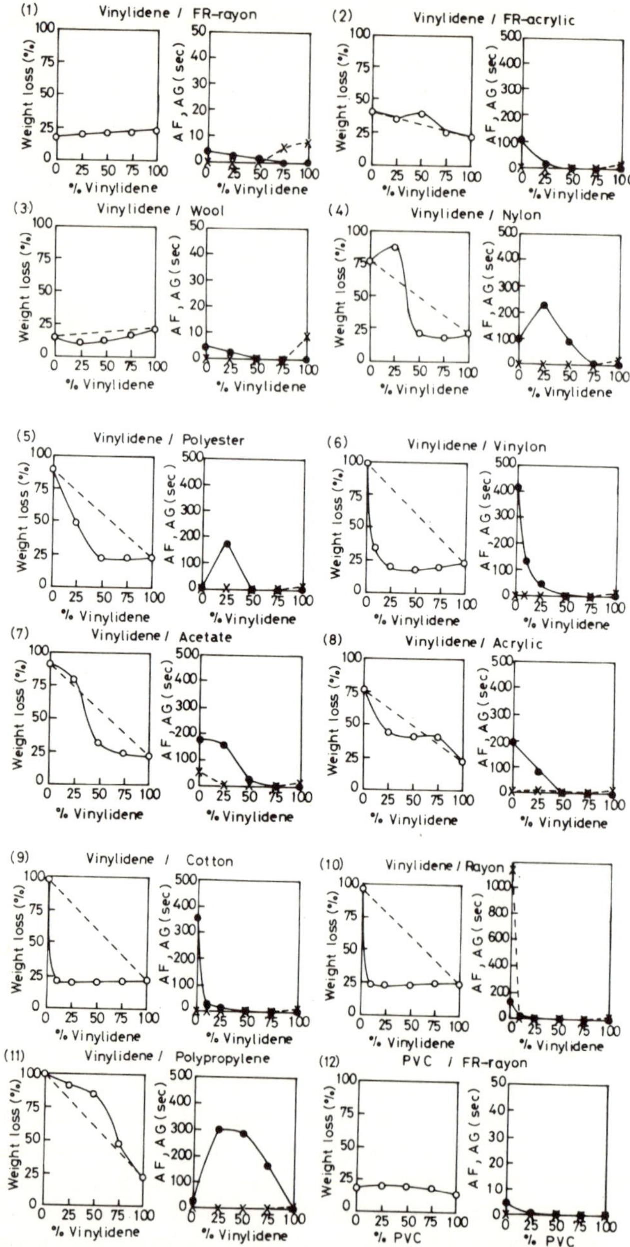

Figure 3. *Flammability of Two-component Spherical Assemblies (72 fiber combinations).*
AF = Afterflame ——●——, *AG = Afterglow* – –X– –
Modacrylic (Vinylidene chloride type) = –·–·–·–

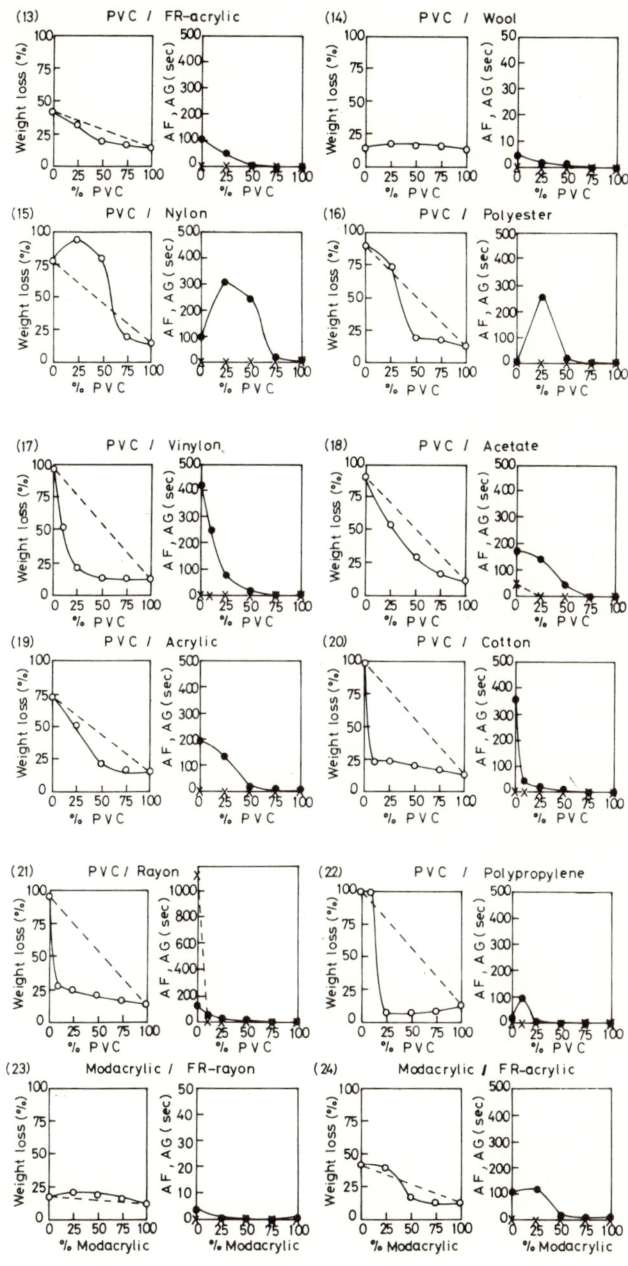

Figure 3. *continued.*

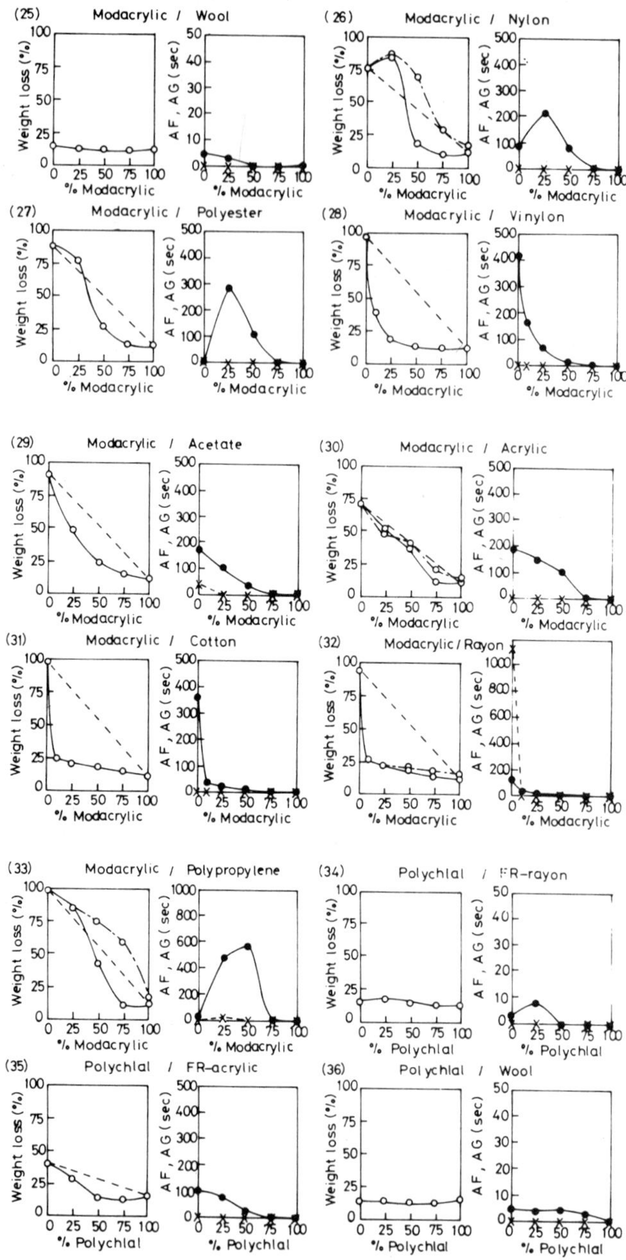

Figure 3. *continued.*

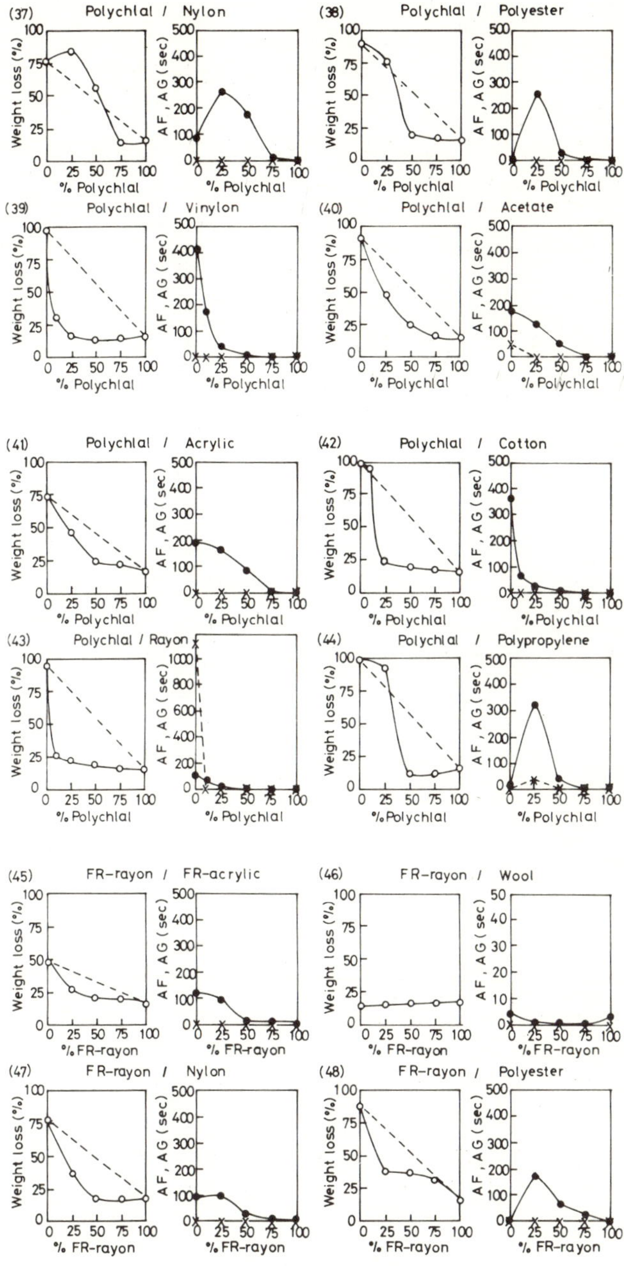

Figure 3. *continued.*

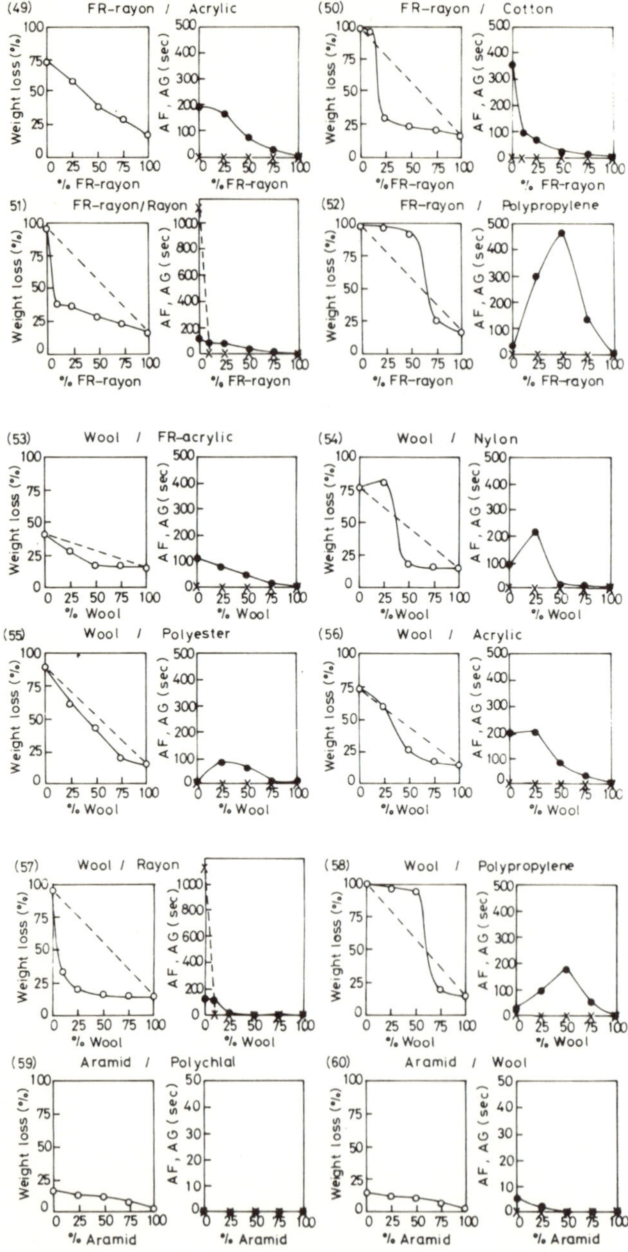

Figure 3. *continued.*

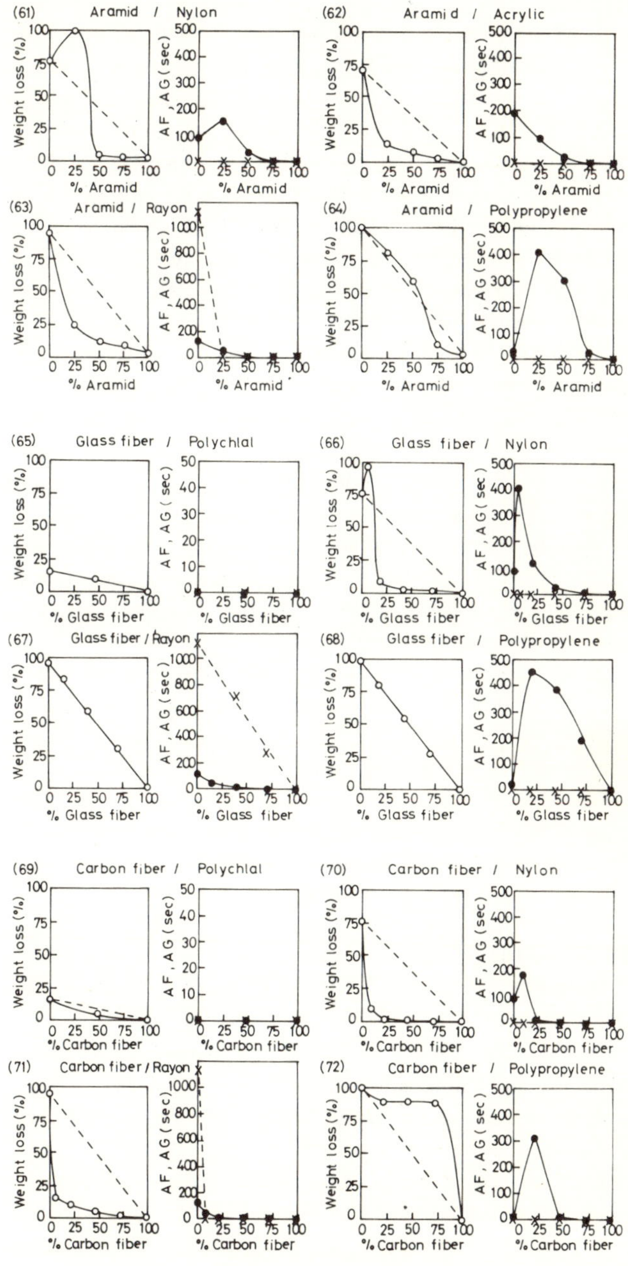

Figure 3. *continued.*

The additive property that flammability of two-component fiber systems is indicated as average flammability of the two component fiber was not found in almost all systems.

As regards the loss of weight, the following four types were seen: (1) positive deviation type, (2) negative deviation type, (3) transition type (transisting from positive to negative deviation according to the blend ratio), (4) average type (Linear type).

Vinylidene/polypropylene or modacrylic(vinylidene chloride type)/polypropylene blend is of a typical positive deviation type. As combustion heat of Polypropylene is very high (10.5 kcal/g) [9] and therefore high calorie is fed back from its combustion, those blends of flame retardant fiber are easy to burn. Moreover, they have a tendency to generate porous solid residue (char) after burning and their burning are accelerated by scaffolding effect of the residue [10].

Negative deviation type is remarkably found in the blends of flame-retardant fiber containing chlorine with and cotton, rayon or vinylon. Cellulosic fiber cannot hold its burning unless it has considerably great porosity, compared with acetate and acrylic [11]. And vinylon decreases in porosity because of contractability when heated. When the surface of spherical assemblies burns, in its inside, pyrolysis zone is created in them due to heat conduction from the burning.

In the zone, the burning is repressed by hydrogen chloride gas generated from pyrolysis of the fiber containing chlorine and prevented to advance to the inside (extinguishes, generating white smoke). Creation of pyrolysis zone is effective for the generation of combustion preventing gas.

For the woven fabrics, it is so thin that pyrolysis zone can't be created, and it's surface area is so large that good flame retardancy cannot be obtained unless much flame-retardant fiber is blended.

Transition type was frequent in blend of PVC or poly chlal and polypropylene or nylon. These systems are hard to be ignited, if flame-retardant fibers are blended to more than a certain degree, and indicates a behavior near to self-extinguishing. However, when ignited with the blend ratio insufficient, they show rapid increase in the flammability and burn almost entirely.

Flammability of two-component fiber systems of combustible and flame-retardant fibers is mostly influenced by the results of competitive effects between the combustion effect of combustible fiber and the depression effect of flame-retardant fiber.

In the blends of this type it is key point whether combustion effect proceeds depression effect and combustible fiber is ignited at first. These blends, once ignited, burn so furiously as supported by high calorific value of polypropylene and nylon that the effect of flame-retardant fiber is almost lost.

Average type was found only in the blends of glass fiber and rayon or polypropylene. Polypropylene component burned out in flaming combustion. Rayon presented flaming combustion at first and then, presenting non-flaming combustion, burned out after long term of its burning.

Both glass fiber and carbon fiber are incombustible, but they presented quite different inclination when blended in similar combination. For example, glass fiber/rayon and glass fiber/polypropylene were proved to be of average type, meanwhile carbon fiber/rayon indicated negative deviation and carbon fiber/polypropylene showed positive deviation.

In the case of carbon fiber/nylon blend, nylon on its surface burned at first and then the combustion was interrupted when the surface was coated by carbon fiber. Carbon or char is not effective fuel and further has flame-proof to act as thermal insulator, insulating an internal combustible component from the flame and control feedback of heat. These are also supposed to act as barrier to prevent oxygen from scattering to the inside and escaping pyrolysis gas to the surface. As stated above, creation of char zone is effective to fire retardation. When the fibers which are easy to create char by heating, such as FR rayon, wool and aromid, were blended in a high ratio, similar inclinations were observed. It is reported that the compound material of phenol resin and nylon fabrics is useful as an ablation material and it seems to have a common problem to this [12].

Meanwhile, carbon fiber/polypropylene blend burns well, because melted polypropylene exudes to the surface of assemblies and act as fuel. Carbon fiber acts as scaffold to accelerate burning rather than as thermal insulator and burns itself, in blends with fiber which has high melting property and high burning heat, such as polypropylene. From these results, the flammability of spherical assembly blends depends upon such factors as solubility, combustion heat, formation of char and pyrolysis zone, and scaffolding effect of each component fiber.

Especially, combustion heat is a critical factor for determining self-extinguishing or not. The self-extinguishing varies on the conditions of exposuring flame. The results obtained here should be relative.

Flammability of Blend Fabrics

As model fabric of two-component fiber systems, various union fabrics were woven for trial and their flammability were examined.

The flammability on single component fabrics is shown in Table 3, and that of two-component blend fabrics is in Table 4–Table 7. And relations between their fiber content and OI values are shown in Figure 4. If OI values of blend fabrics are obtained as arithmetical average values of single component fabrics, they should have been on the dotted line in the Figure.

Most of blends, however, deviate from average to positive or negative. This is because pyrolysis and combustion of each component fiber of two-component fiber systems reacts each other thermally and chemically. Even if PVC with high OI value is blended with nylon having low OI value by the same amount, the OI value of the blends is almost equal to that of nylon. It is thought that pyrolysis of PVC is accelerated chemically by aliphatic polyamide [13] as well-known, and also by combustion of co-existing nylon because of its high combustion heat.

Table 3. Flammability of Single Component Fabrics

Component	Weight (g/m²)	OI	45° Test		Vertical Test				
					DOC FF 3-71		AATCC 34-1969		
			Char Area (cm²)	Afterflame (sec)	Char Length (cm)	Afterflame (sec)	Char Length (cm)	Afterflame (sec)	Afterglow (sec)
PVC	179	36.0	6.1	0	7.0	0	12.1	0	0
Modacrylic	193	28.0	4.5	0	2.1	0	9.6	0	0
Polychlal	210	31.5	6.7	0	1.7	0	6.7	0.2	0
FR-rayon	216	25.5	110.9	0	BEL (5/5)	15.5	BEL (1/3)	8.2	0
Wool	210	23.0	227.8	0	BEL (5/5)	27.9	BEL (3/3)	16.8	0
Nylon	192	22.5	4.2	0	4.7	3.4	BEL (3/3)	122.7	0
Polyester	192	23.0	4.6	0	5.8	10.0	BEL (3/3)	66.4	0
Vinylon	216	20.5	230.1	57	BEL (5/5)	67.1	BEL (3/3)	43.2	0
Acrylic	192	18.5	237.7	0	BEL (5/5)	23.8	BEL (3/3)	17.4	0
Cotton	164	17.5	BEL	25	BEL (5/5)	21.5	BEL (3/3)	12.2	22.7
Rayon	222	19.0	BEL	117	BEL (5/5)	33.6	BEL (3/3)	23.7	461.3
Polypropylene	178	18.5	BEL	98	BEL (5/5)	43.7	BEL (3/3)	69.3	0
Aramid	295	30.5	3.4	0	0.5	0	0.9	0	0

Table 4. Flammability of Blend Fabrics Composed of PVC and Other Fibers

Component	Weight	OI			45° Test				Vertical Test DOC FF 3-71				Vertical Test AATCC 34-1969					
					Warp Direction		Filling Direction		Warp Direction		Filling Direction		Warp Direction			Filling Direction		
(Warp/Filling)	(g/m²)	Warp Direction	Filling Direction	Deviation from Average	Char Area (cm²)	After-flame (sec)	Char Area (cm²)	After-flame (sec)	Char Length (cm)	After-flame (sec)	Char Length (cm)	After-flame (sec)	Char Length (cm)	After-flame (sec)	After-glow (sec)	Char Length (cm)	After-flame (sec)	After-glow (sec)
PVC/PVC	179	36.0*			6.1	0			7.0	0			12.1	0	0			
PVC/Modacrylic	200	33.5	33.0	+	6.6	0	6.0	0	3.2	2.8	6.4	0	8.8	0	0	9.9	0	0
PVC/Polychlal	219	33.5	34.0	+	7.6	0	5.9	0	0.9	0	3.4	0	8.8	5.5	0	9.8	0	0
PVC/FR-rayon	199	28.0	28.0	-	10.6	0	10.7	0	3.9	0.6	6.2	1.5	13.2	0	0	10.2	0	0
PVC/Wool	170	26.5	25.0	-	9.1	0	6.6	0	3.8	0	4.6	0	14.2	4.0	0	11.8	1.4	0.8
PVC/Nylon	186	22.0*	22.5*	-	5.3	0	5.5	0	5.5	0	6.1	0	BEL (4/6)	53.8	0	BEL (4/6)	64.8	0
PVC/Polyester	190	24.5*	24.5*	-	4.7	0	5.6	0	5.5	0	6.5	0	14.2	26.5	0	9.8	0	0
PVC/Vinylon	187	27.0	22.0	-	6.8	0	8.7	0	BEL (2/5)	85.0	4.7	0.8	BEL (3/5)	72.9	0	9.7	0	0
PVC/Acrylic	196	23.5	22.0	-	5.7	0	13.7	0	9.5	12.0	6.9	0	9.2	6.1	0	BEL (1/5)	25.2	0
PVC/Cotton	192	22.5	22.0	-	232.4	0	279.9	0	BEL (5/5)	23.0	BEL (5/5)	27.8	BEL (3/3)	10.9	119.2	BEL (3/3)	7.8	0
PVC/Rayon	196	24.0	22.0	-	228.8	0	231.0	0	BEL (5/5)	21.5	BEL (5/5)	19.3	BEL (3/3)	9.0	0	BEL (3/3)	9.5	0
PVC/Polypropylene	184	25.0*	25.0*	-	5.7	0	5.6	0	6.3	0	5.5	0	BEL (1/6)	18.6	0	14.3	8.5	0
PVC/Aramid	182	33.5	34.0	+	9.2	0	9.5	0	4.0	5.5	5.8	0	5.2	0	2.0	8.5	0	2.2

BEL : Burns Entire Length

*Stitched with cotton yarns

Table 5. Flammability of Blend Fabrics Composed of Modacrylic and Other Fibers

Component	Weight	OI			45° Test				Vertical Test									
									DOC FF 3-71				AATCC 34 -1969					
					Warp Direction		Filling Direction		Warp Direction		Filling Direction		Warp Direction			Filling Direction		
(Warp/Filling)	(g/m²)	Warp Direction	Filling Direction	Deviation from Average	Char Area (cm²)	After flame (sec)	Char Area (cm²)	After flame (sec)	Char Length (cm)	After flame (sec)	Char Length (cm)	After flame (sec)	Char Length (cm)	After flame (sec)	After glow (sec)	Char Length (cm)	After flame (sec)	After glow (sec)
Modacrylic/ Modacrylic	193	28.0			4.5	0			2.1	0			9.6	0	0			
Modacrylic/PVC	200	33.0	33.5	+	6.0	0	6.6	0	6.4	0	3.2	2.8	9.9	0	0	3.8	0	0
Modacrylic/ Polychlal	200	30.5	32.5	+	4.3	0	4.8	0	1.5	0.6	2.4	0	7.7	1.7	0	7.4	0	0
Modacrylic/ FR-rayon	197	26.5	26.0	-	7.5	0	7.2	0	3.6	0.4	1.6	2.3	11.0	0	0.3	7.1	0	0
Modacrylic/Wool	179	26.5	26.5	+	8.2	0	7.8	0	3.9	0	1.2	0.5	8.6	0	0	8.0	0	0
Modacrylic/Nylon	198	23.7	26.0	+	5.0	0	6.8	0	4.2	0	1.6	0	12.1	3.1	0	12.3	13.6	0
Modacrylic/ Polyester	201	27.5	30.0	+	5.4	0	5.2	0	3.6	0	4.5	4.7	BEL (1/6)	19.0	0	9.5	6.0	0
Modacrylic/ Vinylon	194	24.0	24.0	Av	7.0	0	7.4	0	1.8	1.5	3.1	0	BEL (3/3)	36.9	0	11.9	3.8	0
Modacrylic/ Acrylic	216	23.0	23.0	Av	9[illegible].1	25.5	4.2	0	BEL (4/5)	46.2	BEL (5/5)	50.9	BEL (2/4)	47.8	0	BEL (3/5)	31.9	0
Modacrylic/ Cotton	198	24.5	24.5	+	256.3	22.5	34.0	25	BEL (5/5)	41.3	BEL (5/5)	28.4	BEL (2/3)	28.2	0	BEL (3/3)	28.3	0
Modacrylic/Rayon	207	25.0	24.0	+	144.9	0	23.0	0	4.0	2.1	BEL (5/5)	27.9	13.6	0.8	0	8.6	0.5	2.0
Modacrylic/ Polypropylene	206	26.5	29.0	+	5.3	0	5.3	0	3.6	0	5.0	0	10.9	3.2	0	10.4	8.4	0
Modacrylic/ Aramid	194	31.0	30.5	+	4.6	0	4.9	0	2.5	2.9	BEL (1/5)	13.0	5.6	0	6.5	8.3	0	6.2

Table 6. Flammability of Blend Fabrics Composed of Polychlal and Other Fibers

Component	Weight	OI			45° Test				Vertical Test									
									DOC FF 3-71				AATCC 34-1969					
					Warp Direction		Filling Direction		Warp Direction		Filling Direction		Warp Direction			Filling Direction		
(Warp/Filling)	(g/m²)	Warp Direction	Filling Direction	Deviation from Average	Char Area (cm²)	After flame (sec)	Char Area (cm²)	After flame (sec)	Char Length (cm)	After-flame (sec)	Char Length (cm)	After-flame (sec)	Char Length (cm)	After-flame (sec)	After-glow (sec)	Char Length (cm)	After-flame (sec)	After-glow (sec)
Polychlal/ Polychlal	210	31.5			6.7	0			1.7	0			6.7	0.2	0			
Polychlal/ FR-rayon	210	25.5	25.0	-	16.9	0	15.2	0	4.3	0.3	1.0	1.7	BEL (2/5)	9.8	0	7.9	2.3	0
Polychlal/Wool	199	26.5	25.0	-	10.8	0	6.8	0	4.9	0	4.6	0	10.1	0	0	11.8	7.9	0
Polychlal/Nylon	213	23.5	24.0	-	6.7	0	6.4	0	4.5	0	1.4	0.9	BEL (2/6)	30.9	0	BEL (3/3)	51.3	0
Polychlal/ Polyester	215	24.0	28.0	±	5.4	0	6.7	0	5.4	0	1.6	0.6	BEL (4/6)	52.5	0	BEL (4/4)	53.8	0
Polychlal/ Vinylon	202	24.0	25.0	-	55.3	0	104.5	0	BEL (5/5)	44.8	BEL (4/5)	39.7	BEL (2/3)	32.7	0	BEL (3/3)	36.8	0
Polychlal/ Acrylic	230	23.0	21.5	-	269.0	69.0	265.2	91.0	BEL (5/5)	36.5	BEL (5/5)	33.4	BEL (3/3)	27.6	0	BEL (3/3)	24.7	0
Polychlal/Cotton	212	23.0	23.0	-	266.1	0	281.6	0	BEL (5/5)	28.2	BEL (5/5)	29.6	BEL (3/3)	16.6	0.4	BEL (3/3)	23.4	0
Polychlal/Rayon	212	23.0	23.0	-	247.2	32.0	239.0	8.0	BEL (5/5)	21.5	BEL (5/5)	31.0	BEL (3/3)	17.2	0	BEL (3/3)	29.2	0
Polychlal/ Polypropylene	206	24.0	26.5	-	7.3	0	3.2	0	5.5	0	1.9	1.0	14.6	7.4	0	BEL (5/6)	76.8	0
Polychlal/Aramid	220	30.0	30.5	-	6.2	0	7.4	0	1.7	0	1.3	0	7.8	0	0	8.1	0	1.3

Table 7. Flammability of Other Blend Fabrics

Component	Weight	OI			45° Test				Vertical Test									
									DOC FF 3-71				AATCC 34-1969					
					Warp Direction		Filling Direction		Warp Direction		Filling Direction		Warp Direction			Filling Direction		
(Warp/Filling)	(g/m²)	Warp Direction	Filling Direction	Deviation from Average	Char Area (cm²)	After flame (sec)	Char Area (cm²)	After flame (sec)	Char Length (cm)	After flame (sec)	Char Length (cm)	After flame (sec)	Char Length (cm)	After flame (sec)	After glow (sec)	Char Length (cm)	After flame (sec)	After glow (sec)
FR-rayon/Wool	204	25.5	25.5	+	69.8	0	58.6	0	BEL (5/5)	18.4	BEL (5/5)	17.0	BEL (1/3)	4.5	0	BEL (1/3)	2.7	0
FR-rayon/ Polyester	225	22.0	21.0	-	346.6 (BEL)	123	290.3	38	BEL (5/5)	45.5	BEL (5/5)	35.5	BEL (3/3)	34.1	0	BEL (3/3)	23.8	0
FR-rayon/Aramid	213	28.5	28.5	+	110.8	0	111.0	0	3.8	5.5	BEL (3/5)	10.6	2.8	0.4	0.1	BEL (1/3)	2.3	0
Wool/Polyester	184	22.5	23.0	Av	82.9	1	92.7	0	BEL (2/5)	21.2	BEL (2/5)	21.7	15.1	5.6	0	18.9	15.0	0
Aramid/Polyester	194	23.0	24.5	-	113.5	23	37.3	4	BEL (5/5)	47.7	BEL (5/5)	44.0	BEL (3/3)	31.9	0	BEL (3/3)	37.3	0
Vinylidene/ Polyester	184	27.0	27.0	-	5.5	0	7.0	0	BEL (5/5)	44.7	3.9	0	BEL (3/3)	51.1	0	15.8	0	0
Modacrylic*/ Polyester	190	24.5	25.5	-	4.7	0	[illegible].6	0	5.7	0.4	5.6	0	BEL (1/3)	35.3	0	BEL (1/3)	26.7	0

* Acrylonitrile-Vinylidene Chloride Copolymer

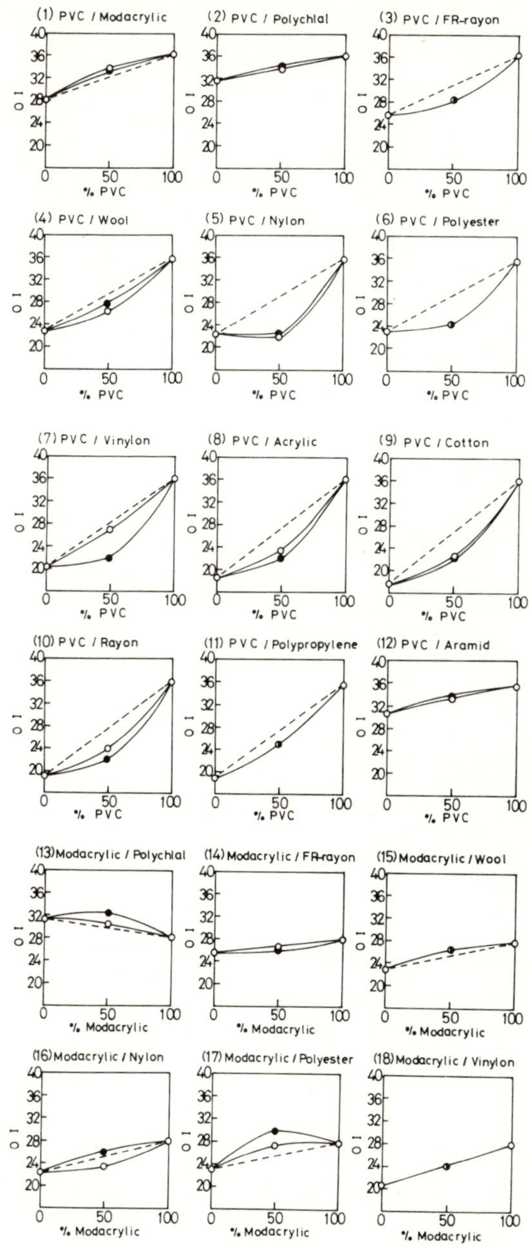

Figure 4. *Relation between Blend Ratio and Oxygen Index (38 fiber combinations).*
○ = *Warp direction,* ● = *Filling direction*

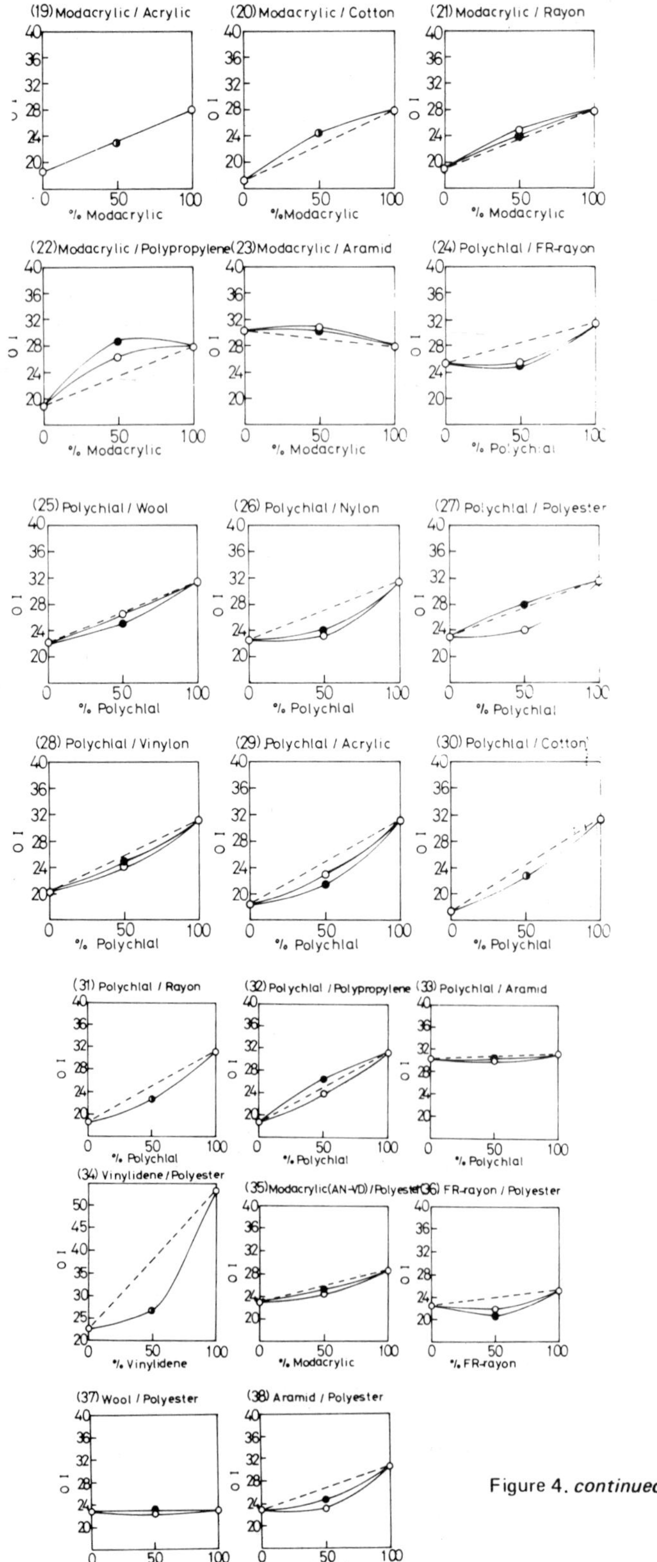

Figure 4. *continued.*

45° angle test is prescribed in Japan Fire Deffence Law and its standards are less than 30 cm^2, in char area and less than 3 sec in residual flame time.

Char area of flame-retardant fibers blends is small, but most blends of flame-retardant fiber and inflammable fiber are easy to burn. The blends with cellulosic fiber are insufficient in flame retardation, different from the case of spherical assembly. This is because depressing reaction is weakened by hydrogen chloride gas due to no generation of pyrolysis zone, although combustion effect is accelerated according to the increase of effective surface area for combustion. Generally, as long as flame-retardant fiber blend ratio of thin fabrics should be higher than that of thick fabrics, its effectiveness cannot be expected.

Char area of blend fabrics of melting fiber and flame-retardant fiber like nylon, polyester and polypropylene is small. This is the so-called dripping effect that these fabrics melt move away from fire source and drop with the flame to prevent the spread of fire.

In the vertical test, a lot of blends burn out completely and the flammability can be examined under severe conditions. Especially, blend fabrics of flame-retardant fiber and aromid are noted to be very excellent.

Foremost difference between Doc. FF3-71 and AATCC 34-1969 is exposure time; 3 sec in the former, and 12 sec in the latter. The results generally showed longer char length in AATCC method which describes longer exposure time, and showed that a lot of specimens completely burnt out. But, there are some blends like modacrylic/rayon and wool/polyester showed longer char length in Doc. method which described shorter exposure time.

This seems to have relation to the phenomena that when exposure time is equal to the shortest time necessary for ignition, the char length is the longest [14].

And the flammability varies also to the testing direction of specimen. It can't be absolutely said that flame-retardant fiber must be woven vertically or horizontally because the results are changeable depending upon the combination of blends. How to reflect a matter of direction in the practical goods should be considered in the future.

Correlation between the values measured in 45° and vertical test and OI values could be found to some degree. For example, in AATCC method, all the samples which have the OI more than 28.5 presented self-extinguishing (less than 12 cm or char length) all the samples which have the OI less than 22 burnt out and the other samples which have the OI between them showed various result case by case. But the blend of soluble and non-soluble fiber tends to prove that char length is longer because of scaffolding effect and the high-soluble blends tends to show the shorter char length because of dripping effect, even if they have the same OI value as others, the flammability were measured on union fabric as model of blend fabrics.

Blend effect of flame-retardant fiber varies according to constructional factors [15] such as knits, wovens, yarn, compactness or weave and also composite type such as blend or union fabric even if the blend ratio is the same. Further consideration should be required about these matters.

Smoke Generation of Blend Fabrics

Smoking of textile goods in burning varies greatly upon burning conditions, for example, temperature level, concentration of oxygen, flaming or non-flaming etc. Various methods for smoking measurement have been sofar used [16]. In this study, smoking by oxygen index method is adopted. This method may matter about the difference from the atmosphere of actual fire. However, the examination on various high polymer materials conducted by Uehara [17, 18] shows their smoking properties correspond with those under high-temperature burning in other measurement methods and they are considered to be equivalent to the smoke generation in exposure of big flame source or after flashover of fire. Stepniczka and others [19] have measured smoking of cotton and polyester in the same method.

The union fabrics as shown in Table 3–Table 7 were also used as the specimens for smoking measurement. Typical smoking curves are shown in Figures 5–9. As smoke density meter is set near the burning column, delicate smoking conditions are being caught minutely.

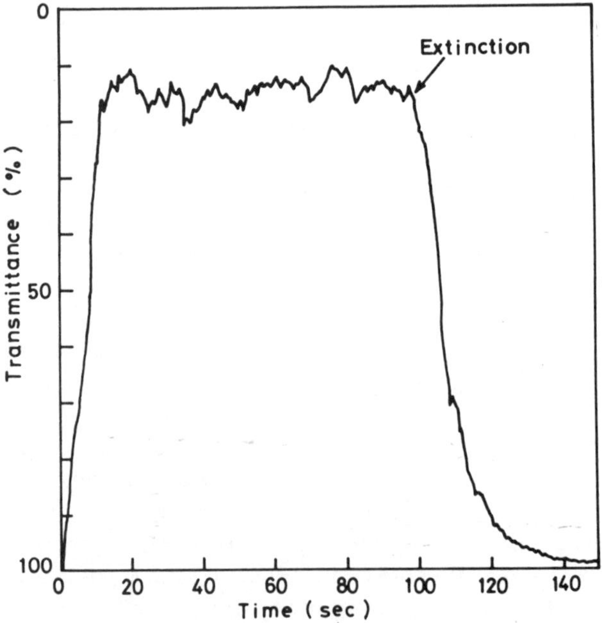

Figure 5. *Smoke Generation Curve for Modacrylic. (Oxygen concn. = 28.5%)*

Burning times from flame exposure to the specimen till its burning out and extinguishing are about 1.5–2 min though there was a little difference according to the kind of blend fabrics. Figure 5 shows an example for generation of dense smoke

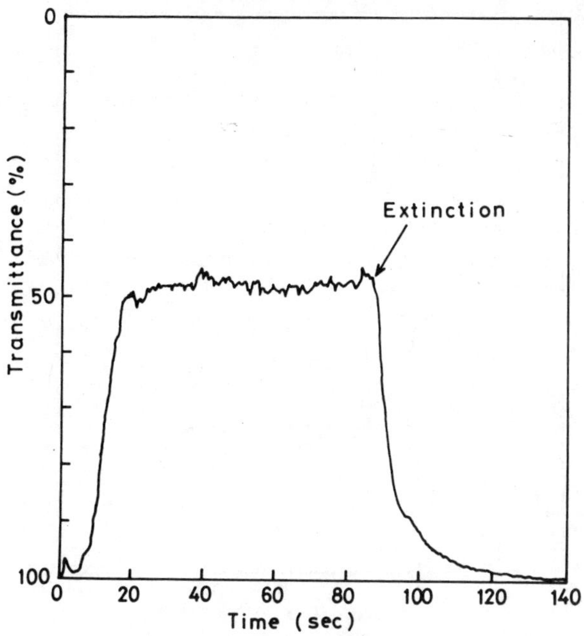

Figure 6. *Smoke Generation Curve for FR-rayon. (Oxygen concn. = 26.0%)*

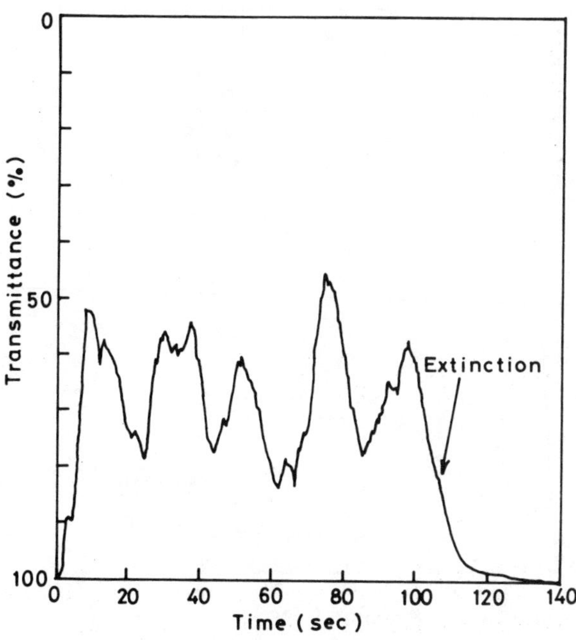

Figure 7. *Smoke Generation Curve for Acrylic. (Oxygen concn. = 19.0%)*

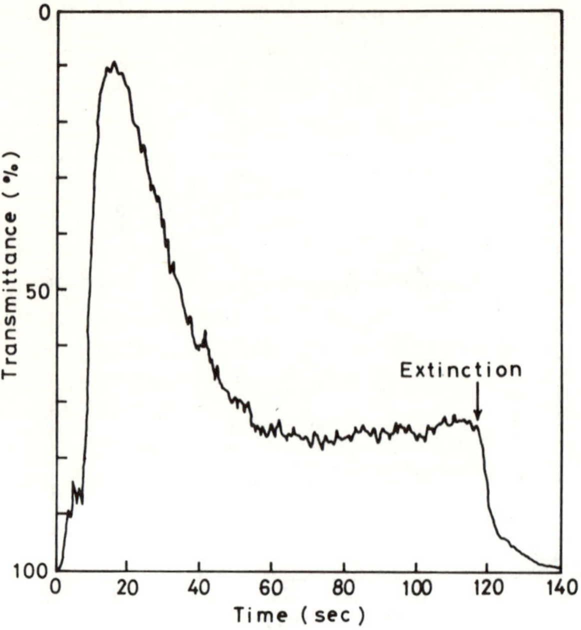

Figure 8. *Smoke Generation Curve for Polychlal/Cotton. (Oxygen concn. = 23.5%)*

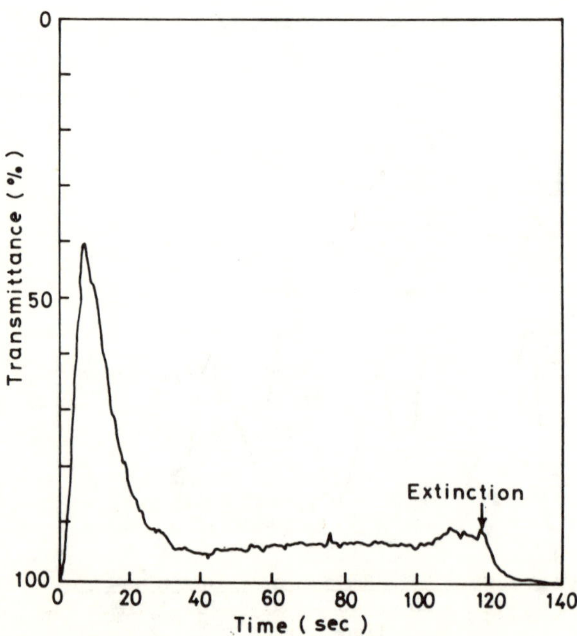

Figure 9. *Smoke Generation Curve for Polychlal/Aramid. (Oxygen concn. = 31.0%)*

and much soot. Figure 6 indicates a constant speed of smoking and typical steady burning and smoking state. In Figure 7, both burning and smoking progress irregularly. And in Figure 8 and Figure 9, smoke density is high at first and rapidly decreases, to stabilize and continue under the burning condition similar to the steady burning. For example, in the case of polychlal/aramid, at first the dense smoke with soot seems to generate from polychlal and, once carbonaceous residue or char of aramid creates, the amount of smoke (generation) exhausted seems to decrease because of soot sticking to the residue or char.

The results for smoking measurement of single component fabrics are shown in Table 8. Flame-retardant fibers containing chlorine, such as PVC, modacrylic and polychlal, generated dense smoke with soot and gave large values of Cs max and smoke generation coefficient. While for the combustible fiber, smoke generation of nylon, vinylon or cotton etc. is small except acrylic generates smoke a little more. And flame-retardant fiber containing chlorine shows an eminent self-extinguishing under the small flame source to give small amount of burning, and therefore shows less smoke generation.

Table 8. Smoke Generation of Single Component Fabrics

Component	Oxygen Concentration (%)	Cs max	$C \times 10^3$ ($Cs \cdot m^3$)	$K \times 10^3$ ($Cs \cdot m^3/g$)
PVC	37.0	4.27	40.8	75.5
Modacrylic	28.5	4.75	69.9	120.6
Modacrylic*	30.0	2.10	18.8	32.4
Polychlal	32.5	3.31	34.4	54.6
FR-rayon	26.0	1.60	23.8	36.6
Wool	26.0	0.50	5.5	8.8
Nylon	23.0	0.05	0.5	0.9
Polyester	24.0	0.51	3.1	5.3
Vinylon	21.0	0.10	1.3	2.0
Acrylic	19.0	2.07	18.9	32.6
Cotton	18.0	0.09	1.5	3.1
Rayon	19.5	0.39	4.1	6.2
Polypropylene	19.5	0.34	4.3	8.1
Aramid	32.5	0.69	6.8	8.5

C: Total amount of smoke K: Smoke generation coefficient

The results for two-component blend fabrics are shown in Table 9–Table 11. PVC blends fabrics combined with nylon, polyester or polypropylene generated considerably large amount of smoke.

Table 9. Smoke Generation of Blend Fabrics Composed of PVC and Other Fibers

Component (Warp/Filling)	Oxygen Concentration (%)	Cs max	C × 10^3 (Cs·m³)	K × 10^3 (Cs·m³/g)
PVC/PVC	37.0	4.27	40.8	75.5
PVC/Modacrylic	35.5	3.43	38.7	64.5
PVC/Polychlal	35.0	4.77	43.5	65.9
PVC/FR-rayon	28.5	6.34	23.8	39.7
PVC/Wool	28.0	4.33	35.4	69.5
PVC/Nylon	23.0	7.02	87.7	156.6
PVC/Polyester	25.5	6.29	73.5	129.0
PVC/Vinylon	27.5	2.94	35.4	63.3
PVC/Acrylic	25.5	3.15	37.5	63.6
PVC/Cotton	23.0	6.29	31.8	54.8
PVC/Rayon	24.5	7.83	39.5	67.0
PVC/Polypropylene	25.5	7.64	82.3	149.6
PVC/Aramid	34.0	1.55	9.6	17.5

Table 10. Smoke Generation of Blend Fabrics Composed of Modacrylic and Other Fibers

Component (Warp/Filling)	Oxygen Concentration (%)	Cs max	C × 10^3 (Cs·m³)	K × 10^3 (Cs·m³/g)
Modacrylic/Modacrylic	28.5	4.75	69.9	120.6
Modacrylic/Polychlal	31.5	4.24	43.4	72.4
Modacrylic/FR-rayon	27.0	4.42	25.6	43.3
Modacrylic/Wool	27.5	4.08	47.3	87.5
Modacrylic/Nylon	25.0	7.33	102.8	174.2
Modacrylic/Polyester	27.5	9.21	81.9	136.5
Modacrylic/Vinylon	25.5	5.99	87.6	151.0
Modacrylic/Acrylic	24.0	4.94	74.4	114.5
Modacrylic/Cotton	25.5	3.46	12.4	21.0
Modacrylic/Rayon	26.0	3.30	17.3	29.4
Modacrylic/Polypropylene	27.0	4.24	23.8	38.4
Modacrylic/Aramid	31.5	1.71	10.1	17.4

Table 11. Smoke Generation of Blend Fabrics Composed of Polychlal and Other Fibers

Component (Warp/Filling)	Oxygen Concentration (%)	Cs max	$C \times 10^3$ ($Cs \cdot m^3$)	$K \times 10^3$ ($Cs \cdot m^3/g$)
Polychlal/Polychlal	32.5	3.31	34.4	54.6
Polychlal/FR-rayon	26.5	4.71	46.7	74.2
Polychlal/Wool	27.0	3.22	55.4	92.3
Polychlal/Nylon	25.0	7.61	104.4	163.1
Polychlal/Polyester	25.0	6.34	74.8	115.1
Polychlal/Vinylon	25.0	5.80	89.5	146.8
Polychlal/Acrylic	24.0	4.33	77.1	111.7
Polychlal/Cotton	23.5	4.61	26.4	41.3
Polychlal/Rayon	24.0	4.24	29.9	46.8
Polychlal/Polypropylene	25.5	6.64	45.5	73.4
Polychlal/Aramid	31.0	1.53	6.9	10.5

And modacrylic and polychlal blends with nylon, polyester, vinylon and acrylic generated also large amount of smoke similarly. But those blends combined with aromid showed less amount of smoke generation. Among them, relations between the fiber composition and smoke generation coefficient of the blends with nylon and aromid together with their OI valves are shown in Figure 10. From the above results, smoke generation of two-component fiber systems hardly depends upon the additive property of the component. This is because the reflections in the prolysis and burning of one component are influenced by other component inevitably. For instance, nylon burns perfectly in itself and so generates a few amount of smoke, but once existed together with PVC, burns imperfectly and the amount of its smoke generation increases because of hydrogen chloride gas generated from the pyrolysis. On the other hand, PVC burns only under the considerable high concentration of oxygen, but once existed together with nylon, comes to burn even under lower concentration of oxygen so that it shows more imperfect burning and the amount of its smoke generation increases.

It is supposed that pyrolysis and burning reflections of each component in PVC/nylon blends shows synergistic effect to interact towards increasing the other's smoking as described above. Blends of aromid with PVC, modacrylic or polychlel present a decrease of smoke generation from the system because much soot generated in the burning of these flame-retardant fibers are sticked fast to carbonaceous residue or char created by burning of aromid. As char is not created yet just after ignition as shown in Figure 9, dense smoke is emmitted. Generally, inflammable fiber blend showed an increase of smoking, when combined with flame-retardant fiber containing chlorine. If flame-retardant fiber is blended so sufficiently that the blend fabric shows self-extinguishing, the blend is meritorious, but if the blend ratio is insufficient, the demerit that smoke generation increases will be caused. Therefore, the blending of flame-retardant fiber like this should be avoided.

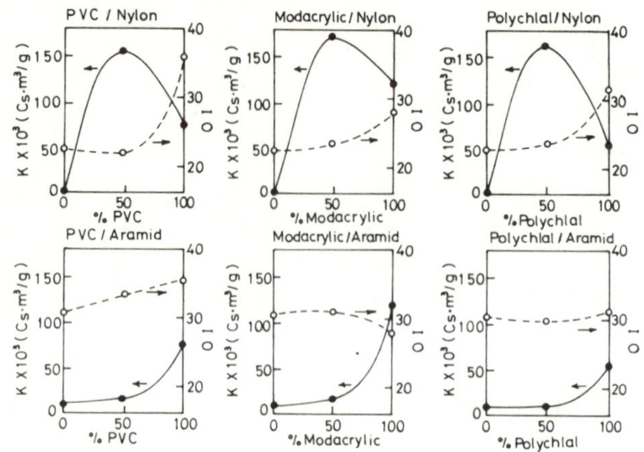

Figure 10. *Relation between Blend Ratio and Smoke Generation Coefficient.*

CONCLUSIONS

Flammability of two-component fiber systems blending flame retardant fiber was discussed using two models of spherical assembly and union fabric.

In almost all cases, flammability of two-component fiber blends is different from the additive property ("average") of the components. This is because the pyrolysis and burning action of a component is influenced by the other component inevitably. Phenomena of the burning-accelerating effect of polypropylene with high calorific value, the scaffolding effect between melting and non-melting fibers, the insulating effect of thick char zone in spherical assembly and the effective flame retardation due to the creation of pyrolysis zone were observed.

As regards smoking, flame-retardant fiber containing chlorive generates dense smoke with soot and when nylon of polyester is blended to the fiber, the total amount of smoke increases more remarkably.

Therefore, two-component fiber systems should be recognized not as a mixture of the components but as a new composite material. And these data of flammability on these experimental blends will contribute to the development of flame retardant two-component fiber materials.

ACKNOWLEDGMENT

The authors wish to thank Dr. Y. Uehara, Yokohama National University for his helpful advice.

ciently that the blend fabric shows self-extinguishing, the blend is meritorious, but if the blend ratio is insufficient, the demerit that smoke generation increases will be caused. Therefore, the blending of flame-retardant fiber like this should be avoided.

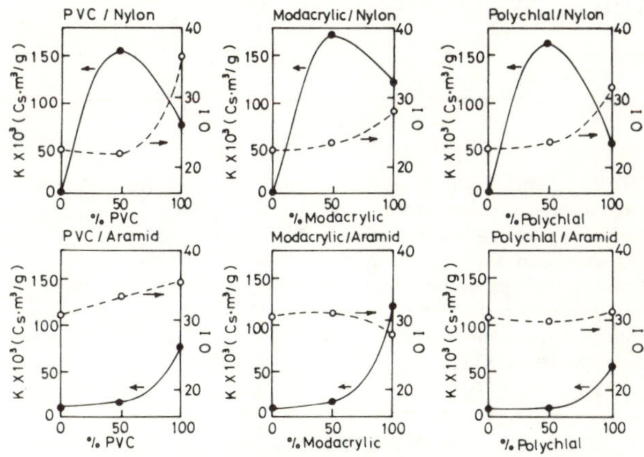

Figure 10. ***Relation between Blend Ratio and Smoke Generation Coefficient.***

CONCLUSIONS

Flammability of two-component fiber systems blending flame retardant fiber was discussed using two models of spherical assembly and union fabric.

In almost all cases, flammability of two-component fiber blends is different from the additive property ("average") of the components. This is because the pyrolysis and burning action of a component is influenced by the other component inevitably. Phenomena of the burning-accelerating effect of polypropylene with high calorific value, the scaffolding effect between melting and non-melting fibers, the insulating effect of thick char zone in spherical assembly and the effective flame retardation due to the creation of pyrolysis zone were observed.

As regards smoking, flame-retardant fiber containing chlorive generates dense smoke with soot and when nylon of polyester is blended to the fiber, the total amount of smoke increases more remarkably.

Therefore, two-component fiber systems should be recognized not as a mixture of the components but as a new composite material. And these data of flammability on these experimental blends will contribute to the development of flame retardant two-component fiber materials.

ACKNOWLEDGMENT

The authors wish to thank Dr. Y. Uehara, Yokohama National University for his helpful advice.

E. L. Finley, Frances G. McDermott and W. H. Carter

Louisiana State University
P.O. Box 18107
Baton Rouge, Louisiana 70803

FIRE BEHAVIOR OF GARMENTS ON MANNEQUINS

(Received February 2, 1974)

ABSTRACT: Fabrics which were alike in all aspects of their physical structure except fiber content were tested for fire behavior, in garment form representing an article of outer apparel, on a life-size mannequin. The basic garment was used as a free-hanging form and as a fitted-in form at the waistline. The effect of fabric in the two types of garment form on the mannequin were studied for maximum combustion temperature, surface heat, burning rate, and radiation. Total heat content by fiber content of the fabric was determined.

INTRODUCTION

Testing the fire behavior of fabric used in a normal size garment is the nearest possible situation to an accidental clothing fire involving a human subject. The garment method requires more fabric, and the testing requires more time to complete than testing small specimens of fabric. Study of the fire behavior of garments on life-size mannequins appeared feasible since scant information was found in the literature on fiber content of fabric and garment behavior in a clothing fire.

The work herein reported was developed on two general hypotheses:

1. The fire behavior of a simple type garment as tested on a mannequin would be basically changed when the fabrics differed only in fiber content.

2. The fire behavior of a simple garment would be changed by modifying its conformation, or shape, on a mannequin.

DISCUSSION AND RESULTS

The conformation or drape of the garment on a mannequin (body form) affects the fire behavior of the garment [5]. Basically, apparel garments for outer apparel are supported on the human form at the shoulder which is a horizontal line of the body, or designed to be supported at the waist and held in place by a belt or tie. Therefore, garments may "swing" with varying amounts of fullness from either of these two positions on the human figure, or the garments may be more form fitting and supported from either position, Figure 1 and Figure 2. The fire behavior of garments on mannequins include such characteristics as propagation of combustion,

Figure 1. A-line dress.

Figure 2. Belted dress.

flame temperature, surface heat generated, potential heat content of fabric, radiation rate, and heat transfer to a surface such as the skin, and the potential of these factors for inflicting serious body injury.

The selection of fabrics was determined primarily according to fabrics widely used by consumers in substantial amounts for various types of apparel and on the basis of which fibers are blended together in yardage [6]. These fabrics were all-cotton and cotton-polyester blends and both have many applications for utilitarian-service, or high-style purposes. The cotton-polyester fabrics were in two different blend ratios. One fabric was an equal blend in a ratio of 50/50 cotton-polyester, and the other fabric was a blend of more cotton and less polyester in a ratio of 70/30. Cotton-polyester blend fabrics are widely used in many types of clothing items for men, women, and children, as well as for household textiles. These apparel items include shirts, trousers, pajamas, blouses, skirts, dresses, slips, gowns, and many items for children.

The fire behavior of these particular fabrics were investigated before and after treating with fire-retardant finishes. The equipment that was used for testing fire behavior of the garments included a fireproof mannequin with surface thermocouples and/or thermoplates at 18 locations, an ignition system using bunsen-type burners, C.P. butane for fuel, a fuel supply system, and individual data recording instrument for each thermocouple lead from the mannequin, Figure 3 and Figure 4 [3].

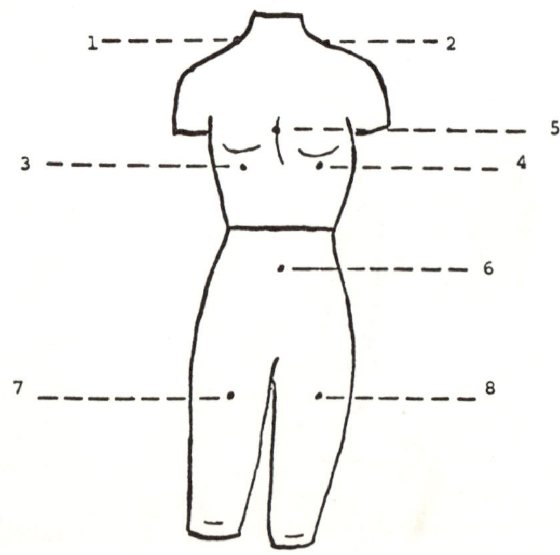

Figure 3. Location of thermocouples on front mannequin.

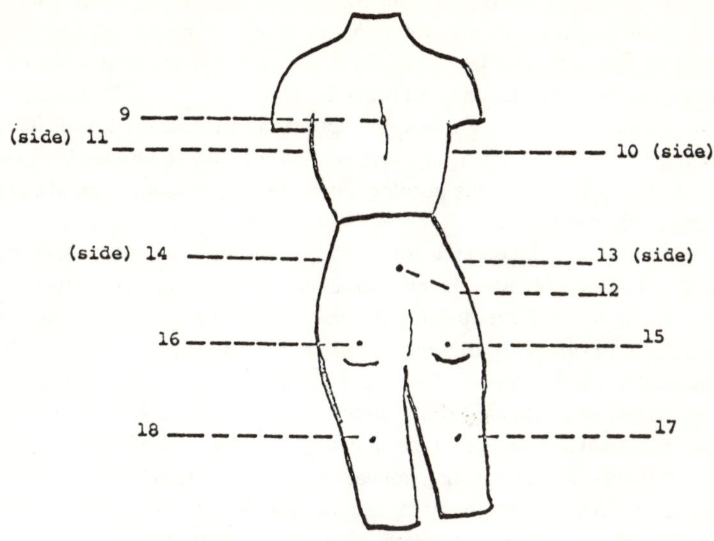

Figure 4. Location of thermocouples on back mannequin.

When flame from each burner was impinged onto the all-cotton fabric at the hem edge, the fabric was ignited immediately, supported combustion, and the flame spread over the garment surface, causing the fabric surface to char. Surface flame then mostly self-extinguished. Only minor penetration of flame occurred through the fabric in this stage. The effect of surface flame over the fabric caused the fabric to shrink to the mannequin. The second stage burning involved flameless combustion or edge burning. The flameless combustion stage caused a slower burning rate and this was recorded as a longer time to produce the maximum flame temperature (Table 1). Flameless combustion of the fabric was self-extinguished on the all-cotton fabric across the upper back shoulder, which was a slightly convex surface. Likewise, flame was self-extinguished on small fragments of fabric just below the shoulder and on the breast area across the front of the garment. These areas were not convex as the surface of the fabric and the mannequin lay flat together. A small amount of fine ash or residue remained after burning the cotton garment with very minimal emission of smoke occurring. None of the residual ash adhered to the mannequin surface. Immediate flame spread over the surface of the cotton garment, followed by flameless combustion, did not set up convection currents to move the fabric away from the mannequin. The garment, therefore, shrank toward the mannequin. Actual flame temperature was more completely measured at the thermocouple points for the cotton garment because less heat was radiated away from the mannequin. In the testing room environment very little smoke was produced from the decomposition of the cotton fabric into carbon dioxide and water. The relative humidity increased one to two percent after an experiment. However, the testing room was not equipped to control temperature and relative humidity conditions. Differences occurred between maximum flame temperatures for all-cotton fabric and the two cotton-polyester fabrics as these were used in the simple A-line dress burned on a fireproof mannequin [4] (Table 2).

In a fire situation involving a single, unsupported garment, the slower flameless, or second stage burning, might offer some element of time to remove the garment. However, the fire behavior of the burning cotton-polyester fabrics would not allow such advantage for the person.

Cotton-polyester blend fabrics were as easily ignited as all-cotton, and burned more rapidly. Cotton-polyester blend fabrics in the same garment style, and of comparable fabric weight, produced a different fire behavior on the mannequin. Both blend fabrics burned rapidly with a strong, vertical flame advance with complete combustion of the fabric. The rapid flame advance was accompanied by emission of fumes with flame on the upper-fabric surface as well as between the fabric and the mannequin. Total combustion of the blend fabric, as flames advanced vertically, created strong convection currents between the garment and the mannequin. This caused the free-hanging garment to move away from the mannequin. As the rapid and complete combustion of the blend fabric occurred, sections of fabric were separated by flame from the garment and these sections dropped to the platform below and continued to burn until consumed.

Sections of fabric joined together at garment seams remained attached to the garment for a few seconds longer than single layer of fabric without a seam. Zippers made of cotton-tape with metal gripper teeth were used as the type of closure on all of the garments. The zipper tape burned slower than the adjacent dress fabric to which it was attached. The metal parts disintegrated and fell as red-hot bits onto the platform surface below. The garments made of cotton-polyester blend fabrics produced a molten residue which hardened and adhered to the mannequin surface. More melt residue remained from the 50/50 fabric than from the 70/30 cotton-polyester fabric. The flaming blend fabrics produced noxious fumes and black smoke; whereas, the amount of smoke produced by burning cotton was negligible. The shrinking-in performance did not occur with the garments made of cotton-polyester blend fabric.

Burning time, or rate, was more rapid for the blend fabrics, and maximum flame temperature reached and flame duration occurred in less time. However, it was apparent to observers conducting testing that as combustion occurred, the blend fabrics produced more intense heat than the all-cotton fabric. Burning rate, as recorded in seconds, decreased as the percent of polyester fiber content increased in the blend. Maximum radiation rate increased as polyester fiber content increased. Radiation time decreased as the percent of polyester fiber content increased [2].

Table 3 shows a mean summary for temperature of untreated fabrics by thermocouples on the mannequin. The flame temperature was lower on top of the shoulder for all fabrics and all garments. Flame was generally extinguished before it passed over the mid-shoulder point for an all-cotton garment. The flames reached

12 to 18 inches above the top of the shoulder for the blend fabrics. Thermocouple positions 3 and 4 were centered under the breast prominence where a concentration of heat occurred, and at these positions the highest flame temperature was recorded above the waist position. Thermocouple 5, located on the mannequin front at approximately the sternum position was in a concave area. This registered the next highest flame temperature for all fabrics. Thermocouple 9 at the mid-center back above the waistline on the mannequin registered lower flame temperature for all fabrics except the shoulder position. The all-cotton fabric temperature was lower than the blends at position 9. The difference in temperature was attributed to secondary stage burning of the all-cotton fabric where the zipper tape burned very slowly and was sometimes self-extinguished. At thermocouples 10 and 11, located 2 to 3 inches below the arm, a ledge effect held some of the heat over these points.

Thermocouple 7 on the mid-thigh front leg position, registered the lowest flame temperature for all fabrics below the waist. This was attributed to the skirt drape in the A-line shape. The mid-center back thermocouple 12, just below the waist, also registered a lower flame temperature. At this point the garment drape allowed space between the fabric and the mannequin due to the concave contour of the mannequin, near the waist position and above the rounded contour of the hips. Thermocouples 15 and 16 located on the hip position registered the highest flame temperature of all positions on the lower mannequin. Thermocouples 13 and 14 on the side hip, just below the waist, were positioned where the mannequin form narrowed in to the waist. Maximum flame temperature was higher for all-cotton fabric and the 70/30 cotton-polyester blend than for the 50/50 cotton-polyester blend.

An Eppley pyrheliometer was used to measure the radiation rate. The surface plane of the pyrheliometer sensor was in a parallel position with the vertical center of the mannequin and set 18 inches distant from the mannequin. This position was an arbitrary selection since there were many positions which might have been selected. Therefore, flame temperature, flame time, radiation time, used as rate, show that these factors were not completely measured by the thermocouples on the mannequin surface because of the characteristic burning behavior, or billowing away, of the cotton-polyester blend fabrics, Table 4 and Table 5.

A Parr oxygen bomb calorimeter was used to determine total heat content for each of the three fabrics. Each fabric sample was rolled into a tight, compact cylinder, weighing approximately 1 g. Five replications were burned for each fabric. Standard procedures were followed for operating the calorimeter. Heat content (cal/g), as determined by the standard Parr calorimeter procedure was used to measure the fiber potential for heat liberation during combustion. The potential total heat content increased in direct ratio to the polyester fiber content in the fabric. When these particular blend fabrics were burned, the polyester fiber increased the potential heat content and radiation rate.

Surface temperature generated by the burning untreated fabrics was measured by copper thermoplates mounted on the mannequin at the same locations as the thermocouples. These data were obtained in degrees F and converted into equivalent cal/cm^2/15 sec. The copper discs were 0.025 inches thick, 1 inch in diameter, and weighed 10 grams. These thermoplates had a heat absorption coefficient similar to human skin. The data in Table 6 show that these fabrics, if used singly next to a body, would produce severe third-degree injury if they were ignited accidently. On the basis of data reported by Behnke [1], these data seem to indicate that a full blister would occur at 0.205 cal/cm^2/15 sec. In terms of mean cal/cm^2/15 sec for all thermoplate locations on the mannequin for these three fabrics, the ascending rank order was 3.827 for the 50/50 cotton-polyester blend; 3.986 for all-cotton; and 4.521 for 70/30 cotton-polyester blend.

The lower surface heat phenomenon, Cal/cm^2/min, for the 50 percent polyester fabric was in agreement with the higher calculated radiation rate, shown to be 1.8; and followed in order of radiated rate of 1.7 for 70/30 cotton-polyester; and 0.7 for all-cotton. Therefore, data for total heat content in cal/g showed that all-cotton ranked lowest followed in order by 70/30 cotton-polyester blend and 50/50 cotton-polyester blend in order of 3952.4, 4398.3, and 4716.6 cal/g, respectively. According to the location of thermoplate position on the mannequin, higher surface heat was registered on the areas below the normal waistline rather than above the normal waistline. This was in general agreement with the thermocouple point data in maximum flame temperature for positions below and above the waist.

Some of each of the three fabrics, all-cotton, and the 70/30 and 50/50 cotton-polyester blends were treated with each of three flame-retardant finishes, THPC-MM-Urea, APO-THPC, and THPOH-NH_3. The treated fabrics did not ignite to support combustion when the direct flame was impinged on the garment at the hem edge for 3 minutes. The fabric was charred at the six flame contact points within 10 seconds after impinging the flame. The flame liberated volatile gases and fumes from the chemically treated fabrics. The volatile gases appeared to collect between the fabric and the mannequin near the waist position. Heat rising from the flames impinged at the garment hem edge below caused some spontaneous ignition of the volatile gases and flames of a few seconds duration would occur between the garment and mannequin at these spots. When this happened, isolated char spots appeared on the fabric surface. This momentary flaming was visible through the fabric.

The shape of char areas at the bottom of the garment was somewhat like an isosceles triangle rounded at the apex. The treated fabrics flamed at the hem edge for several inches above the tip of the burner flame for less than 10 seconds. The gases and fumes produced color staining on the fabric which ranged from grey to yellow and deep orange. The color of the stains differed according to the combination of flame-retardant finish and fiber content. Evidences of polyester fiber residue were observed in the char on the 50/50 treated cotton-polyester fabric. The melt residue was visible as minute shiny particles.

In these studies, the effect of drape or shape of the A-line dress, as a single layer of fabric on the mannequin, was changed by adding a narrow self-fabric tie-belt at the waist position. When the belt was added, the dress was held closely to the mannequin at the waistline, and just above and below the waistline the fabric was formed in folds. The effect of this alteration in garment shape caused a difference in burning behavior between the A-line and belted garments with no flame-retardant finish. The mean for maximum flame temperature at thermocouple points was higher for the untreated fabrics where the dress fabric formed in folds than where the dress fabric fitted loosely or lay flat against the mannequin surface at the side, upper bodice area across the front, above breast area, and across the upper back shoulder (Table 7). The burning time in seconds for peak flame temperature was longer for the blend fabrics than for the all-cotton fabric. The 50/50 blend fabric had a longer burning time or rate than the 70/30 blend. Vigorous flame-up occurred on the fabric when the belted-in constriction was released at the waist. Flame was impinged for 5 seconds at the garment hem and the self-supported flame advanced rapidly in the upward direction to the belt where the flame advance was halted until the belt was burned through. Characteristic char behavior of the all-cotton belted dress with a flame-retardant finish was shown to be similar to the char characteristics of the fabrics in the A-line dress.

SUMMARY

All-cotton fabric produced a higher combustion temperature than the cotton-polyester blend fabrics in the A-line garment. This was due to the combined effect of a first and secondary stage burning, an extended time of burning and because the garment shrank against the mannequin. The cotton-polyester fabrics burned more rapidly and produced a higher radiation rate because the garment billowed away from the mannequin due to convection currents generated by rapid burning of the fabric. The total heat content, in Cal/g, was higher for the cotton-polyester fabrics than for all-cotton.

The burning rate for the belted dress was different from the A-line dress. The belted dress produced a higher combustion temperature for all fabrics than the A-line dress. Surface heat produced by the combustion of fabrics would produce potentially serious burn injury. All fabrics treated with flame-retardant finishes did not support combustion. The fiber content of fabrics caused a difference in the char patterns and emission of volatile gases.

Table 1. Mean Data for Temperature — Time and Surface Heat for Fabrics with No Flame-Retardant Finish

	Temperature	Time		Surface Heat
Fabric	Peak (°F)	To Peak (Sec)	Over Peak (Sec)	$Cal/Cm^2/15$ sec
All cotton	399.59	37.02	14.07	3.986
70/30, C-P	386.96	30.22	9.01	4.521
50/50, C-P	360.59	25.28	7.74	3.827

C-P, Cotton Polyester

Table 2. Calculated F Values for Flame Temperature as Measured on Mannequin Form

		F Value		
Source	d.f.	Flame	To Peak	Over Peak
Fabric	2	21.8**	4.5**	5.4*
Thermocouple	17	15.6**	18.4**	11.0**
Fabric x thermocouple	33	2.4	1.1	2.4

*Significant at 5% level
**Significant at 1% level

F test is the ratio of two mean squares used for testing the null hypothesis of no treatment differences.

Table 3. Mean Summary of Temperature °F for Untreated Fabrics by Thermocouples on the Mannequin

Thermocouple Position on Mannequin	Fabric x Thermocouple	
	All Cotton	Cotton-Polyester 70/30 + 50/50
Above Waist		
Front		
1	214.4	296.4
3 & 4	458.4	410.8
5	390.2	348.3
Mean	354.4	351.8
Back		
9	226.6	318.6
Under Arm		
10 & 11	346.7	352.0
Overall Mean	309.2	340.8
Below Waist		
Front		
6	509.4	423.0
7	263.6	255.6
Mean	386.5	339.3
Back		
12	387.0	321.3
15 & 16	560.9	483.8
17 & 18	455.0	301.3
Mean	467.6	368.8
Side Hip		
13 & 14	435.6	460.9
Overall Mean	429.9	389.7

Table 4. Mean Values of Flame Temperature, Radiation and Heat Content as Measured for Three Fabrics

	Mannequin			Pyrheliometer			Parr Calorimeter
	Flame (°F)	To Peak (Sec)	Over Peak (Sec)	Time During Radiation (Sec)	Recorder Gain (M/v)	Calculated Radiation Rate (Cal/cm²/min)	Heat Content (Cal/Gm)
All-cotton	522.8	39.0	16.6	106.0	1.6	0.7	3952.4
70/30, C-P	426.1	31.5	12.8	74.4	3.9	1.7	4398.3
50/50, C-P	380.5	31.5	16.5	74.6	4.2	1.8	4716.6

C-P Percent Cotton/Polyester Fiber Content

Table 5. Calculated F Values for Radiation Rate Measured by Pyrheliometer and Parr Calorimeter

		Parr Cal.	Pyrheliometer		
Source	d.f.	Heat Content (Cal/Gm)	Time During Radiation (Sec)	Recorder Gain (M/v)	Radiation Rate (Cal/Cm²/Min)
Fabric	2	131.7**	9.2*	110.8**	90.7**
Residual	12	–	–	–	–

*Significant at 5% level
**Significant at 1% level

Table 6. Summary Mean Cal/cm²/15 sec for Three Fabrics with No Flame-Retardant Finish

	Cal/cm²/15 sec		
		Garment Position	
Fabric	All Areas	Upper	Lower
All-cotton	3.986	3.672	4.176
70/30, C-P	4.521	4.386	4.908
50/50, C-P	3.827	3.514	4.233

Table 7. Mean Peak Flame Temperature and Mean Over-Peak Time for Three Fabrics in Belted Garments

Belted	Fabrics		
	All-Cotton	70/30, C-P	50/50, C-P
Temperature (degrees F)	608.1	531.1	487.3
Burning rate (seconds)			
To-peak	33.02	28.16	24.43
Over-peak	11.96	13.43	16.34

C-P, Cotton-Polyester Blend

REFERENCES

1. W. P. Behnke, and R. E. Seaman. Development of Clothing for Protection from Thermal Hazards. Paper Delivered to the ASTM Committee, D-13, Belmont Plaza Hotel, New York, March, 1969.
2. W. H. Carter, E. L. Finley, and B. R. Farthing. Flame Temperature, Heat Radiation and Heat Content Measured for All-Cotton and Cotton-Polyester Fabrics. *J. Fire & Flammability*, Vol. 4/April, 1973, pp. 106–107.
3. E. L. Finley, and W. H. Carter. Description of a Fireproof Torso-Form and Ignition System Used as a Testing Technique for Studying Garment Flammability. *J. Fire & Flammability*, Vol. 1/April, 1970, pp. 166–174.
4. E. L. Finley, and W. H. Carter. Temperature and Heat-Flux Measurement on Life-Size Garments Ignited by Flame Contact. *J. Fire & Flammability*, Vol. 2/October, 1971, pp. 298–319.
5. E. L. Finley, and C. T. Butts. Garment Conformation on a Mannequin Changes Flammable Characteristics. *J. Fire & Flammability*, Vol. 4/July 1973, pp. 145–155.
6. *Textile Organon*. Vol. XLIII, No. 5, May, 1972, p. 60.

E. L. Finley

Etta Lucille Finley is a Professor on the teaching staff, and a Project Leader in textile research, in the School of Home Economics at Louisiana State University and Agricultural and Mechanical College in Baton Rouge, Louisiana. She received the B.S. and M.A. degrees from Texas Woman's University and has done sabbatical study at the University of Tennessee. She is the author and co-author of 26 technical papers. She has been an invited lecturer at the University of Utah, American Home Economics Association, Research Section; Association of College Professors of Textiles and Clothing; and the AATCC, Gulf Coast Section.

F. G. McDermott

Frances G. McDermott is an Instructor assigned to textile research in the School of Home Economics at Louisiana State University and Agricultural and Mechanical College. She received the B.S. degree from Mississippi State College for Women, and the M.S. degree from Louisiana State University.

W. H. Carter

William Hobart Carter is a Professor of Agricultural Engineering at Louisiana State University and Agricultural and Mechanical College in Baton Rouge Louisiana. He received his B.S. and M.S. degrees in Agricultural Engineering from Iowa State University in 1929 and 1934, respectively. Before joining the faculty of Louisiana State University in 1930, he was employed by the U.S.D.A. Division of Agricultural Engineering. He has experience in farm power and machinery, farm and home utilities, and household equipment.

CARLOS J. HILADO

Research and Development Department
Union Carbide Corporation
South Charleston, West Virginia 25303

FIRE STUDIES OF BEDDING MATERIALS

(Received July 26, 1973)

ABSTRACT: Increased concern with the flammability of bedding materials prompted a limited investigation of the fire behavior of readily available materials, to obtain guidance regarding the proper use of existing materials and the development of new and improved materials. A variety of tests, ranging in scale from small specimens to full mattresses with sheets and pillows, were performed.

Currently available materials and technology have been shown capable of giving innerspring and urethane foam core type mattresses adequate resistance to small ignition sources. The presence and involvement of sheets, pillows, and pillowcases increases the severity of exposure and introduces many variables affecting the resistance of the mattress to ignition and involvement.

It is the opinion of the author that a bedding system cannot be considered to have a given level of resistance to ignition unless that level of resistance is possessed by all the components of the system: mattresses, sheets, pillows, pillowcases, blankets, and sleepwear.

Fire experience indicates that population habits such as smoking in bed may be the major factor in bedding fires. The extent to which bedding materials must be improved to attempt to offset human carelessness is an issue which has not been completely resolved.

INTRODUCTION

INCREASING CONCERN FOR public safety has resulted in greater attention to the flammability behavior of a variety of consumer products. The amendment of the Flammable Fabrics Act has enlarged the scope of this scrutiny from clothing to other products such as blankets and floor coverings.

Bedding materials have been involved in a significant number of fires year after year, primarily because of the widespread habit of smoking in bed. A study of 3,145 single-fatality fires reported to the National Fire Protection Association from February 1966 through January 1968 [1] showed that 525 were clothing fires and 2,620 were nonclothing fires, and of the nonclothing fires 307 were caused by smoking in bed and 237 were caused by smoking on upholstered furniture.

Cigarettes and matches are among the principal sources of ignition in building fires. Of the 996,600 building fires in the United States in 1971, 118,400, or 12 percent, were attributed to smoking and matches, and 70,400, or 7 percent, were attributed to children and matches [2]. Of the 992,000 building fires in the United States in 1970, 107,200, or 11 percent, were attributed to smoking and matches, and 63,800, or 6 percent, were attributed to children and matches [3]. Of the 973,000 building fires in the United States in 1969, 111,800, or 11 percent, were attributed to smoking and matches, and 79,400, or 8 percent, were attributed to children and matches [4].

Bedding materials can be exposed to two types of fire exposure: exposure to a small ignition source such as a lighted match or cigarette, as in cases of occupants smoking in bed and children playing with matches; and exposure to a developing or fully developed room fire involving other articles in the room or the contents of adjacent rooms in the structure, with initial ignition coming from any of a variety of causes.

In the first type of exposure, the behavior of the bedding materials has an important bearing on life safety in that resistance to ignition can prevent a fire, resistance to flame spread can hinder its development, and limitation of combustion products can reduce physical injury and problems of escape. In the second type of exposure, the behavior of the bedding materials is much less significant to the safety of the occupant of the bed, because the temperatures required for ignition tend to be well above lethal levels. The combustion products would be the major concern of occupants of adjacent rooms and of persons combatting the fire.

Bedding materials have traditionally been combustible materials, and requirements of comfort, convenience, and cost will continue to call for substantial use of combustible materials. The flammability characteristics of these materials have generally been considered to present no hazard than could be considered an acceptable risk by the majority of consumers. Increases in population density, especially in urban areas, and changes in population habits have resulted in both increased probability of fire exposure in terms of frequency, and increased probability of fire injury in terms of number of persons affected. The person who smokes in bed probably has in a given radius not only more neighbors but also more neighbors with the same habit. The resultant need to reduce the probability and consequences of fires has been reflected in growing government concern.

Fire tests of bedding and upholstery materials have been performed by the Southwest Research Institute under a contract awarded by the National Bureau of Standards [5, 6]. As a result of these tests and investigations of fire records, the Department of Commerce has issued a flammability standard for mattresses [7] and found a need for a flammability standard for upholstered furniture [8]. In these standards and in earlier standards for carpets and rugs and children's sleepwear [9, 10, 11], the government approach has been to reduce fire hazard by reducing the probability of ignition by small ignition sources.

Pursuing the philosophy that manufacturers should, as a matter of public responsibility, explore methods of improving the safety of their products without the need for forcible government action, Union Carbide Corporation has undertaken a limited investigation of the flammability characteristics of readily available bedding materials, to obtain guidance as to the proper use of existing materials and the development of new and improved materials. This paper presents the results of this work.

PLAN OF INVESTIGATION

The work was divided into three phases: first, a study of the behavior of various types of mattresses and sofa-bed cushions when subjected to small ignition sources, to determine their relative susceptibility to ignition or progressive smoldering; second, a study of the behavior of various types of mattresses furnished with sheets and pillows of different materials, when exposed to a small ignition source, to determine the relative fire resistance of complete bedding systems; and third, a study of the smoke producing characteristics of various mattress constructions, to determine the relative sight obscuration which results. Mattress behavior was emphasized because mattresses are the major product of the bedding industry and its most identifiable one.

The investigation had several limitations which were recognized at the outset. The lack of room-scale fire test facilities prevented the acquisition of information on the effect of placement, ventilation, and interaction with other materials, and limited thermal and optical measurements to the vicinity of the materials involved. The almost complete absence of published work on bedding fire tests resulted in the dependence of the procedures of each test on the results of previous tests, to at least some extent. Information was almost entirely qualitative at the start of the test program, and became more quantitative as experience was accumulated.

The articles tested were limited to those which were readily available, or which could be constructed by existing methods from readily available materials. The fire-retardant samples of cotton cover and urethane foam were commercially available to the bedding industry, and are not necessarily representative of the best technology available at the time this paper is read. Modacrylic and aromatic polyamide sheets and pillowcases were tested, but there were no fire-retardant cotton analogs which had the permanent fire retardance and esthetics of these two types of materials.

TESTS ON MATTRESS SECTIONS

The purpose of this series of tests was to determine the relative susceptibility of representative types of mattress constructions to ignition or smoldering when subjected to small ignition sources. Three types of small ignition sources were selected: a cigarette, a match, and a methenamine pill; the first two are prominent causes of bedding fires, and the third is a standard ignition source specified in government standards for carpets [9, 10].

Mattresses were tested without sheets and pillows in order to isolate the effect of mattress construction and materials from interaction with the behavior of sheets and pillows. The samples were tested in the bone-dry condition to provide the most unfavorable conditions and to avoid the effects of humidity variations. These conditions were not unrealistic, since they can be approached by humidity levels in the winter months.

Test Procedure

Sections one foot wide, each the full width of the mattress, were cut from each mattress sample. The cutting of sections from the innerspring constructions presented special problems, because the special saw required to cut through the metal innerspring generated enough heat and sparks to scorch sisal padding in the vicinity of the cut. The use of water on the saw blade provided sufficient cooling to prevent ignition in most cases, but one smoldering fire had to be extinguished with water. As a result of this experience, spark and hot-surface ignition methods were considered for inclusion in the study. These types of ignition are widely used in fire research on liquids and vapors, but were judged not to be representative of enough actual fire experiences.

The test sections were conditioned for one hour at 105°C in a circulating-air oven and then sealed in polyethylene containers until the time of test. A sheet of asbestos-cement board, 1/4-inch thick, supported the test section in a large laboratory hood. Three ignition sources were placed on top of each sample: a lighted non-filter cigarette (regular length), a lighted methenamine pill (Eli Lilly, No. 1588), and a lighted wooden match. Figure 1 shows a typical arrangement.

Figure 1. *Typical arrangement for tests on mattress sections. Ignition sources, left to right: cigarette, methenamine pill, match. Test No. 7, time 15 seconds.*

Test Results

Fifteen tests were performed on mattress sections. These are summarized in

Table 1. Four tests involved innerspring constructions, two of these with fire-retardant covers; one test involved individually wrapped coil springs with padding and quilting; four tests involved latex foam rubber core constructions, two of these with fire-retardant covers; and six tests involved urethane foam core constructions, one of these with fire-retardant cover and one with both fire-retardant cover and fire-retardant core.

The lighted cigarette failed to ignite any mattresses of the innerspring and urethane foam core types, and damage was limited to a small charred area with a little darkening or melting out of the underlying material. The lighted cigarette produced progressive internal smoldering in all the latex foam rubber core mattresses, and the affected area attained a 12-inch diameter after 22 minutes when the cover was not fire-retardant, and after 35 to 44 minutes when the cover was fire-retardant cotton. The use of cut sections to show internal construction was particularly illustrative in this case since it permitted detection of smoldering behavior, but burning or smoldering rate increased when the area affected by flame or smoldering reached the cut edge with its greater access of oxygen. The presence of cut edges may provide no more access of air than the "breathing" action caused by normal movement of an occupant sleeping on the mattress, but this point would be difficult to resolve. For the purposes of this study, the valid portion of the test was considered to end when the affected area reached the cut edge, which corresponded to a diameter of 12 inches.

In several tests, the lighted methenamine pill and lighted match started spreading fires which reached the cigarette before the cigarette could show any serious effects. In such cases, the test was repeated with only a lighted cigarette as the ignition source.

The internal smoldering of the latex foam rubber produced dense and irritating smoke, grayish white at first, changing to yellowish brown as smoldering progressed and turning to black when the smoldering changed to burning. Two samples of latex foam rubber mattresses ignited upon contact with air after being carried out of the laboratory to alleviate smoke conditions; the ignition delay time was about 30 seconds, and the smoldering rubber was deliberately ignited with a match in subsequent tests as a safer method of disposal.

The lighted wooden match was quickly extinguished when in contact with the fire-retardant cotton cover used in six of the tests, and with the plastic-coated cover used in one test. Spreading fires resulted in all other tests, or progressive smoldering in latex foam rubber.

Damage from a lighted methenamine pill was limited to small charred areas in the four tests employing fire-retardant cotton covers on innerspring and urethane foam core type mattresses. Progressive smoldering occurred in latex foam rubber despite the use of a fire-retardant cover, which increased the time required to involve a 12-inch diameter area from 20 minutes to 31–32 minutes.

Some of the differences in behavior described are shown by Figures 2 to 16.

Table 1. Tests on Mattress Sections

Test No.	Mattress Construction: Core	Padding	Cover	Ignition Source	Result	Time, minutes for Affected Area to Reach 12 inches Diameter
1	Innerspring unit	Latex sisal pad, felt batt	Spun rayon, quilted 3/4 inch urethane foam	Cigarette	Small charred area	
				Methenamine pill	Spreading fire	3.5
				Wooden match	Spreading fire	3.5
3	Innerspring unit	Latex sisal pad, 1 inch urethane foam	Spun rayon, quilted 1/2 inch urethane foam	Cigarette	Small charred area	
				Methenamine pill	Spreading fire	3.5
				Wooden match	Spreading fire	3.5
8	Innerspring unit	1/4 inch latex sisal pad, 3/4 inch urethane foam	Cotton FR	Cigarette	Small charred area	
				Methenamine pill	Small charred area	
				Wooden match	Small charred area	
7	Innerspring unit	1/4 inch tufflex FR, 3/4 inch urethane foam	Cotton FR	Cigarette	Small charred area	
				Methenamine pill	Small charred area	
				Wooden match	Small charred area	
9	Coil springs, individually wrapped	1 inch urethane foam	Rayon, quilted two 1/4 inch layers urethane foam	Cigarette	Small charred area	
				Methenamine pill	Spreading fire	4
				Wooden match	Spreading fire	4+
2	Latex foam rubber, 4 inch	None	Spun rayon, quilted felt	Cigarette	Progressive smoldering	22
				Methenamine pill	Progressive smoldering	20
				Wooden match	Progressive smoldering	15
4	Latex foam rubber, 4 inch	None	Spun rayon, quilted 1/2 inch urethane foam	Cigarette	Progressive smoldering	22
				Methenamine pill	Spreading fire	3
				Wooden match	Spreading fire	3
11	Latex foam rubber, 4 inch	None	Cotton FR	Cigarette	Progressive smoldering	44
				Methenamine pill	Progressive smoldering	32
				Wooden match	Small charred area	

Table 1. Tests on Mattress Sections (continued)

Test No.	Mattress Construction			Ignition Source	Result	Time, minutes for Affected Area to Reach 12 inches Diameter
	Core	Padding	Cover			
15	Latex foam rubber, 4 inch	None	Cotton FR	Cigarette	Progressive smoldering	35
				Methenamine pill	Progressive smoldering	31
				Wooden match	Small charred area	
5	Urethane foam, 4 inch	None	Rayon, quilted 1/4 inch urethane foam	Cigarette	Small charred area	
				Methenamine pill	Spreading fire	5
				Wooden match	Spreading fire	5
6	Urethane foam, 5.5 inch	None	Rayon	Cigarette	Small charred area	
				Methenamine pill	Spreading fire	6
				Wooden match	Spreading fire	6
14	Urethane foam, 1.5 pcf, 4 inch	None	Cotton, plastic coated	Cigarette	Small charred area	
				Methenamine pill	Small charred area	
				Wooden match	Small charred area	
13	Urethane foam, 1.5 pcf, 4 inch	None	Cotton	Cigarette	Small charred area	
				Methenamine pill	Spreading fire	6
				Wooden match	Small charred area	
12	Urethane foam, 1.5 pcf, 4 inch	None	Cotton FR	Cigarette	Small charred area	
				Methenamine pill	Small charred area	
				Wooden match	Small charred area	
10	Urethane foam FR, 2.0 pcf, 4 inch	None	Cotton FR	Cigarette	Small charred area	
				Methenamine pill	Small charred area	
				Wooden match	Small charred area	

Figure 2. *Test No. 1, time 3 minutes. Regular cover, padded innerspring core.*

Figure 3. *Test No. 3, time 3 minutes. Regular cover, padded innerspring core.*

Figure 4. *Test No. 4, time 3 minutes. Regular cover, latex foam rubber core.*

Figure 5. *Test No. 5, time 5 minutes. Regular cover, urethane foam core.*

Figure 6. *Test No. 6, time 5 minutes. Regular cover, urethane foam core.*

Figure 7. *Test No. 9, time 5 minutes. Regular cover, individually wrapped coil springs.*

Figure 8. *Test No. 13, time 5 minutes. Regular cover, urethane foam core.*

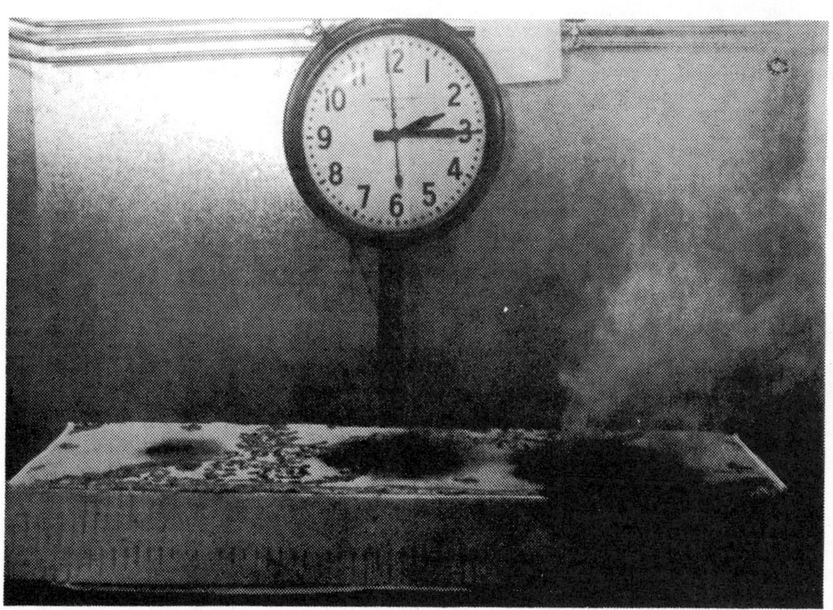

Figure 9. *Test No. 2, time 15 minutes. Regular cover, latex foam rubber core.*

Figure 10. *Test No. 11, time 40 minutes. Fire retardant cover, latex foam rubber core.*

Figure 11. *Test No. 15, time 35 minutes. Fire retardant cover, latex foam rubber core.*

Figure 12. *Test No. 7, cut-away view after test. Fire retardant cover, padded innerspring core.*

Figure 13. *Test No. 8, cut-away view after test. Fire retardant cover, padded innerspring core.*

Figure 14. *Test No. 10, cut-away view after test. Fire retardant cover, fire retardant urethane foam core.*

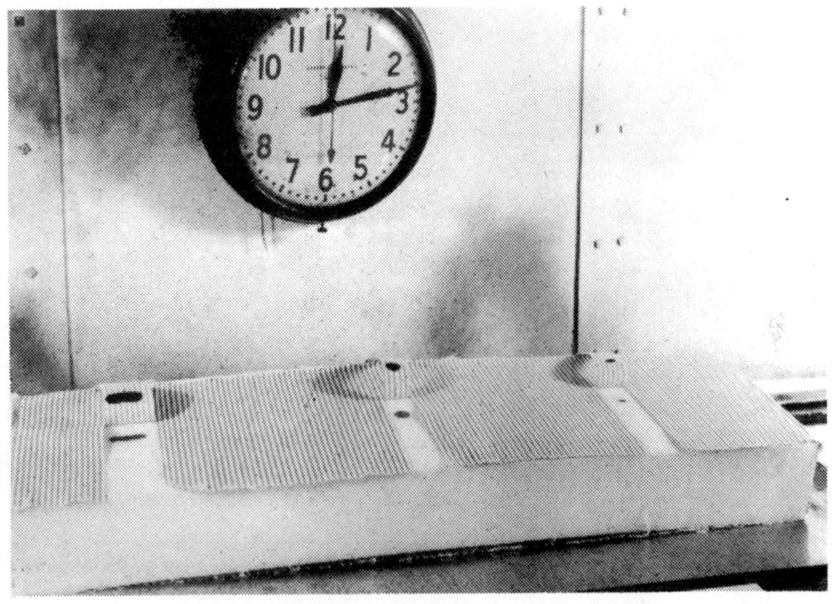

Figure 15. *Test No. 12, cut-away view after test. Fire retardant cover, urethane foam core.*

Figure 16. *Test No. 14, cut-away view after test. Plastic coated cover, urethane foam core.*

Type of Ignition

Spreading fires or progressive smoldering were produced by a lighted methenamine pill in 10 of the 15 tests, by a lighted match in 7 of the 15 tests, and by a lighted cigarette in 4 of the 15 tests. The methenamine pill averaged out as the most severe ignition source, and the lighted cigarette the least severe. If only spreading fires are considered, the methenamine pill produced these results in 7 of the 15 tests, and the match in 6 of the 15 tests. These two ignition sources tend to be similar in severity of exposure. For test purposes the methenamine pill is preferred because it is on the average more severe and seems more reproducible. The reproducibility of this type of exposure is shown in Table 2.

Table 2. Reproducibility of Methenamine Pill Exposure

Test No.	Burning Time, sec.
7	90
8	100
10	90
11	95
12	85
13	100
14	94
15	90

Ease of Extinguishment

Fires in rayon covering and in urethane foam were relatively easy to extinguish with water or carbon dioxide, but fires in felt batt and in latex foam rubber were difficult to extinguish with water and almost impossible to extinguish with carbon dioxide. Smoldering latex foam rubber produced large quantities of smoke when hit by water.

TESTS ON SOFA-BED CUSHIONS

The purpose of this brief series of tests was to determine the relative susceptibility of some types of sofa-bed cushions to ignition when exposed to small ignition sources. The smaller size of the cushions relative to mattresses made it impossible to use more than two ignition sources simultaneously. On the basis of experience with mattress sections, only the cigarette and methenamine pill were used.

Test Procedure

Each sample cushion was cut in half to expose the entire cross section, and a half-section was used as the test specimen. Conditioning for one hour at 105°C in a circulating-air oven was followed by storage in a polyethylene container until the time of test. The specimen was supported by 1/4-inch thick, asbestos-cement board. Two ignition sources were placed on top of each specimen: a lighted non-filter cigarette (regular length) and a lighted methenamine pill (Eli Lilly, No. 1588). Figure 17 shows a typical arrangement.

Test Results

Three tests were performed on sofa-bed cushion sections. These are summarized in Table 3. The lighted cigarette produced no ignition in any of the tests. The open-mesh fabric covered cushions were ignited by the methenamine pill and eventually consumed, but the vinyl fabric covered cushion did not ignite. The presence of polyester fiberfill between the fabric covering and the urethane foam cushion seemed to accelerate flame spread, probably due to greater access of air. The resistance of the vinyl fabric to ignition is attributed to the substantial chlorine content of polyvinyl chloride.

Some of the described differences in behavior are shown in Figures 18 to 20.

TESTS ON MATTRESSES WITH SHEETS AND PILLOWS

The purpose of this series of tests was to determine the effect of fire involvement of sheets and pillows on the behavior of various types of mattress constructions. The choice of sheets and pillows is the prerogative of the mattress owner, and beyond the control of the mattress manufacturer. The exposure severity of burning

Figure 17. *Typical arrangement for tests on sofa-bed cushion sections. Ignition sources: cigarette, methenamine pill. Test No. 1, time 15 seconds.*

Table 3. Tests on Sofa-Bed Cushion Sections

Test No.	Cushion	Cover	Ignition Source	Result
1	Urethane foam, 5-inch	Vinyl fabric	Methenamine pill Cigarette	Small charred area Small charred area
2	Urethane foam, 4-inch Polyester fiberfill	Open-mesh fabric	Methenamine pill Cigarette	Spreading fire Small charred area
3	Urethane foam, 5-inch	Open-mesh fabric	Methenamine pill Cigarette	Spreading fire Small charred area

sheets and pillows is greater than that of the small ignition sources employed in the preceding tests, but the response of the mattress under this exposure had not been previously studied.

Because of the tendency exhibited by latex foam rubber core mattresses to undergo progressive smoldering and produce large quantities of dense smoke, this type of mattress was not used in these tests. The majority of tests in this series involved mattress constructions which exhibited satisfactory resistance to small ignition sources.

Figure 18. *Test No. 1, time 10 minutes. Vinyl fabric cover.*

Figure 19. *Test No. 2, time 5 minutes. Open mesh fabric cover.*

Figure 20. *Test No. 3, time 5 minutes. Open mesh fabric cover.*

Three types of pillows were used: one with 100 percent latex foam rubber filling, one with 100 percent polyester fiberfill filling, and one filled with 50 percent down and 50 percent crushed duck feathers. Down-and-feather pillows were available with 25/75, 50/50, and 75/25 ratios of down to feathers, and the 50/50 ratio was selected as representative. All pillows had 100 percent cotton covering. These three types of pillows together are believed to constitute the large majority of the pillow market.

In the majority of tests in this series, the sheets and pillowcases were 100 percent cotton, fine comb percale. None of the cotton sheets, pillowcases, or pillow covers were fire retardant. The fire-retardant cotton materials generally available to the bedding industry at the time did not have the permanent fire retardance and esthetics of the modacrylic and polyamide materials which were used in several tests.

Test Procedure

The size of the hood available for the full-scale tests limited sample size to twin bed size mattresses (38 by 75 inches). In most of these tests, the mattress was covered with two cotton sheets, the upper sheet turned down at the end where a single pillow, covered with a pillowcase, was placed. Only single sheets of modacrylic and polyamide were used due to material limitations. The entire assembly was supported by a 1/4-inch thick sheet of asbestos-cement board, with a similar sheet of asbestos-cement board along the back wall of the hood to prevent damage to the structure.

To obtain some measure of incident radiation from the burning pillow and pillowcase initially, and from the burning mattress and sheets subsequently, a radiometer purchased from Heat Technology Laboratories was positioned at various distances along the top surface of the sheets covering the mattress. The radiometer was moved from location to location along the centerline, with the receiving surface at right angles to the top surface of the mattress and 1.5 inches above it.

To obtain a measure of the temperatures reached in the vicinity of the pillow during the test, thermocouples were positioned on the top surface as shown in Figure 21. In cases where burning was expected in the interior of innerspring type mattresses, thermocouples were positioned in the interior of the mattress as shown in Figure 22.

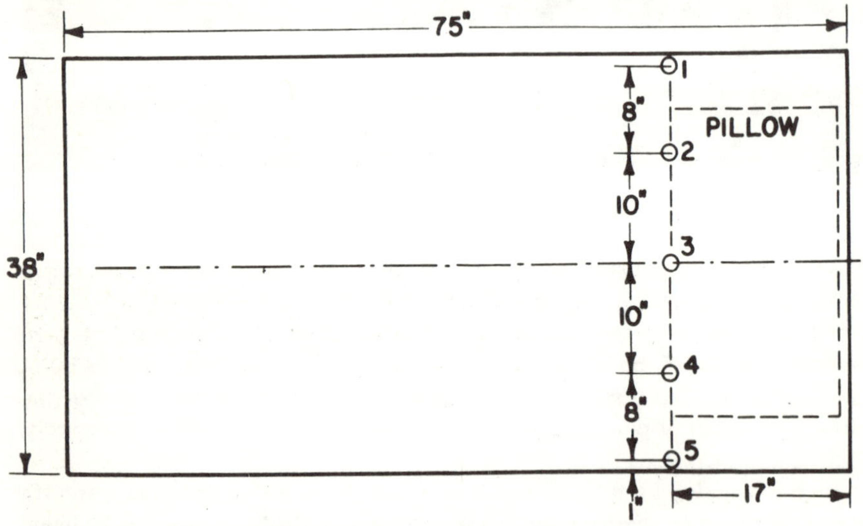

Figure 21. *Locations of thermocouples on top surface of mattress.*

Because the test procedure gradually developed in sophistication as the series progressed, heat flux and temperature measurements were employed only in the later tests.

The mattresses were not given special conditioning because of the lack of available conditioning facilities for articles of this size. Review of the environmental conditions prevailing during storage, transfer, and testing indicated that relative humidity was 30 ± 20 percent; this was less severe than the bone-dry conditioning used in the preceding phase but more severe than typical laboratory conditioning to 50 or 65 percent relative humidity.

Ignition was accomplished by placing a lighted methenamine pill (Eli Lilly, No.

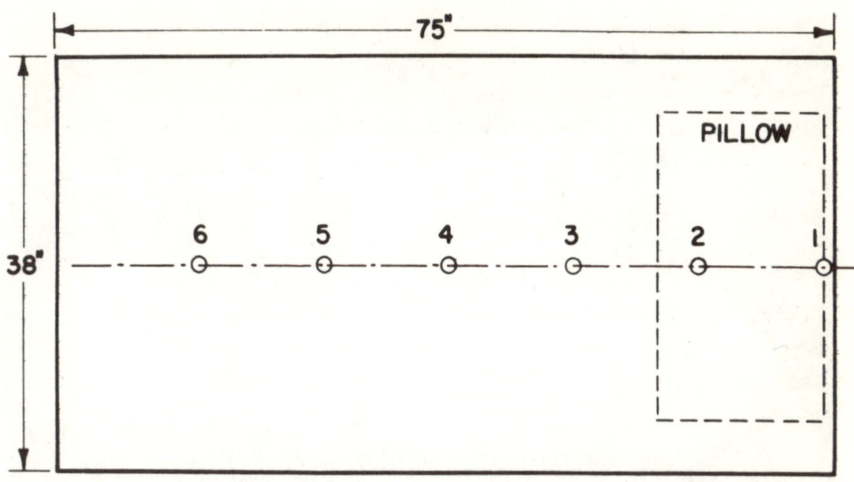

Figure 22. *Locations of thermocouples in core of innerspring mattress.*

1588) on top of the cased pillow. Flame generally progressed from pillowcase to pillow covering to pillow filling, and from pillowcase to sheets to mattress.

Typical test arrangements are shown by Figures 23 to 25.

Figure 23. *Typical arrangement for tests on mattresses with sheets and pillows. Test No. 4.*

Figure 24. *Typical arrangement for tests with thermocouples on top surface. Test No. 10.*

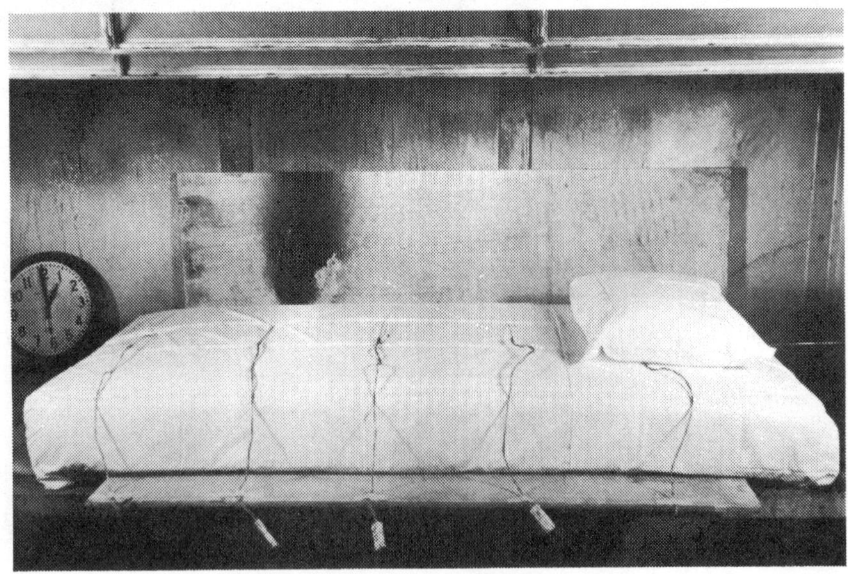

Figure 25. *Typical arrangement for tests with thermocouples in core. Test No. 13.*

Test Results

Thirteen tests were performed on mattresses with sheets and pillows. These are summarized in Table 4. Eight tests involved innerspring type mattresses and five tests involved urethane foam-core type mattresses.

Pillows with latex foam rubber filling were used in most of the tests because they provided the most severe fire exposure. The latex foam rubber filling burned steadily and generated considerable heat. Polyester fiberfill filling melted back gradually and the thick liquid was ignited by the burning pillowcase and pillow covering. Down-and-feather filling was scorched by the burning pillowcase and pillow covering but did not burn, and most of the filling was intact after the test. Because the down-and-feather pillow kept the upper half of the pillowcase and pillow cover away from the mattress while they were burning and tended to smother the lower half, the net fire exposure is believed to be less severe than the burning of an empty pillow cover and pillowcase laid on the mattress.

In all tests with cotton sheets and cotton pillowcases, the mattress was eventually ignited, and the time to ignition was a function of exposure severity. One mattress construction employing fire-retardant urethane foam became involved after 9 minutes exposure to a burning latex foam rubber pillow, 12 minutes exposure to a burning polyester fiberfill pillow, or 23 minutes exposure to a burning down-and-feather pillow. In the last case there was a significant induction period between exposure and ignition, and in that case the mattress ignited in the area next to the far wall where the asbestos-cement boards increased exposure intensity by geometric configuration and increased exposure time by their high heat capacity.

Mattresses with innerspring cores exhibited sudden releases of gases which had accumulated in the essentially hollow interior. These gases seemed to be combustible vapors generated by thermal decomposition of the mattress materials, and internal burning was observed after the gas release. It is not known at this time whether the evolved combustibles mixed with the air already present in the innerspring core to form a flammable composition which was then brought to autoignition temperature; or whether rapid pressure build-up within the mattress, inadequately relieved by diffusion through the padding and cover, resulted in a rupture in the fire-weakened mattress, through which the flame front moved into the interior. The sensing devices used did not have sufficient sensitivity to resolve this point.

Despite the use of fire-retardant materials, the exposure severity of burning sheets and pillows was sufficient to overcome the resistance of all the mattresses tested. Two approaches to reducing severity were to reduce the fraction of incident heat flux absorbed by the mattress by using sheets for fire protection, and to reduce the incident heat flux by minimizing involvement of the pillow.

Using the first approach, the use of a polyamide sheet on top of the mattress prevented ignition of one mattress and significantly delayed the ignition of another;

Table 4. Tests on Mattresses with Sheets and Pillows

Test No.	Mattress Construction				Pillow			Involvement		Time, minutes to Event							
	Core	Padding	Cover	Sheet	Filling	Cover	Pillowcase	Pillow	Mattress	A	B	C	D	E	F	G	H
4	Innerspring unit	68% blended cotton felt, 32% sisal pads	Rayon	Cotton	100% Latex foam rubber	Cotton	Cotton	Yes	Yes	5	7	8	12	14			
13	Innerspring unit	68% blended cotton felt, 32% sisal pads	Rayon	Cotton	100% Latex foam rubber	Cotton	Cotton	Yes	Yes								
6	Innerspring unit	3/4 inch urethane foam FR, 1/4 inch tufflex	Cotton FR	Cotton	100% Latex foam rubber	Cotton	Cotton	Yes	Yes	6	6	8	12	15	20	25	28
7	Innerspring unit	3/4 inch urethane foam FR, 1/4 inch tufflex	Cotton FR	Cotton	100% Latex foam rubber	Cotton	Cotton	Yes	Yes	4	5	6	11	14	20		25
10	Innerspring unit	3/4 inch urethane foam FR, 1/4 inch tufflex	Cotton FR	Modacrylic	100% Latex foam rubber	Cotton	Cotton	Yes	Yes	5		8		25			
11	Innerspring unit	3/4 inch urethane foam FR, 1/4 inch tufflex	Cotton FR	Polyamide	100% Latex foam rubber	Cotton	Cotton	Yes	No	7							
9	Innerspring unit	3/4 inch urethane foam FR, 1/4 inch tufflex	Cotton FR	Modacrylic	100% Latex foam rubber	Cotton	Modacrylic	No	No								
8	Innerspring unit	3/4 inch urethane foam FR, 1/4 inch tufflex	Cotton FR	Polyamide	100% Latex foam rubber	Cotton	Polyamide	No	No								
5	100% urethane foam	None	Rayon	Cotton	100% Latex foam rubber	Cotton	Cotton	Yes	Yes	7	8	9	12	14	16		20
12	100% urethane foam	None	Rayon	Polyamide	100% Latex foam rubber	Cotton	Cotton	Yes	Yes*	5			10				
1	100% urethane foam FR	None	Cotton FR	Cotton	100% Latex foam rubber	Cotton	Cotton	Yes	Yes	4	5	7	12	15	20		30
2	100% urethane foam FR	None	Cotton FR	Cotton	100% Polyester fiberfill	Cotton	Cotton	Yes	Yes	4	5	8	10	15	20		
3	100% urethane foam FR	None	Cotton FR	Cotton	50% down, 50% feathers	Cotton	Cotton	Yes	Yes*	6	8	12		20	36		40

Event A — fire spreads over edge of pillow
B — fire spreads to sheets

Event C — fire spreads to top edge of mattress
D — pillow ceases burning
E — fire reaches 50 percent of mattress length

Event F — fire reaches 60 percent of mattress length
G — fire reaches 70 percent of mattress length
H — fire reaches 80 percent of mattress length

*Significant induction period before involvement of mattress.

the first mattress contained fire-retardant materials, and the second did not. The performance of the polyamide sheet is ascribed to its high strain release temperature of about 750°F [12] which enabled it to maintain integrity of the fabric structure even in the process of charring. Modacrylic tended to shrink under fire exposure, leaving areas of the substrate material exposed. Any fire resistant fabric which chars in place without distortion would probably provide the protection desired.

Using the second approach, modacrylic and polyamide pillowcases prevented ignition of the latex foam rubber pillow by a lighted methenamine pill. The latex foam rubber pillow did not appear susceptible to progressive internal smoldering when covered with a fire-retardant pillowcase.

Figures 26 to 38 show some of these results in photographs.

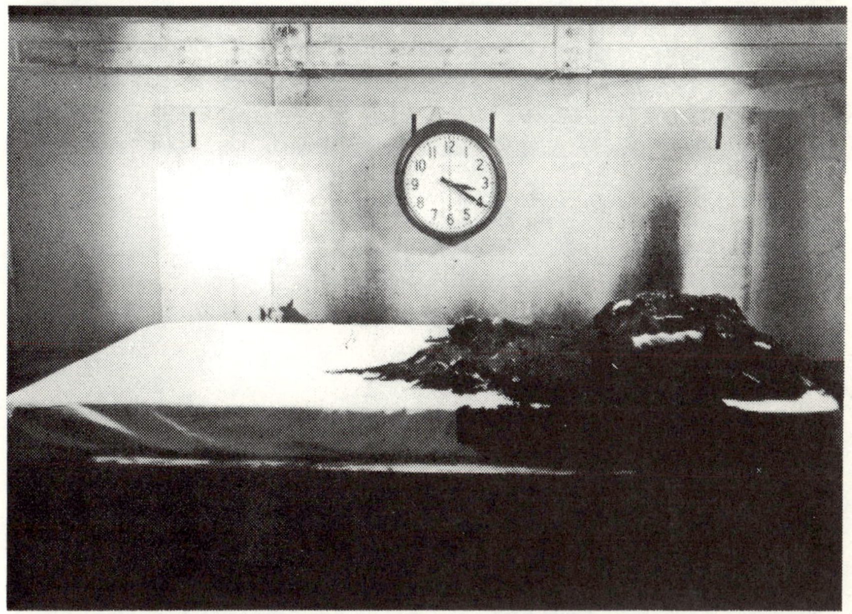

Figure 26. *Test No. 3, time 20 minutes. Down-and-feather pillow, fire retardant mattress.*

Reproducibility of Fire Exposure

The selection of a latex foam rubber pillow with cotton pillow cover and pillowcase, ignited by a methenamine pill, as a standard fire source in full-scale tests gave good reproducibility, considering the variations possible in the manner in which a pillow is stuffed into a pillowcase and laid on a mattress. This source of variation was minimized by having the same technician make the bed for each test. The precision with which the various events in the burning process were repeated is shown by Table 5.

Figure 27. *Test No. 2, time 20 minutes. Polyester fiberfill pillow, fire retardant mattress.*

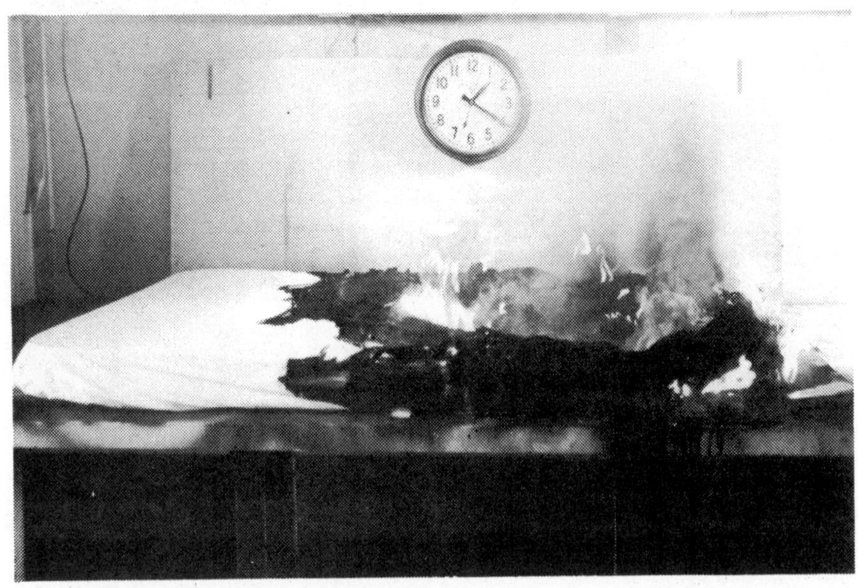

Figure 28. *Test No. 1, time 20 minutes. Latex foam rubber pillow, fire retardant mattress.*

Figure 29. *Test No. 5, time 20 minutes. Cotton sheet, cotton pillowcase, regular mattress.*

Figure 30. *Test No. 12, time 20 minutes. Polyamide sheet, cotton pillowcase, regular mattress.*

Figure 31. *Test No. 4, time 20 minutes. Cotton sheet, cotton pillowcase, regular mattress.*

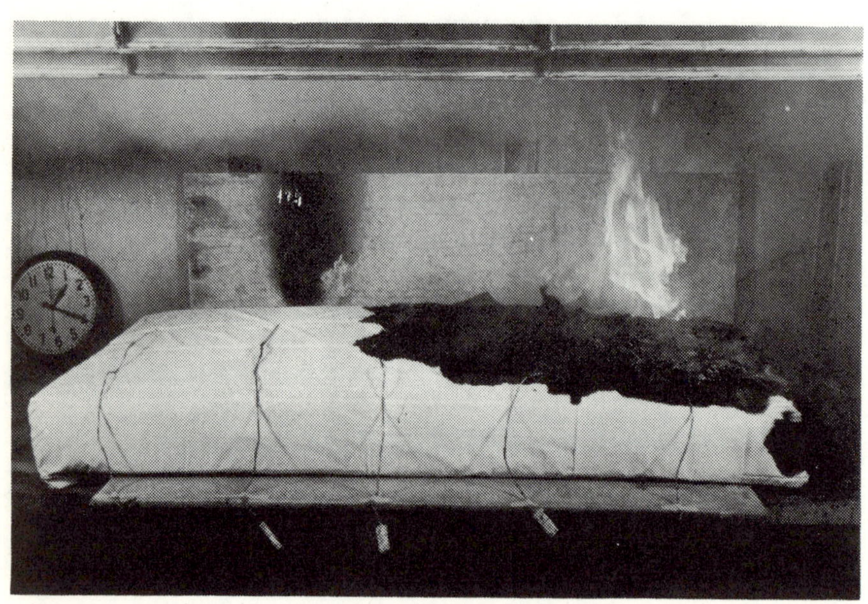

Figure 32. *Test No. 13, time 20 minutes. Cotton sheet, cotton pillowcase, regular mattress.*

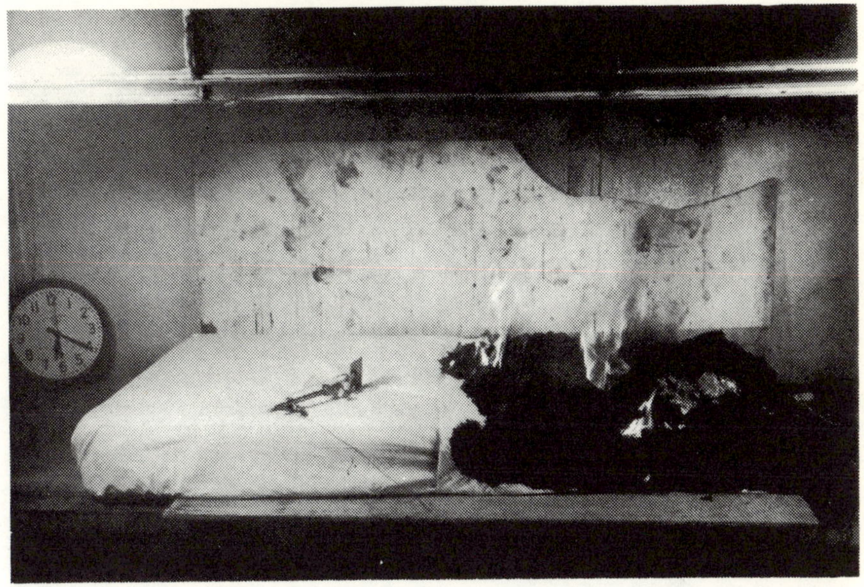

Figure 33. *Test No. 6, time 20 minutes. Cotton sheet, cotton pillowcase, fire retardant mattress.*

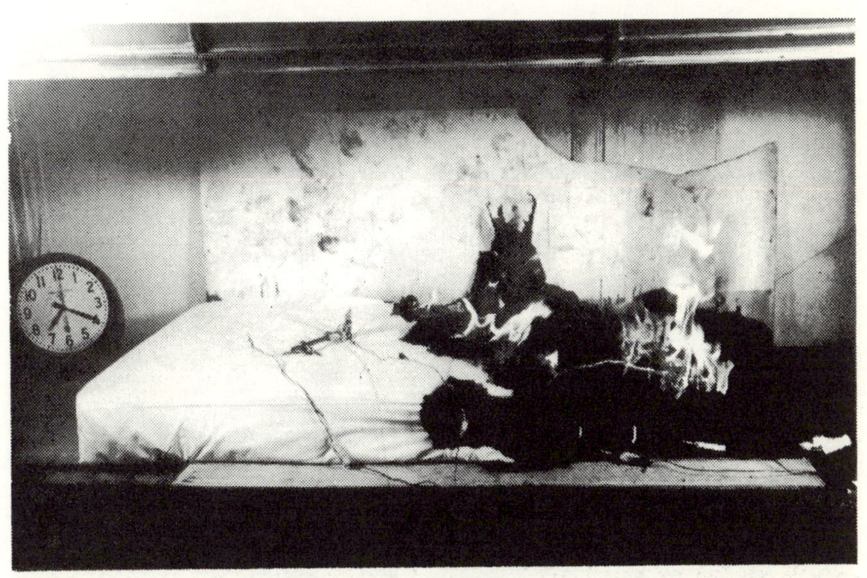

Figure 34. *Test No. 7, time 20 minutes. Cotton sheet, cotton pillowcase, fire retardant mattress.*

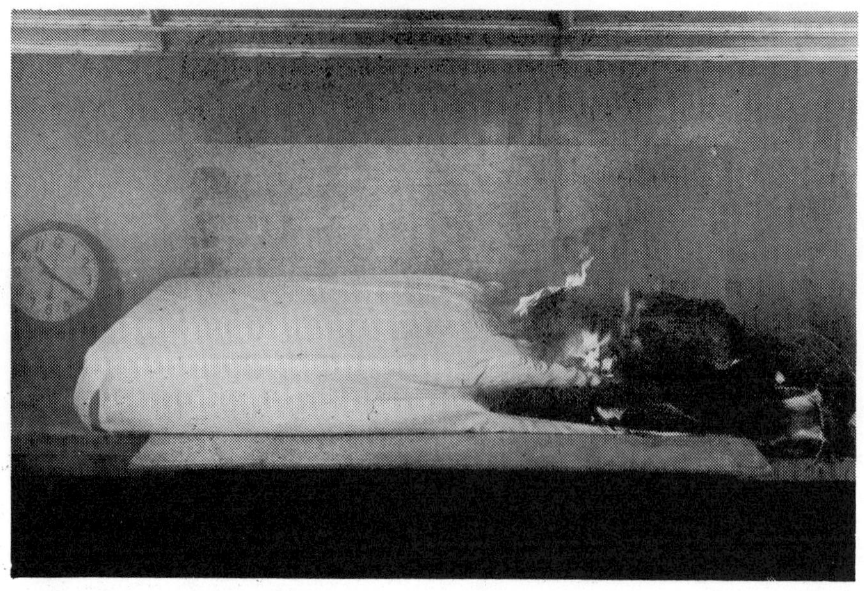

Figure 35. *Test No. 10, time 20 minutes. Modacrylic sheet, cotton pillowcase, fire retardant mattress.*

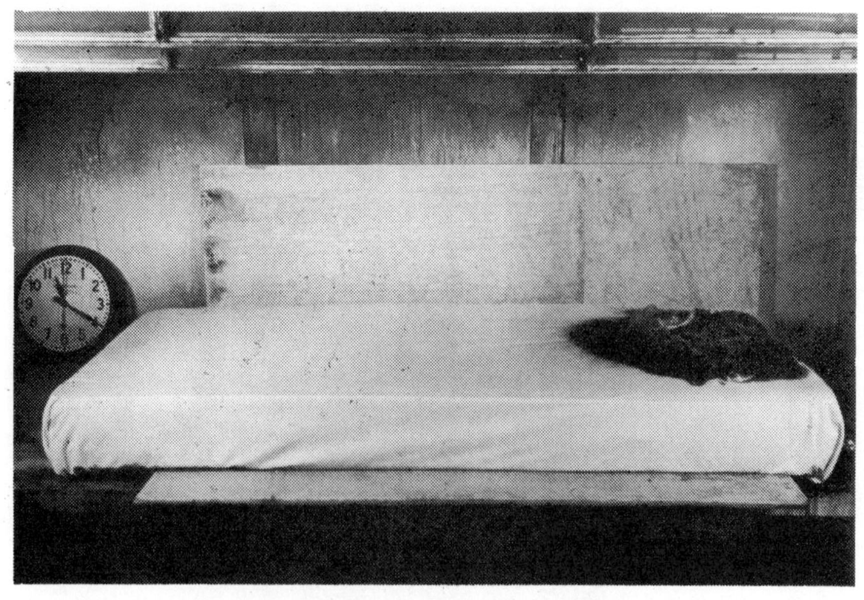

Figure 36. *Test No. 11, time 20 minutes. Polyamide sheet, cotton pillowcase, fire retardant mattress.*

Figure 37. *Test No. 9, time 5 minutes. Modacrylic sheet, modacrylic pillowcase, fire retardant mattress.*

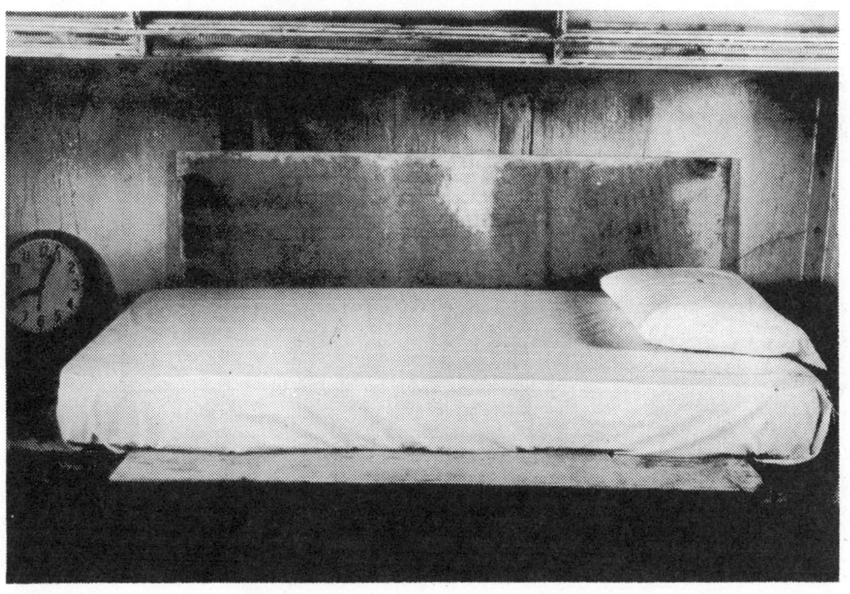

Figure 38. *Test No. 8, time 4 minutes. Polyamide sheet, polyamide pillowcase, fire retardant mattress.*

Table 5. Reproducibility of Fire Exposure
Burning Latex Foam Rubber Pillow in Cotton Pillowcase
Ignited by Methenamine Pill

Test No.	Time, minutes, for: Fire Spread Over Edge of Pillow	Fire Spread to Sheets	Fire Spread to Top Edge of Mattress	Pillow to Cease Burning
1	4	5	7	12
4	5	7	8	12
5	7	8	9	12
6	6	6	8	12
7	4	5	6	11

The temperatures recorded in the vicinity of the burning pillow are shown in Figures 39 to 42. The differences between the tests indicate some interaction with the mattress, with some mattresses contributing more combustible volatiles or conducting away more heat. The use of fire retardant materials in or on top of the mattress appears to reduce exposure severity.

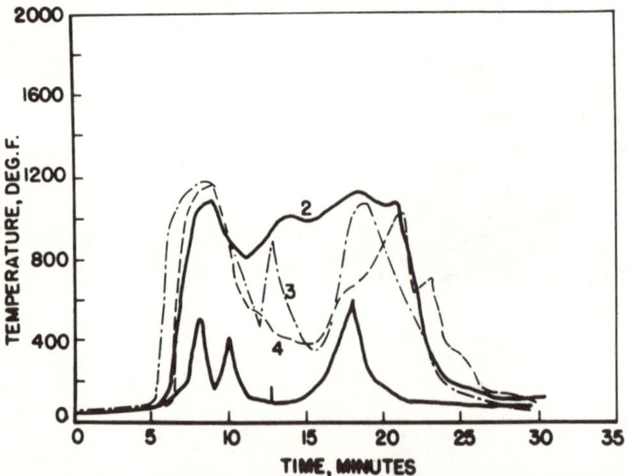

Figure 39. *Temperatures at surface near pillow, Test No. 6. Latex foam rubber pillow, cotton pillowcase, cotton sheet, fire retardant mattress. Thermocouple locations shown in Figure 21.*

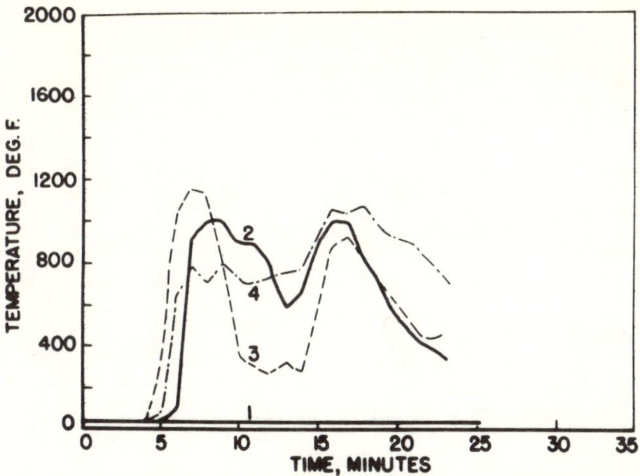

Figure 40. *Temperatures at surface near pillow, Test No. 10. Latex foam rubber pillow, cotton pillowcase, modacrylic sheet, fire retardant mattress. Thermocouple locations shown in Figure 21.*

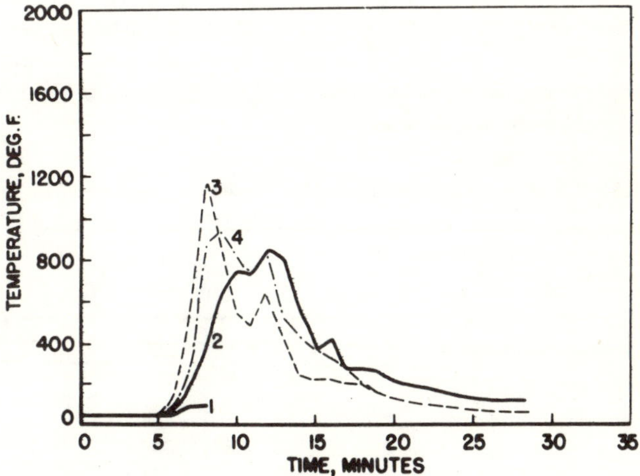

Figure 41. *Temperatures at surface near pillow, Test No. 11. Latex foam rubber pillow, cotton pillowcase, polyamide sheet, fire retardant mattress. Thermocouple locations shown in Figure 21.*

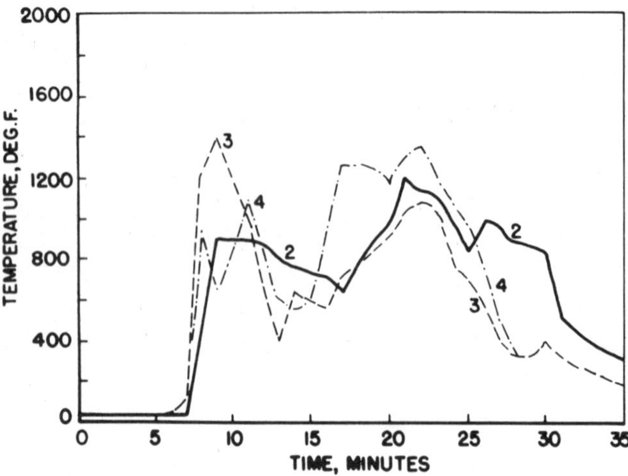

Figure 42. *Temperatures at surface near pillow, Test No. 12. Latex foam rubber pillow, cotton pillowcase, polyamide sheet, standard mattress. Thermocouple locations shown in Figure 21.*

Surface Temperatures

The temperatures reached at the mattress surface once a pillow becomes fully involved are in the vicinity of 1000°F, as shown by Figures 39 to 42. These temperature levels explain why supposedly fire-retardant mattress materials exhibit insufficient resistance to ignition. The ignition temperatures of most available padding and core materials are below 1000°F. Table 6 lists some ignition temperatures obtained in a Setchkin apparatus[13].

Table 6. Flash-Ignition Temperatures of Some Cushioning Materials

Material		Flash-Ignition Temperature, °C
Flexible urethane foam		280
Flexible urethane foam, fire-retardant, Sample	1	405
	2	364
	3	394
	4	407
	5	397
	6	390
	7	416
Styrene-butadiene rubber foam		270
Neoprene rubber foam, fire-retardant		395

The value of protective sheets may lie in keeping the flame front from penetrating to the mattress itself, so that any flammable gas mixtures produced would have to be heated to the autoignition temperature, rather than ignited by a flame front at the lower flash-ignition temperature.

Core Temperatures

Once burning occurs in the interior of an innerspring type mattress, temperatures in the vicinity of 1200°F can be developed by combustion in this relatively well-insulated enclosure. Figures 43 to 44 present the temperatures recorded inside two mattresses, one with and one without fire-retardant components. The fire-retardant mattress exhibited internal combustion earlier in the test; this may be

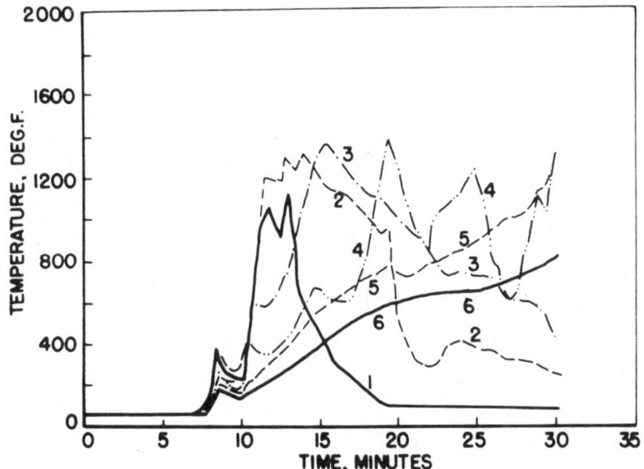

Figure 43. *Temperatures in core, Test No. 7. Fire retardant mattress. Thermocouple locations shown in Figure 22.*

related to earlier occurrence of the gas release phenomenon described in a preceding section, and thus be due to differences in structural strength, rather than inherent flammability of the materials. The fire-retardant mattress produced lower peak temperatures and smaller areas under the temperature-time plots, indicating substantially less heat produced during burning.

Heat Flux

Heat flux data were difficult to obtain with continuity because the radiometer had to be moved frequently to escape the advancing flame front. Most measurements were less than 1.0 Btu/ft^{2}sec, and reached 2.0 Btu/ft^{2}sec only when the flames were almost in contact with the radiometer. From the data obtained, it

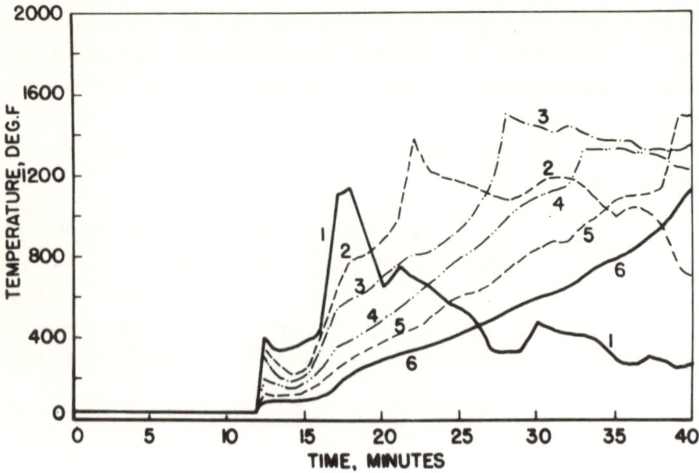

Figure 44. *Temperatures in core, Test No. 13. Standard mattress. Thermocouple locations shown in Figure 22.*

would appear that to attain the irradiance level of 2.2 Btu/ft^{2}sec used in the National Bureau of Standards smoke density test [14], a material would have to be positioned 6 to 12 inches from a burning latex foam rubber pillow at the stage of full involvement. The period of full involvement of the pillow and maximum heat flux is less than five minutes.

The particular radiometer used for these measurements had previously been checked in the NBS apparatus and found to agree within 2 percent with the radiometer used to calibrate the apparatus.

SMOKE DENSITY TESTS

The lack of room-scale test facilities prevented the measurement of smoke produced by burning complete mattresses under controlled conditions, and limited smoke density tests to those which could be performed in laboratory test chambers. Two test chambers were available: the Rohm and Haas XP2 smoke density chamber [15] and the National Bureau of Standards smoke density chamber [14].

It was recognized that the data obtained in small-scale laboratory tests do not necessarily predict behavior in actual fire situations. This series of tests was intended to provide guidance as to relative smoke-producing characteristics, using the means at hand, and to indicate the relative usefulness of different laboratory techniques in simulating the smoke conditions observed in the larger-scale tests described in earlier sections.

Apparatus

The Rohm and Haas XP2 smoke chamber [15] is a cabinet measuring 30 inches high, 12 inches wide, and 12 inches deep, completely enclosed except for 1-inch high ventilating openings around the bottom. For standard tests, the heat source is a propane-air flame from a pencil-tip burner, applied at a 45-degree angle to the bottom of a horizontal specimen. A horizontal photometer path, 20 inches above the bottom of the chamber, is used for measuring light absorption. The instrument purchased from George Eysenbach of Bristol, Pennsylvania, was equipped with a photometer providing an overall sensitivity range of 100 to 1.

The National Bureau of Standards smoke chamber [14] is a completely enclosed cabinet, measuring 3 feet wide, 3 feet high, and 2 feet deep, in which a specimen 3 inches square is supported vertically in a frame such that 2-9/16 inches square is exposed to heat under either flaming or nonflaming (smoldering) conditions. The heat source is an electric furnace, adjusted with the help of a circular foil radiometer to give a heat flux of 2.2 Btu/ft^{2}sec (2.5 watts/cm^2) at the specimen surface. A vertical photometer path for measuring light absorption is employed to minimize measurement differences due to smoke stratification which could occur with a horizontal photometer path at a fixed height; the full three-foot height of the chamber is used to provide an overall average for the entire chamber. The instrument purchased from American Instrument Company of Silver Spring, Maryland, was equipped with a photomultiplier providing an overall sensitivity range of 1000 to 1.

Test Procedure

In an effort to establish procedures most relevant to the burning of complete bedding systems, three groups of tests were conducted on mattress materials. In the first series of tests, one-inch cubes of core materials, and one-inch squares of cover and pad materials, were subjected to the standard Rohm and Haas XP2 test procedure involving a four-minute exposure to a burner flame fueled by propane at 40 psi pressure.

In the second series of tests, sections of mattress constructions, each 2 inches square and the depth of the constructions, were exposed to a lighted methenamine pill (Eli Lilly, No. 1588) in the XP2 smoke chamber with the burner turned off and no other ignition source present. For innerspring constructions, the innerspring was replaced by the existing specimen support screen, because the metal springs would not burn and would merely complicate specimen preparation. For solid foam core constructions, the thickness of core material was reduced to 2 inches, so that the total thickness of material on the specimen support screen would not exceed 3 inches in any case. By this approach, total specimen height was restricted to 2 ± 1 inches, and the variation in distance from the photometer path was correspondingly reduced.

In the third series of tests, sections of mattress constructions, each 3 inches square and with the same depth as in the second series of tests, were subjected to the standard NBS test, involving 20 minutes exposure to an electric furnace delivering a heat flux of 2.2 Btu/ft^2 sec, under both flaming and nonflaming conditions.

Test Results

Smoke density data obtained in these three series of tests are presented in Tables 7 to 9. In the first series of tests, dealing with individual materials, the cover materials, with the exception of the plastic coated cover, produced little smoke. Among the pad materials, two urethane foam samples had moderate smoke and the rest little or none. Latex foam rubber produced very dense smoke. There was moderate smoke from fire-retardant urethane foam and little smoke from non-fire-retardant urethane foam.

Table 7. Smoke Produced from Mattress Components
Rohm and Haas XP2 Test

Material		Light Absorption, percent	
		Maximum	Final
Core Materials (Specimens 1 x 1 x 1 inch)			
Latex foam rubber, Sample	A	96.3	50.0
	B	99.7	64.3
	C	97.7	59.7
Flexible urethane foam, Sample	A	5.0	0
	B	4.3	1.0
	C (1.5 pcf)	8.7	0.3
	D (1.5 pcf)	7.3	1.0
Flexible urethane foam, fire-retardant, 2.0 pcf		27.0	4.3
Pad Materials (Specimens 1 x 1 inch)			
Latex sisal pad, 1/4 inch, Sample	A	0.7	0
	B	6.0	0.3
	C	1.3	0
Felt batt, 1 inch		0	0
Tufflex, 1/4 inch		0	0
Flexible urethane foam, Sample	A, 1 inch	6.7	0.3
	B, 3/4 inch	20	4.5
	C, 3/4 inch	24.3	3
Cover Materials (Specimens 1 x 1 inch)			
Rayon, quilted with 3/4 inch urethane foam		4.3	0
1/2 inch urethane foam		1.7	0.3
1/2 inch urethane foam		1.3	0
1/4 inch urethane foam		1.3	0
felt		0	0
Cotton, fire-retardant, Samples A, B, C		0	0
Cotton		0	0
Cotton, plastic coated		24.3	3

Table 8. Smoke Produced from Mattress Constructions Modified Rohm and Haas Test

Material	Ignition	Absorption, percent		Specific Optical Density		Time to Ds = 16 min.
		Max.	Res.	Maximum, Dm	Final, Df	
Spun rayon cover, urethane foam 3/4" (a), felt batt, latex sisal pad	Yes	81	2	64.9	0.8	3.17
Spun rayon cover, felt (a), latex foam rubber 2"	No	86	0	76.8	0	2.58
Spun rayon cover, urethane foam 1/2" (a), urethane foam 1", latex sisal pad	Yes	18	4	7.7	1.6	–
Spun rayon cover, urethane foam 1/2" (a), latex foam rubber 2"	Yes	99	64	180	39.9	0.92
Cover, urethane foam 1/4" (a), urethane foam 2"	Yes	32	3	15.1	1.2	–
Cover, urethane foam 2"	Yes	32	6	15.1	2.4	–
Cotton cover FR, urethane foam 3/4", latex sisal pad 1/4"	No	0	0	0	0	–
Cotton cover FR, urethane foam 3/4", latex sisal pad 1/4"	No	4	0	1.6	0	–
Cotton cover FR, urethane foam FR 2"	No	4	0	1.6	0	–
Cotton cover FR, latex foam rubber 2"	No	3	0	1.2	0	–
Cotton cover FR, urethane foam 2"	No	3	0	1.2	0	–
Cotton cover, urethane foam 2"	Yes	35	7	16.8	2.8	1.92
Plastic coated cover, urethane foam 2"	No	3	0	1.2	0	–

(a) Quilted with cover

Table 9. Smoke Produced from Mattress Constructions
National Bureau of Standards Test

Material	Exposure		Min. Trans. pct	Specific Optical Density Maximum, Dm	Specific Optical Density Final, Df	Time to Ds = 16 min.
Spun rayon cover, urethane foam 3/4" (a), felt batt, latex sisal pad	F		7.62	147.1	5.0	0.46
	NF		0.062	422.1	53.9	0.37
	NF		0.105	392.0	62.0	0.40
Spun rayon cover, felt (a), latex foam rubber 2"	F		0.0145	504.8	43.7	0.82
	NF		0	780	68.8	0.22
Spun rayon cover, urethane foam 1/2" (a), urethane foam 1", latex sisal pad	F		0.147	372.7	8.2	2.75
	NF		0.00906	532.0	60.6	0.49
Spun rayon cover, urethane foam 1/2" (a), latex foam rubber 2"	F		0	780	107.3	0.35
	NF		0	780	131.6	0.26
Cover, urethane foam 1/4" (a), urethane foam 2"	F	(b)	0.920	267.2	12.5	4.50
	F		4.98	171.3	7.3	3.95
	NF	(b)	0	780	84.5	0.30
	NF		0.014	506.8	50.1	0.37
Cover, urethane foam 2"	F		11.8	122.0	6.6	2.87
	NF		0.110	389.0	38.5	0.40
Cotton cover FR, urethane foam 3/4", paper synthetic pad 1/4"	F		6.80	153.6	4.0	0.78
	NF		0.0381	449.9	20.1	0.33
	NF		0.0435	442.3	18.6	0.31
Cotton cover FR, urethane foam 3/4", latex sisal pad 1/4"	F		4.9	172.4	6.0	1.05
	NF		0.0201	486.4	21.4	0.34
Cotton cover FR, urethane foam FR 2"	F		0.0650	419.1	15.7	0.50
	NF		0.0150	502.8	16.8	0.35
Cotton cover FR, latex foam rubber 2"	F	(b)	0	780	74.4	0.20
	F		0	780	61.0	0.13
	NF	(b)	0	780	9.1	0.30
	NF		0	780	20.0	0.37
Cotton cover FR, urethane foam 2"	F		14.2	111.5	4.6	2.65
	NF		0.0150	502.8	20.9	0.36
Cotton cover, urethane foam 2"	F		2.30	215.4	9.8	1.91
	NF		0.0688	415.8	30.8	0.52
Plastic coated cover, urethane foam 2"	F	(b)	0.410	314.0	16.4	0.14
	F		1.71	232.4	17.8	0.15
	NF	(b)	0.460	307.4	27.8	0.50
	NF		0.0010	657.5	103.0	0.33

(a) Quilted with cover
(b) Not wrapped in aluminum foil

The results of the second and third series of tests tended to confirm expectations based on the results of the first series, with the effect of core materials predominating due to their higher weight or volume fraction of the composite. Composites containing latex foam rubber generally exhibited the highest smoke densities.

It is the opinion of the author that the second series of tests, employing methenamine pill ignition in the XP2 smoke chamber, gave the results which most closely approximated the smoke observed in the larger-scale tests. The principal advantage of this group of data is that it tends to incorporate in the results the effect of relative ease of ignition.

The NBS test procedure involves a heat flux level which, by both direct measurement and observation of results, constitutes substantially more severe fire exposure than was developed in the larger-scale tests. For fully developed room fires, the NBS test may be the most appropriate, but data such as time to reach a critical density level become irrelevant, since the occupant of the bed would probably have been killed several times over by lethal levels of temperature and carbon monoxide.

CONCLUSIONS

The use of fire-retardant cotton covering in innerspring and urethane foam core type mattresses has been shown to provide adequate resistance to ignition by small fire sources applied to the mattresses themselves. Use of fire-retardant cotton covering did not prevent the same ignition sources from causing progressive smoldering in latex foam rubber core mattresses.

The presence of sheets, pillows and pillowcases on the mattress, and the involvement of these articles in fire, increases exposure severity, and introduces many interacting variables all affecting the resistance of the mattress to ignition and involvement. Each of the following changes has been shown to improve this resistance:

1. Fire-retardant cotton covering instead of non-fire-retardant cotton or rayon covering in the mattress.
2. Fire-retardant urethane foam instead of non-fire-retardant urethane foam in the mattress.
3. Down-and-feather filling instead of polyester fiberfill in the pillow, and polyester fiberfill instead of latex foam rubber filling in the pillow.
4. Aromatic polyamide sheets instead of non-fire-retardant cotton sheets.
5. Modacrylic or polyamide pillowcases instead of non-fire-retardant cotton pillowcases.

Latex foam rubber cores in mattresses exhibited the characteristics of progressive internal smoldering, vigorous burning when ignited, and production of large quantities of dense smoke.

Currently available materials and technology are capable of giving innerspring and urethane foam core type mattresses adequate resistance to small ignition sources. Resistance to burning sheets and pillows can be achieved in varying degrees by selection of materials.

Fire experience indicates that population habits may be the major factor in bedding fires. The extent to which bedding materials must be improved to attempt to offset human carelessness is an issue which has not been resolved.

It is the opinion of the author that a bedding system cannot be considered resistant to ignition by fire exposure of a given severity unless the necessary level of resistance is possessed by all the components of the system: mattresses, sheets, pillows, pillowcases, blankets, and sleepwear.

REFERENCES

1. Anon., "The Single-Fatality Fire," *Fire Journal*, Vol. 63, No. 1, 34–35 (January 1969).
2. Anon., "Fires and Fire Losses Classified, 1971," *Fire Journal*, Vol. 66, No. 5, 65–69 (September 1972).
3. Anon., "Fires and Fire Losses Classified, 1970," *Fire Journal*, Vol. 65, No. 5, 21–25, 28 (September 1971).
4. Anon., "Fires and Fire Losses Classified, 1969," *Fire Journal*, Vol. 64, No. 5, 65–69 (September 1970).
5. C. A. Hafer, C. H. Yuill, "Characterization of Bedding and Upholstery Fires," Final Report, NBS Contract No. CST-792-5-69, SWRI Project No. 3-2610, Southwest Research Institute, San Antonio, Texas (March 31, 1970).
6. C. H. Yuill, "The Life Hazard of Bedding and Upholstery Fires," *Journal of Fire and Flammability*, Vol. 1, No. 4, 312–323 (October 1970).
7. U.S. Department of Commerce, "Standard for the Flammability of Mattresses," DOC FF 4–72, *Federal Register*, Vol. 37, No. 110, 11362–11367 (June 7, 1972).
8. U.S. Department of Commerce, "Notice of Finding That Flammability Standard for Upholstered Furniture May Be Needed," *Federal Register*, Vol. 37, No. 230, 25239–25240 (November 29, 1972).
9. U.S. Department of Commerce, "Standard for the Surface Flammability of Carpets and Rugs," DOC FF 1–70, *Federal Register*, Vol. 35, No. 74, 6211–6214 (April 16, 1970).
10. U.S. Department of Commerce, "Standard for the Surface Flammability of Small Carpets and Rugs," DOC FF 2–70, *Federal Register*, Vol. 35, No. 251 (December 29, 1970).
11. U.S. Department of Commerce, "Standard for the Flammability of Children's Sleepwear," DOC FF 3–71, *Federal Register*, Vol. 36, No. 141, 14624–14632 (July 21, 1971).
12. W. H. Gloor, "Heat Resistant Textile Fibers," *American Dyestuff Reporter*, (June 17, 1968).
13. N. P. Setchkin, "A Method and Apparatus for Determining the Ignition Characteristics of Plastics," *Journal of Research of the National Bureau of Standards*, Vol. 43, 591–608 (December 1949).
14. D. Gross, J. J. Loftus, A. F. Robertson, "Method for Measuring Smoke from Burning Materials," ASTM Special Technical Publication No. 422, 166–204 (1967).
15. Rohm and Haas Co., NFPA Quarterly, Vol. 57, 276–287 (January 1964).

Carlos J. Hilado

Carlos J. Hilado is a Project Scientist in the Fire Research Laboratory of the Chemicals and Plastics Research and Development Department of Union Carbide Corporation at South Charleston, West Virginia, where he is concerned with evaluating the flammability characteristics and fire safety aspects of materials. He received his B.S. degree in chemical engineering from De La Salle College in Manila, Philippines in 1954, and his M.S. degree in Chemical Engineering Practice from Massachusetts Institute of Technology in 1956. His M.I.T. Practice School experience included the pulp and paper mill of Eastern Corporation at South Brewer, Maine, the chemical plant of Hercules Inc. at Parlin, New Jersey, and the steel mill of Bethlehem Steel Co. at Lackawanna, New York. Since 1956 he has been with Union Carbide Corporation at South Charleston, West Virginia, where he has worked in the fields of vinyl and urethane foams, vinyl polymerization, testing and evaluation, industrial insulation, and fire research.

A registered Professional Engineer in the state of West Virginia, he is a member of the American Chemical Society, American Institute of Chemical Engineers, American Society for Testing and Materials, Society of the Plastics Industry, National Fire Protection Association, and Combustion Institute. He represents Union Carbide Corp. on several ASTM and SPI committees, and is a member of the Ad Hoc Committee on Fire Safety Aspects of Polymeric Materials of the National Materials Advisory Board of the National Research Council. In 1970 he received the Distinguished Service Award of the Cellular Plastics Division of the SPI, and in 1973 he received the Educational Service Award of the Plastics Institute of America.

He is the author or co-author of 42 technical papers, author of the "Flammability Handbook for Plastics" and "Handbook of Environmental Management," editor-in-chief of the Journal of Fire and Flammability, and editor of the Fire and Flammability book series. He has been an invited lecturer at the University of Utah, University of Detroit, Wayne State University, Stevens Institute of Technology, and Brooklyn Polytechnic Institute.

Nestor B. Knoepfler, Julius P. Neumeyer and John P. Madacsi

Southern Regional Research Center *
New Orleans, Louisiana 70179

MEETING THE MATTRESS FLAMMABILITY STANDARD FF 4-72 WITH BORON TREATED COTTON BATTING PRODUCTS**

(Received July 5, 1974)

ABSTRACT: Mattresses which comply with the Mattress Flammability Standard FF 4-72 can be produced by using cotton batting which has been treated with any of a number of boron compounds. The boron compounds provide resistance to the initiation of smoldering combustion from a cigarette source by forming glass-like boric oxides which are thermally stable to temperatures above 700° C. The most effective of the boric oxide donors is boric acid. Borax-boric acid mixtures are not as effective as boric acid because Na^+ ions catalyze the oxidation of carbonaceous chars. Data are presented comparing the performance of cotton batting treated with different boric oxide donors in preventing the cigarette ignition of mattresses.

Boron-containing compounds have significant vapor pressures and tend to be lost from cotton products over extended periods of time. Data are presented showing the permanence of treatments containing different boric oxide donors during storage at 70, 105, 130 and 160° F at 65% relative humidity. The loss of boric oxide from these systems is dependent upon the temperature of storage and on the boric oxide donor used. Data suggests that the greatest loss of boric oxide occurs within the first 60 days after manufacture and is related to the amount of the donor that is surface deposited.

INTRODUCTION

The promulgation of DOC-PFF 4-71, the Proposed Mattress Flammability Standard, by the Department of Commerce on September 9, 1971 [1] and the publication of the Mattress Flammability Standard FF 4-72 on June 7, 1972 [2], have brought about a significant reassessment of research aimed at the development of flame-retardant cotton batting products.

Research at this Center over a 6-year period has shown that cotton batting can be rendered flame retardant by the application of urea phosphates, borated amido polyphosphates or ammonium phosphates, alone or in combination with thermoplastic or thermosetting resins [3, 4, 5, 6]. Such products pass the AATCC vertical

*One of the facilities of the Southern Region, Agricultural Research Service, U.S. Department of Agriculture.

**This research is being carried out under a Cooperative Agreement between the U.S. Department of Agriculture and the National Cotton Batting Institute, and under a Memorandum of Understanding with the Textile Fibers and By-Products Association and with the National Cottonseed Products Association.

flame test [7], and the modification of that test developed at SRRC for batting products thicker than l/2 inch [8]. The products can readily meet the requirements of Motor Vehicle Safety Standard 302 [9] and exhibit "afterglow" values of less than 5 seconds when tested by either the AATCC or the MVSS302 procedures. When these treated cotton batting products are assembled in mattresses, and the composite structure evaluated by the cigarette test of FF 4-72, smoldering combustion ensues and the mattresses fail to pass the standard [10, 11]. Further analysis of the problem indicates that phosphorus compounds, although considered to be efficient inhibitors of both flaming and glowing in cellulosic fabrics, are not at all effective for preventing smoldering combustion in mattress structures [10, 11, 12, 13].

During the last year and a half research at SRRC to develop smolder resistant cotton batting products has concentrated on the use of boron compounds. This report compares the performance of these products in resisting cigarette ignition in mattress structures. The treatments studied were borax-boric acid mixtures, ammonium pentaborate, boric acid alone, and boric acid-ammonium carbonate systems.

Of concern in the use of the borate salts or boric acid for preventing smoldering combustion is their instability when exposed to elevated temperature and high relative humidity. Boric acid exhibits significant vapor pressure in its solid state as well as in water solutions [14, 15, 16, 17, 18, 19, 20, 21, 22]. It has been reported that fabrics treated with borax-boric acid when exposed to 88% relative humidity and 120°F for 2 weeks show increasing afterglow as the storage proceeds [23]. Loss of glow resistance in fabrics exposed to saturated air at 65°C for 24 hours has been attributed to losses of boric acid from the fabric [23, 24, 25]. Researchers have pointed out that borax-boric acid treatments are not durable and should not be used where service for more than 1 year is required [23, 24, 26].

The Mattress Flammability Standard does not in any way indicate or demand performance after storage or use for extended periods of time. However, the wording of PL 90-189 of December 14, 1967, which amended the Flammable Fabrics Act of 1952 to include mattresses, is such that manufacturers may be liable for failure even after their product has been in service for many years. This report, therefore, will also include the status of an ongoing series of storage tests on treated cotton batting with particular emphasis on the loss of treatment when exposed to steady-state elevated temperatures at 65% relative humidity.

EXPERIMENTAL

The rawstock for this work consisted of a blend of 60% Delta first-cut linters and 40% textile waste fibers from the Piedmont area. The first-cut linters and the textile waste fibers were individually cleaned in conventional textile mill cleaning equipment. After cleaning the textile waste fibers were formed into picker laps. The linters needed for the 60%-40% ratio were spread by hand on the laps, and the mixture repassed through the picker to form laps. These laps were uniform in fiber content and served as feed for subsequent operations.

One- to two-gallon batches of the treating formulations were prepared in the laboratory. The concentration of the formulation was controlled so that the desired add-on could be achieved with about 100% wet pick-up. Where solutions were limited by saturation values, the wet pick-up was controlled to achieve the desired add-on after drying.

Samples of the lap described above were weighed dry, then fed to a laboratory padder. The wet impregnated lap was then reweighed to determine the wet pick-up. Where necessary to obtain the preselected wet pick-up the pressure on the padder squeeze rolls was adjusted, and the wet lap repassed through the padder, and then reweighed. The treated rawstock was dried in a laboratory circulating-air oven at about 220°F. After 24 hours equilibration the treated rawstock was sampled for chemical analysis and then garnetted to yield 3/4 inch thick batts with a nominal density of 2.0 pounds/ft.^{3}.

The treated batts were equilibrated overnight and again sampled for chemical analysis before assembly into a miniature (mini-) mattresses (Figure 1). For the mini-mattresses the base pad used was 3 lb/ft^3 Cotton Flote[1] [27] 1½ in. thick cut to 5 in. X 9 in. The intermediate pad was made of the material to be tested, about 3/4 in. thick cut to size 5 in. X 9 in. The topper pad was also made of the material to be tested, approximately 3/4 in. thick cut to size 8 in. X 9 in. to provide an overhang on the 9 in. sides. A control grey and white pin stripe ticking weighing 8 oz/yd was used for all tests. The ticking used throughout this work had no treatment for flame retardance. It was cut to size 10 in. X 10½ in. To simulate the tape or roll edge of a mattress a strip of treated batting 1/4 in. thick by 1½ in. wide by 9 in. in length was sewn to the ticking as shown in Figure 1. The roll edge was about 1/2 in. high and the center flat between the rolls was approximately 5 in. Note from Figure 1 that the batting being tested extends into the simulated roll edge. The sewn ticking was pulled down over the mock-up structure so that the valley formed at the roll edge was 1½ in. above the base plate, and the crown of the mini-mattress is 2¼ in. above the base plate. The ticking was held in place by six binder clips (3/4 in. size). The ends of the mini-mattress were then sealed with 2 in. wide masking tape.

Previous laboratory evaluations had shown that the most vulnerable place for cigarette ignition is the tape or roll edge [11]. The mini-mattresses were therefore tested by placing a lighted cigarette in the valley formed by the roll edge and the flat surface of the bare mini-mattress. If ignition did not occur the test was repeated diagonally across the sample on the other tape edge with a lighted cigarette being placed between two sheets with the lower sheet being depressed into the valley

---

[1] Trademark registered by National Cotton Batting Institute, Memphis, Tenn. Use of a company or product name by the department does not imply approval or recommendation of the product to the exclusion of others which may also be suitable.

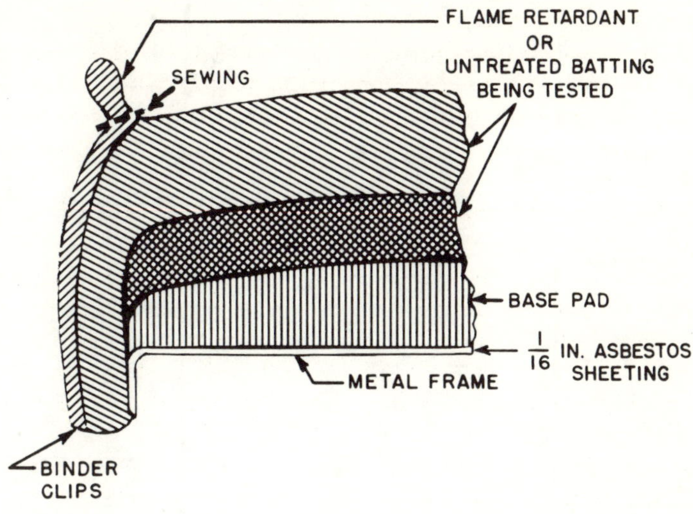

TYPICAL MINI MATTRESS CONSTRUCTION

Figure 1.

between the tape edge and the flat surface. Samples that passed the two sheet test were subjected to a more severe test where two lit cigarettes were placed side by side on the tape edge. Samples that passed all three of the above procedures were considered to exceed the requirements of FF 4-72 sufficiently that a manufacturer could be reasonably sure that his products would consistently comply with the standard.

To test the permanence of borate treatments for flame retardance and smolder resistance, a number of samples, prepared by immersion and drying treatments described earlier, were stored in an environmental chamber. The temperature and humidity could be controlled to $\pm 5^{\circ}$F and ±5% RH. Samples were periodically withdrawn and analysed for boric oxide content.

RESULTS AND DISCUSSION

Cigarette Testing Mini-Mattresses

Many researches over the past 150 years have studied the use of borax and boric acid to obtain flame retardance for cellulosic textile materials [24, 25, 28, 29, 30, 31, 32, 33, 34, 35, 36, 37]. These researchers differed widely in their recommendations as to the optimum ratio of these salts to obtain a balance between flame retardance and glow resistance. Borax confers flame retardance, but appears to promote glowing. On the other hand, boric acid is a good glow retardant but imparts little flame resistance [13, 24, 25, 38].

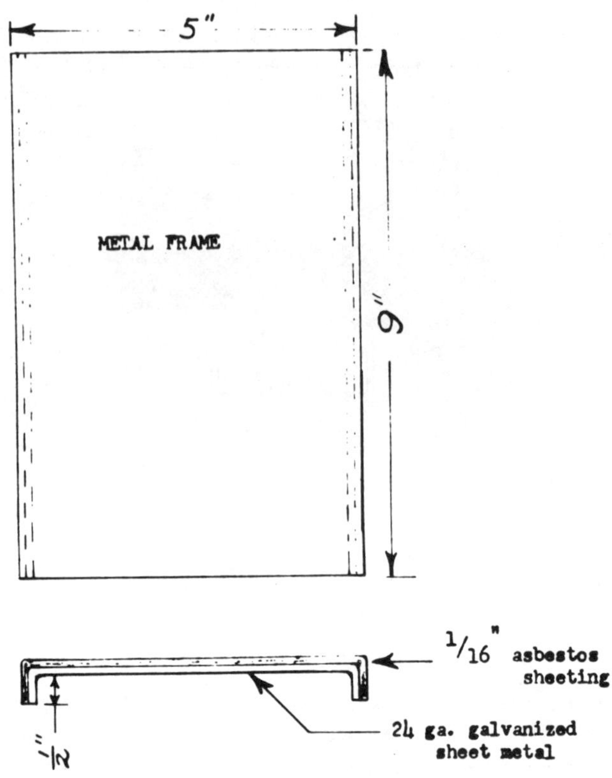

Base Pad: Preferably flame retardant Cotton Flote 3 lb./ft.^{3}
1 1/2 in. thick cut to fit above frame 5 x 9 in.
Place base pad on frame on top of asbestos sheeting.

Intermediate Pad: Material to be tested 3/4 in. thick cut to 5 x 9" size and placed on top of base pad.

Topper Pad: Material to be tested 3/4 in. thick about 2 lb./ft.^{3}
Cut to size 8 x 9" with overlap on 9 in. sides.
Place this pad on top of the intermediate pad.

Ticking: Control or material to be tested. Cut to size 10 x 10 1/2 in. To simulate the tape or roll edge cut two(2) strips of the material to be tested 1/4 in. thick by 1 1/2 x 9 in. Sew this material to the ticking. Size of ticking permits a roll 1/2 in. Roll edges should be sewn so that the center of ticking between rolls is 5 in. Place simulated cover on the assembly, pull down on frame and secure with binder clips. Seal ends with masking tape.

Figure 1A

Borax is a sodium borate containing about 0.45 parts by weight of sodium oxide (Na_2O) to 1.0 part of boric oxide (B_2O_3). Boric acid contains about 56% boric oxide. By varying the ratio of borax to boric acid, the sodium oxide, and boric oxide contents can be varied as shown in Table 1. The data demonstrates that boric acid is more effective in preventing the cigarette initiation of smoldering combustion in mattress structures than are mixtures of borax and boric acid or borax alone.

Table 1. Cigarette Test Data on Tape Edge of Mini-Mattresses Borax/Boric Acid Mixtures on Batting at 15% Add-On

Borax/ Boric Acid Ratio	Wt. Ratio Na_2O/B_2O_3	One Cigarette		Two Cigarettes
		Bare Mattress	Between Two Sheets	Bare Mattress
1/0	0.45	I	I	I
7/3	0.27	N	N	I
6/5	0.20	N	N	I
0/1	0.00	N	N	N

N = No Ignition
I = Ignition

Table 2 shows that an add-on of 20% of the 7:3 ratio borax-boric acid is required to pass the two-cigarette test. Note, however, that products that pass the standard can be obtained with 13.2% add-on.

Data in Table 3 confirm that an add-on as low as 7.98% of borax-boric acid at the 6:5 ratio yields products that pass FF 4-72. An add-on of 25% allowed passage of the two-cigarette test.

Data in Table 4 shows that an add-on as low as 5% of boric acid is sufficient to pass FF 4-72. An add-on of 10% of boric acid yields products that pass the two-cigarette test.

Thermogravimetric Studies

To investigate the apparent differences between the boron and phosphorus systems, purified cellulosic linters treated with borax, boric acid and diammonium phosphate were examined with a thermogravimetric analyzer. Figure 2 shows the weight loss curve (TG) and its derivative (DTG) for the untreated cellulose linters control heated in 8.4% oxygen. The DTG curve is more sensitive to minor weight changes than the original weight-loss curve. This property allows more accurate

Table 2. Cigarette Test Data on Tape Edge of Mini-Mattresses Batting Treated with Borax to Boric Acid Ratio 7:3

$Na_2O/B_2O_3 = 0.27$

	One Cigarette		Two Cigarettes
% Add-on	On Tape Edge of Bare Mattress	Between Two Sheets Tape Edge	On Tape Edge of Bare Mattress
5.3	I	--	--
8.5	N	I	--
13.2	N	N	I
20.0	N	N	N
26.0	N	N	N
31.5	N	N	N

N = No Ignition
I = Ignition

Table 3. Cigarette Test Data on Tape Edge of Mini-Mattresses Batting Treated with Borax to Boric Acid Ratio 6:5

$Na_2O/B_2O_3 = 0.20$

	One Cigarette		Two Cigarettes
% Add-on	On Tape Edge of Bare Mattress	Between Two Sheets Tape Edge	On Tape Edge of Bare Mattress
7.98	N	N	I
15.0	N	N	I
25.0	N	N	N

N = No Ignition
I = Ignition

determination of onset temperature of pyrolysis, and greater resolution of overlapping reactions. The free noncrystalline water is essentially completely removed

Table 4. Cigarette Test Data on Tape Edge of Mini-Mattresses Treated with Boric Acid

Boric Acid Add-On (%)	One Cigarette		Two Cigarettes
	Bare Mattress	Between Two Sheets	Bare Mattress
%5.0	N	N	I
7.5	N	N	I
10.0	N	N	N
15.0	N	N	N

N = No Ignition
I = Ignition

by 120°C. Between 120–220°C no detectable weight loss occurs. The small weight loss between 220° and 280°C was ascribed by Kilzer and Broido [39] to a dehydration process to yield "dehydrocellulose." The major weight loss of cellulose occurs between 280 and 380°C, the region investigated in most thermal studies of cellulose. Between 450 and 580°C the carbonaceous residue, which remains after decomposition, undergoes combustion. The rate of this reaction is much slower than that of the pyrolytic decomposition reaction. Even when the two reactions are nomalized to take into account differences in total weight of materials reacting, the high temperature reaction is much slower in rate. If this reaction is considered to involve only oxidation of the char, then the weight loss is controlled by the rate of diffusion of oxygen down to the surface of the char. On the other hand, the rate of weight loss in the decomposition reaction depends upon the concentration of cellulose and the temperature [40].

Figure 3 compares the DTG curves of purified cellulose in atmospheres of 100% nitrogen and 8.4% oxygen/91.6% nitrogen. In an inert nitrogen atmosphere little weight loss occurs above 450°C. However, in an oxidizing atmosphere a second phase of weight loss occurs between 450° and 580°C. This reaction apparently is associated with oxidation of the carbonaceous residue and is not a simple volatilization process.

The smoldering combustion process can therefore be considered as two distinct reactions: (1) oxidation of volatile products, and (2) oxidation of the char.

Figure 4 shows the DTG curves for the four additives used in this study. Borax (NaO/B_2O_3 = 0.45) loses essentially all of its water of hydration between 50 and 150°C and thereafter is present as anhydrous $Na_2B_4O_7$ ($Na_20.2B_2O_3$) and is stable to over 700°C. Boric acid also undergoes dehydration and is present essentially as the stable oxide, B_2O_3. The amorphous form of boric oxide has no definite melting point, but softens at about 325°C and becomes fully fluid at about 500°C [41].

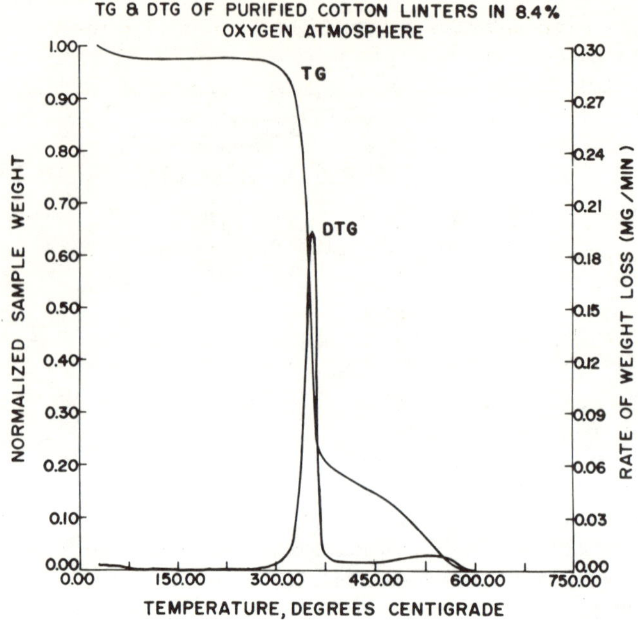

Figure 2.

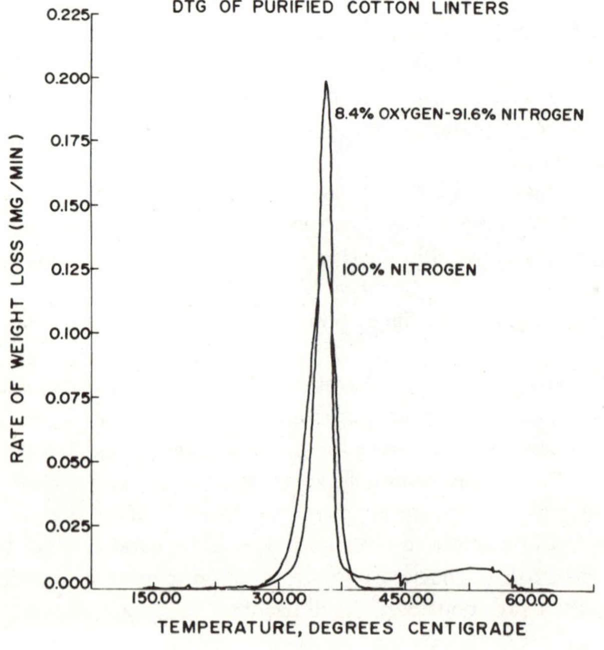

Figure 3.

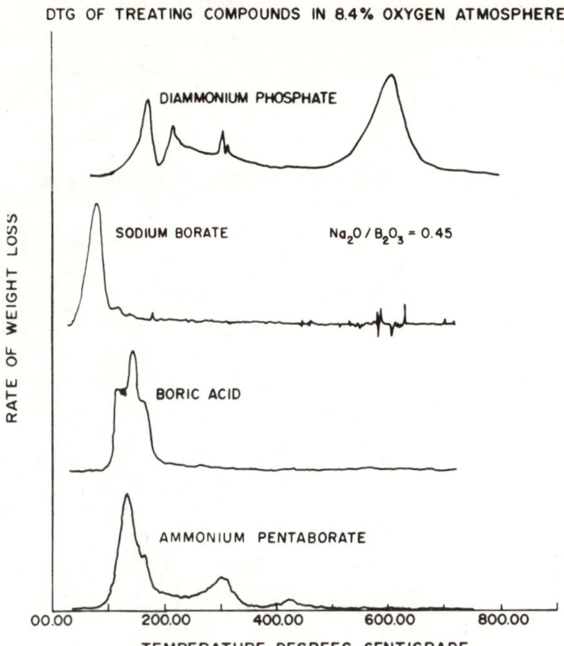

Figure 4.

Ammonium pentaborate loses essentially all of its water of hydration between 75 and 225°C. Two more peaks, at 300°C and 425°C, are associated with the loss of ammonia. Thereafter the compound is present as the stable oxide, B_2O_3. The decomposition of diammonium phosphate is much more complex. The DTG curve reveals four distinct peaks of weight-loss; the first two peaks account for the loss of two moles of ammonia with the subsequent formation of phosphoric acid. The phosphoric acid is further dehydrated between 300–500°C and probably forms the polyphosphoric acid $(HPO_3)_x$ which is a glassy polymer [14]. Above 500°C this glassy polymer apparently breaks down since only a 20% residue remains at 700°C.

The ability of diammonium phosphate to act as a glow retardant for fabrics but not for mattress filling can be explained by thermal data. Visual examination shows that pure diammonium phosphate forms a soft melt at 400°C. At 500°C a harder, more glass-like residue remains. Above 500°C white fumes are released from the residue and rapid weight loss is observed. Cellulose treated with diammonium phosphate is thus protected at temperatures below 500°C but above 500°C, as often occurs locally in a smoldering mattress, this protection is lost when the glassy polyacid decomposes.

Figure 5 shows this effect for samples containing 5 and 15% diammonium phosphate. As the add-on increases, the rate of the reaction involving oxidation of the char, which centers around 540°C, decreases. Then the rate of weight loss increases as the protection of the glassy coating is lost. The more the reaction is initially inhibited by the polyacid, the more dramatic the increase in rate after the coating is destroyed.

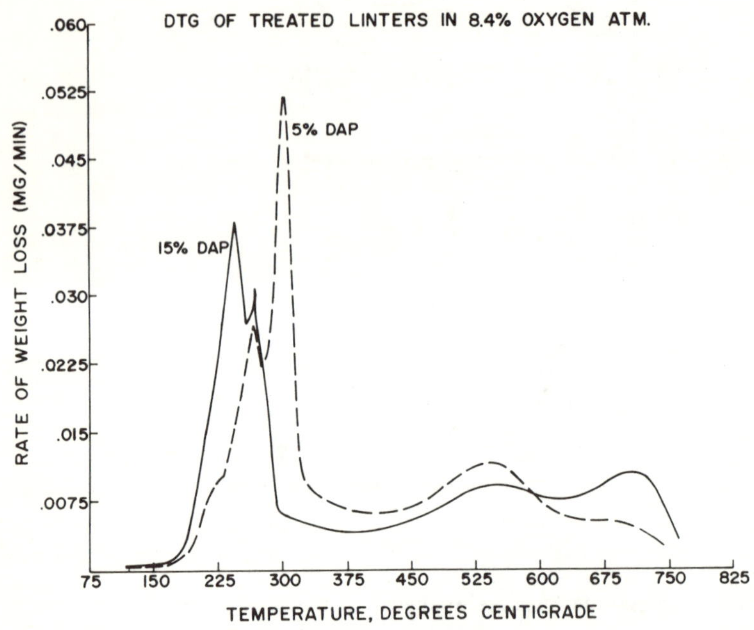

Figure 5.

This effect is illustrated by the enlarged portion of the DTG curves for the temperature interval between 400°C and 725°C (Figure 6). The top curve shows the DTG of purified cotton linters treated to an add-on of 10% by weight of diammonium phosphate; the bottom curve is for pure diammonium phosphate. The dashed curve is the difference between the two which demonstrates the large increase in rate of weight loss once the protection of the polyphosphate glass is lost.

Also important is the decrease in thermal stability (as evidenced by the onset of weight loss) as the concentration of diammonium phosphate is increased. Untreated cellulose begins to lose weight (other than free water) at about 220°C. Samples treated with 5 and 15% diammonium phosphate begin to lose weight at 155 and 135°C, respectively.

The effects of borax on the pyrolytic and char-oxidation reactions are shown in Figure 7. The thermal stability of borax-treated samples is similar to that of untreated cellulose. Unlike diammonium phosphate, an increase in borax concentration does not shift the onset temperature nor the temperature at which the

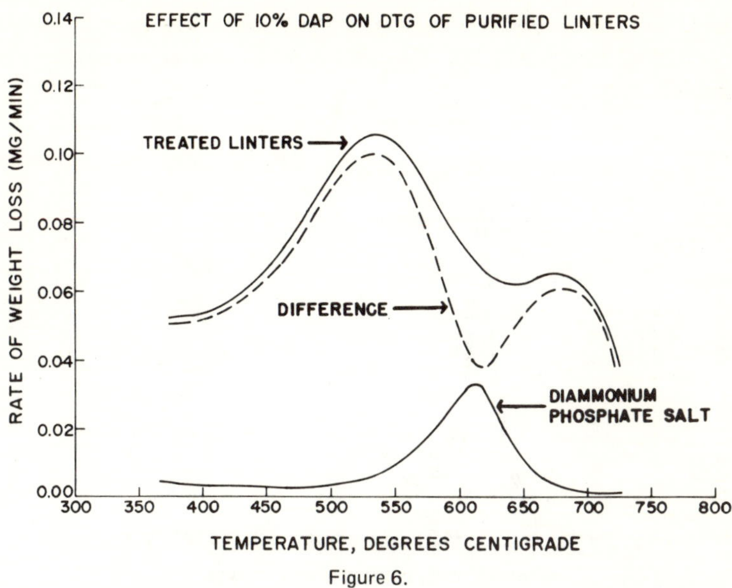

Figure 6.

maximum rate of weight loss occurs. Weight loss associated with the oxidation of carbonaceous residue from borax treated cellulose is more complex than weight loss of untreated cellulose. The reaction proceeds at a rate about equal to that of the control, but at 50°C lower. In all there appear to be three peaks of weight loss during the char oxidation. Differential thermal analysis [12] has shown all three to be exothermic reactions. Alkali metals have been reported to catalyze the oxidation of carbon [42], and apparently the Na^+ ions of borax catalyze the carbonaceous char of cellulose [43].

Thermogravimetric data of cellulose treated with boric acid differ from data with the other two additives (Figure 8). The most important difference is in thermal stability. Although most additives, particularly Lewis acids, lower the thermal stability, boric acid delays the onset of weight loss and exothermic activity associated with the oxidation of volatile pyrolysis products. Boric acid is a weak acid, but at the pyrolysis temperature of cellulose it is not present in acid form, having dehydrated to the oxide, B_2O_3:

$$2H_3BO_3 \xrightarrow[120\text{–}210^\circ C]{-2H_2O} 2HBO_2 \xrightarrow[260\text{–}270^\circ C]{-H_2O} B_2O_3 \quad \text{softens at } 325^\circ C$$

Formation of the boric oxide melt at temperatures above 325°C interferes with diffusion and chemisorption of oxygen on the surface of the residue and thus effectively retards oxidation of the char.

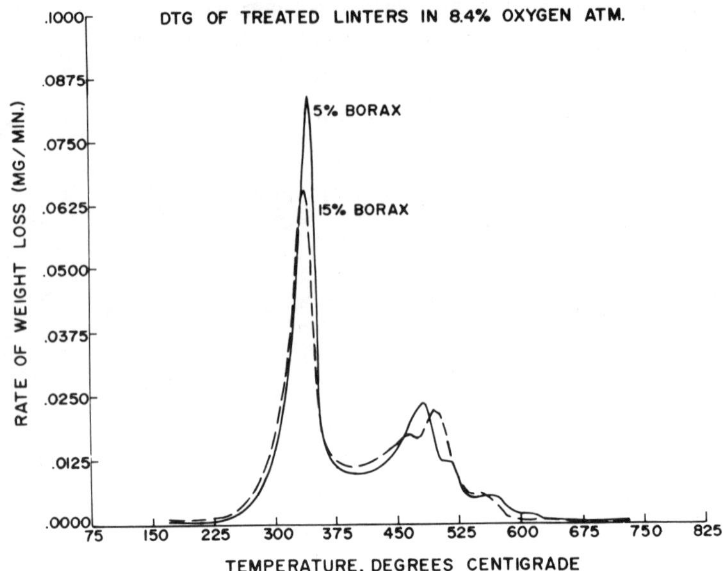

Figure 7.

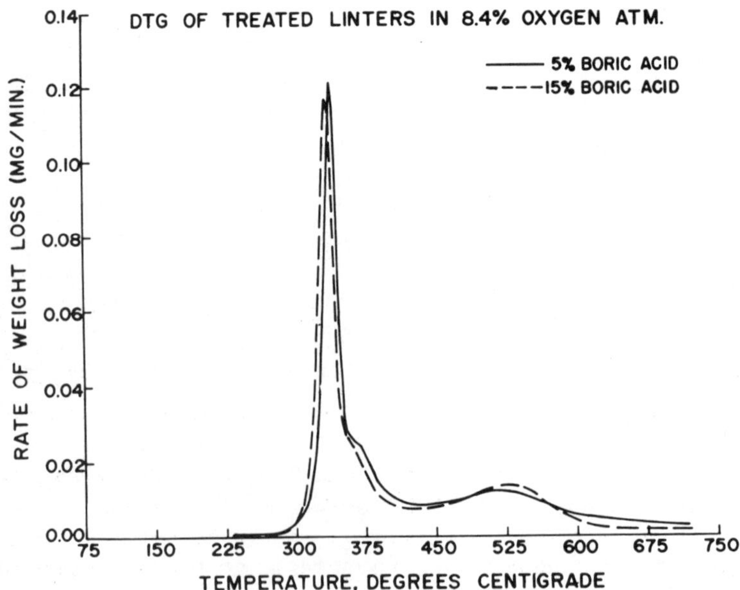

Figure 8.

Figure 9 shows that the effect of treatment with ammonium pentaborate is similar to that with boric acid; thermal stability is not significantly altered and the boric oxide melt is present at temperatures below the char oxidation.

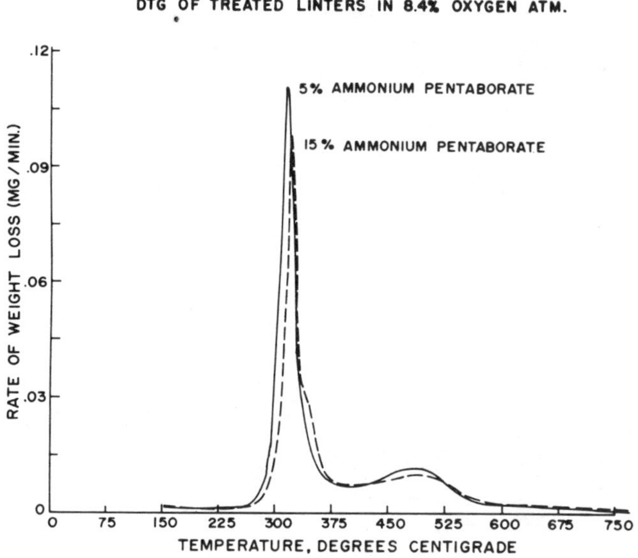

Figure 9.

Table 5 summarizes the thermogravimetric data. The rates shown were normalized by dividing the weight loss for the specific reaction into the maximum rate of that reaction. For diammonium phosphate, which loses weight concurrently with the cellulose, that portion of the weight loss and rate attributable to decomposition of the additive were subtracted before the rate of cellulose weight loss was normalized. As previously noted, diammonium phosphate lowers the thermal stability of cellulose progressively as the add-on increases; borax, ammonium pentaborate and boric acid do not appreciably affect thermal stability. Primary and secondary char oxidations are also listed in Table 5. Borax, a very poor suppressor of afterglow, does not slow down the rate of char oxidation. The fact that the temperature of maximum weight loss is about 50° below that of untreated cellulose definitely suggests some catalytic activity. For diammonium phosphate, which is considered an excellent glow retardant, the maximum rate of reaction for the primary oxidation of the char was between 30 and 40% less than that of the control. However, in what has been termed "secondary oxidation" of the char, above 650°C, the rate of combustion was greater. Boric acid and ammonium pentaborate demonstrate reductions in reaction rate similar to diammonium phosphate in the 450–600°C range. Unlike diammonium phosphate, neither exhibits any secondary oxidation of the char.

Table 5. Normalized Rate of Weight Loss

		Oxidative Pyrolysis (Flaming)	Maximum Rate	Primary Oxidation of Char (Smoldering)	Maximum Rate	Secondary Oxidation of Char	
							Maximum Rate
		°C	mg/min/gm.	°C	mg/min/gm.	°C	mg/min/gm.
Untreated Control Linters		357	0.253	536	0.050	--	--
Diammonium phosphate	5% add-on	305	0.097	540	0.034	670	0.060
	10% add-on	280	0.087	535	0.035	675	0.105
	15% add-on	245	0.087	545	0.032	715	0.060
Borax	5% add-on	343	0.165	485	0.050	--	--
	10% add-on	343	0.125	475	0.050	--	--
	15% add-on	337	0.110	495	0.056	--	--
Boric acid	5% add-on	335	0.175	525	0.042	--	--
	10% add-on	340	0.184	525	0.035	--	--
	15% add-on	340	0.191	520	0.036	--	--
Ammonium pentaborate	5% add-on	315	0.184	483	0.037	--	--
	10% add-on	321	0.201	484	0.034	--	--
	15% add-on	321	0.190	488	0.034	--	--

Stability Studies

Samples of cotton batting treated with ammonium pentaborate; sodium borate ($Na_2O/B_2O_3 = 0.25$); a 1:1 ratio borax-boric acid ($Na_2O/B_2O_3 = 0.18$); boric acid alone; and boric acid-ammonium carbonate were evaluated for durability of treatment to various temperatures at 65% relative humidity.

Table 6 shows ignition data for these samples installed in mini-mattresses and tested within 72 hours after treatment of the fibers. Mini-mattresses constructed with cotton batting having an add-on of 4.3% ammonium pentaborate pass the cigarette test requirements of FF 4-72. An add-on of 9.1% sodium borate ($Na_2O/B_2O_3 = 0.25$) is needed to pass FF 4-72.

Table 6. Performance Data on Cotton Batting Treated with Various Smolder Resist Formulations when Tested Within 72 Hours of Manufacture in a Mini-Mattress Structure

		One Cigarette		Two Cigarettes
	% Add-On	Bare Mattress	Between Two Sheets	Bare Mattress
Ammonium Penta-Borate	4.3	N	N	I
	4.6	N	N	I
	7.0	N	N	I
	7.5	N	N	I
	9.1	N	N	I
	10.0	N	N	N
	14.6	N	N	N
	15.3	N	N	I
Sodium Borate ($Na_2O/B_2O_3=0.25$)	4.0	N	N	I
	4.5	N	I	-
	9.1	N	N	N
	10.9	N	N	N
	15.8	N	N	N
	17.0	N	N	N
Boric Acid	9.3	N	N	N
"	11.4	N	N	I
Boric Acid[1/]	8.9	N	N	I
"	10.0	N	N	N
Borax:Boric Acid ($Na_2O/B_2O_3=0.18$)	8.3	N	I	I
	8.9	N	N	I

1/ 2% Ammonium carbonate added to treating formulation to improve solubility of boric acid.

With boric acid, an add-on of about 10% yields products that pass FF 4-72. To achieve 10% add-on of boric acid, however, wet pick-up must be about 166%, due to the limiting solubility of boric acid in water, or the water must be heated to about 160°F. Neither alternative is desirable because of the high costs of drying and of maintaining the treating bath at 160°F. We found that small amounts of ammonium carbonate increased the solubility of boric acid in water at room temperature. Samples with 10% boric acid add-on prepared with 2% ammonium carbonate in the treating solution, also passed FF 4-72. The sample with 8.9% add-on of borax-boric acid mixture in a 1:1 ratio ($Na_2O/B_2O_3 = 0.18$) passes the single cigarette on bare surface test but does not consistently pass the single cigarette two-sheet test.

When samples of all materials reported in Table 6 were tested with the modified vertical flame test [8], only those with sodium oxide present were resistant to flaming.

Table 7 shows the B_2O_3 content of samples prepared for storage stability testing. The two samples that failed the two-sheet test both contained sodium oxide (Na_2O). These findings also support the belief that Na^+ ions adversely effect smolder resistance. Samples with a B_2O_3 content of 4.11% pass the cigarette tests of FF 4-72. The source of B_2O_3, effects the ability to confer sufficient smolder resistance to pass the two-cigarette test.

Samples of the various products shown in Table 7 were stored at 70, 105, 130, and 160°F at about 65% relative humidity for periods of up to 35 days. At 7-day intervals portions of the stored samples were withdrawn for determination of boric oxide content. Figures 10, 11, 12, and 13 illustrate graphically the loss of B_2O_3 for the respective products plotted as the least squares fit of the data for each temperature. In general, all of the treated products show a loss of B_2O_3 with storage time.

Figures 14 and 15, data from Figures 10, 11, 12, and 13 for the different treatments plotted as isotherms, show that samples from different treatments lose B_2O_3 at different rates. The data, however, are inadequate to predict long-term stability of the treatments to temperature and humidity. We think, however, that at some point in time a steady state will be reached and further losses of B_2O_3 will be at a much slower rate than that for the first 35 days. There are indications that the high initial rate of loss shown by some of the samples might be due to the deposition of the boric acid donor on the surface of the fibers. Once this surface deposition is lost, in 30 to 60 days, any further loss can be expected to be at a much slower rate. Figure 13 indicates that during storage at 105°F and 65% RH the sodium borate ($Na_2O/B_2O_3 = 0.25$) treatment lost boric oxide at a much slower rate than the boric acid and ammonium pentaborate treatments.

Figures 14 and 15 indicate that the boric acid-ammonium carbonate system loses B_2O_3 at a slower rate than boric acid alone, or than ammonium pentaborate.

Figures 16 through 19 show the results after 225 days' storage at 70°F and 65% relative humidity.

In Figures 15 through 18 the experimental curves are the least squares fit to the data assuming a first order irreversible reaction for the loss of boric oxide.

Figure 16 indicates that regardless of the initial add-on of ammonium pentaborate, up to 5.2% B_2O_3 content, essentially all of the B_2O_3 loss occurs in the first

Table 7. Initial B_2O_3 Content of Experimental Samples Prepared for Stability Studies

B_2O_3 Donor	$Na_2O:B_2O_3$ Ratio	Add-On %	B_2O_3 %	Single Cigarette		Two Cigarettes
				Bare Mattress	Two Sheets	Bare Mattress
Ammonium Borate	-	4.3	2.51	N	N	I
Ammonium Borate	-	4.6	3.03	N	N	I
Sodium Borate	1:4	4.0	3.23	N	N	I
Sodium Borate	1:4	4.5	3.35	N	I	-
Boric Acid	-	11.4	3.39	N	N	I
Ammonium Borate	-	7.0	3.48	N	N	I
Borax:Boric Acid (1:1)	-	8.9	3.63	N	N	I
Boric Acid	-	10.0	3.87	N	N	N
Borax:Boric Acid (1:1)	1:5.5	8.3	4.00	N	I	I
Ammonium Borate	-	7.5	4.11	N	N	N
Boric Acid	-	8.9	4.43	N	N	I
Boric Acid	-	9.3	4.68	N	N	N
Ammonium Borate	-	9.1	4.73	N	N	N
Ammonium Borate	-	10.0	4.82	N	N	I
Ammonium Borate	-	15.3	5.09	N	N	I
Ammonium Borate	-	14.6	5.36	N	N	N
Sodium Borate	-	9.1	5.63	N	N	I
Sodium Borate	-	10.9	5.81	N	N	N
Sodium Borate	-	15.8	8.43	N	N	N
Sodium Borate	-	17.0	8.84	N	N	N

N = No Ignition
I = Ignition

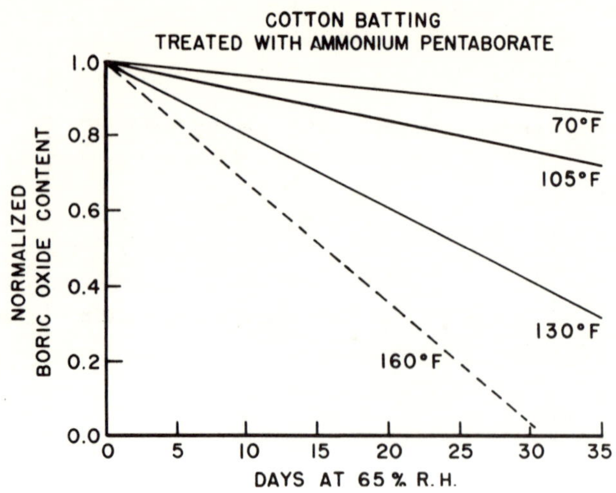

Figure 10.

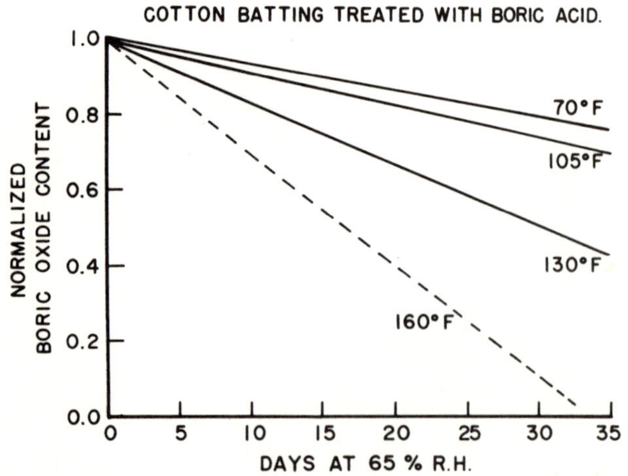

Figure 11.

60 days. Table 7 shows that an add-on as low as 2.51% B_2O_3 derived from ammonium pentaborate provides sufficient smolder resistance to pass FF 4-72.. All of the stored products of Figure 15 could thus be expected to pass FF 4-72.

Figure 17 shows that, for the sodium borate systems, the greatest loss of reagent also takes place within the first 60 days of storage, and that further loss of B_2O_3 with continued storage at 70°F and 65% relative humidity is minimal. Figure 18

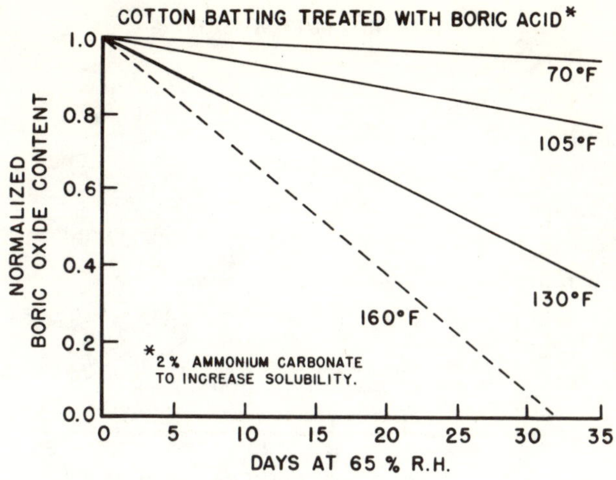

Figure 12.

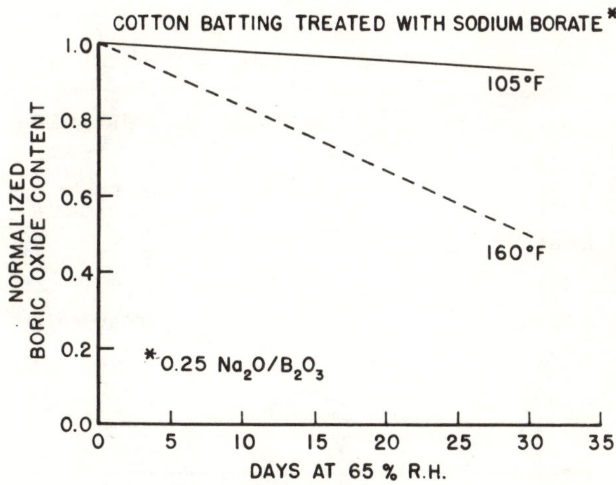

Figure 13.

shows boric oxide loss during 225 days' storage from batting products treated with boric acid. These data also indicate that essentially all of the significant losses took place during the first 60 days of storage.

The data in Figures 15 through 17 can be interpreted to support the hypothesis attributing early losses of B_2O_3 to surface-deposited material. Correlation of the residual B_2O_3 contents of these samples with the B_2O_3 content needed to pass cigarette testing (Table 7) indicates that most of these samples could be expected to

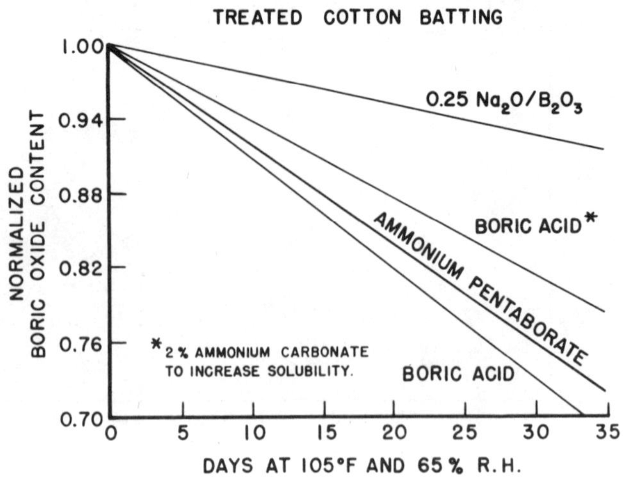

Figure 14.

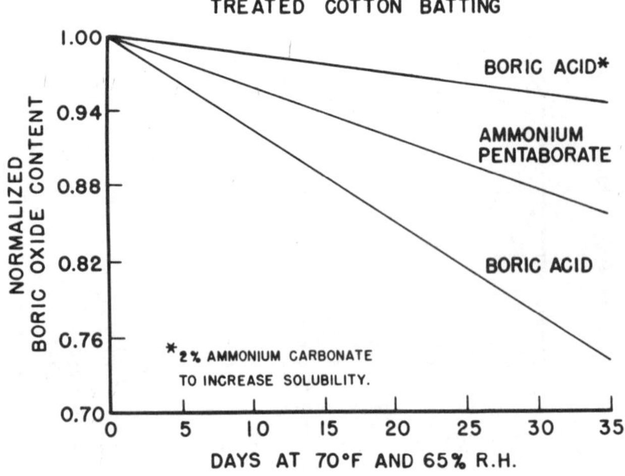

Figure 15.

pass FF 4-72 after extended storage.

Figure 19 shows the data from Figures 16 through 18 after normalization for initial B_2O_3 content. These data indicate that it is possible to determine and predict the eventual residual B_2O_3 content of cotton batting products, treated with any of a number of chemical systems containing the borates, after these products have been stored for extended times at 70°F and 65% relative humidity. About

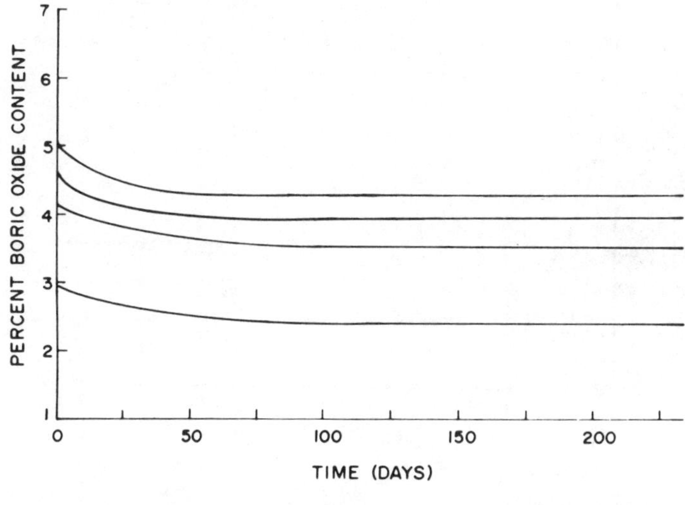

Figure 16.

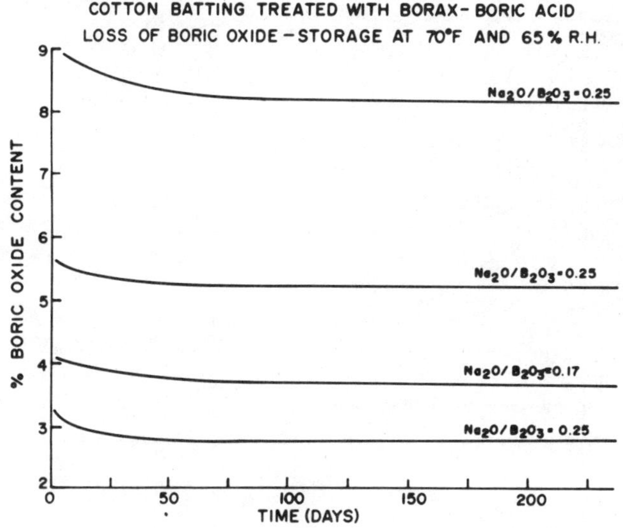

Figure 17.

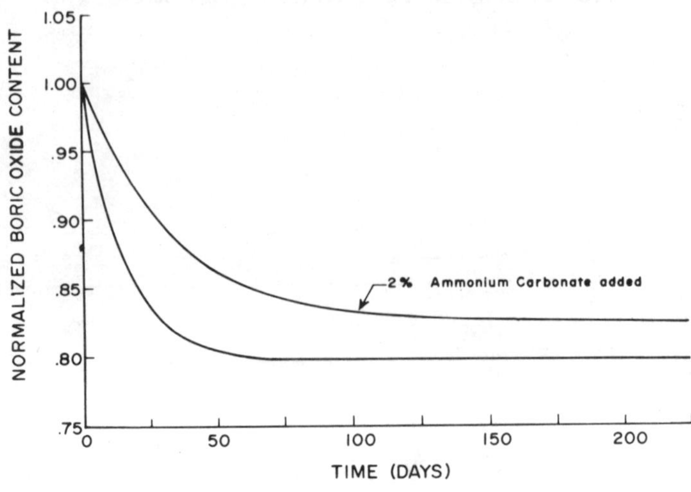

Figure 18.

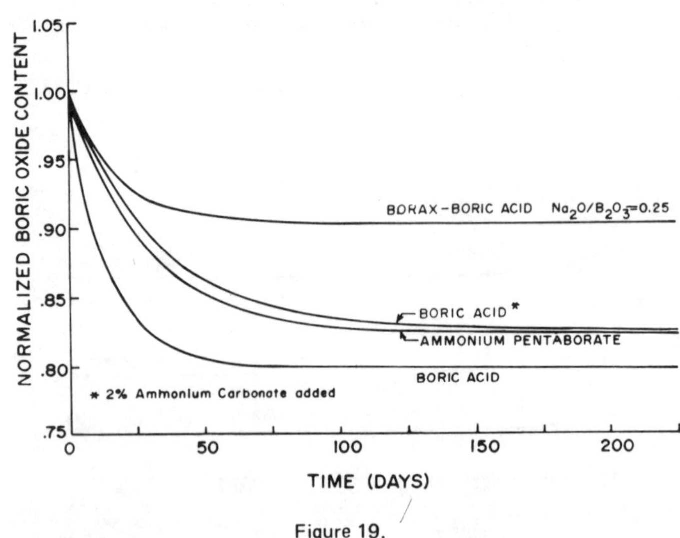

Figure 19.

10% of the sodium borate treatment is lost compared to 17–20% of the boric acid or ammonium pentaborate treatments.

SUMMARY AND CONCLUSIONS

Thermogravimetric data indicate that the formation of boric oxide upon heating is an important factor in the ability of borate systems to resist the initiation of smoldering combustion in cotton filling materials. The presence of sodium ions can be detrimental by promoting smoldering via a catalytic effect on the oxidation of the carbonaceous char.

Data from cigarette ignition tests confirm that requirements of the Mattress Flammability Standards, FF 4-72, can be met by treating cotton batting with a number of chemical systems containing boric oxide donors. A boric oxide content in excess of 2.51% by weight of the cotton is needed to resist the initiation of smoldering by a single cigarette on the bare mattress surface. Maintaining the boric oxide content above 4.1% makes possible passage of the more severe two-sheet test.

There are noticeable losses of the treating compound during the initial period after treatment. The rate of this initial loss varies with the particular boric oxide donor and with the temperature and humidity of the storage conditions. The losses are essentially complete after 60 days storage at 70°F and 65% relative humidity. These data tend to confirm the hypothesis that the early loss of boric oxide results from the volatilization of the boric oxide donors which are surface deposited. The results obtained in this study should be compared with the performance of treated products stored under fluctuating conditions of temperature and relative humidity.

ACKNOWLEDGMENTS

The authors thank Mr. Hollis L. Miller for carrying out the many extractions required in this study, and Mr. W. T. Gentry, Jr., for the B_2O_3 determinations.

REFERENCES

1. Proposed Flammability Standard for Mattresses DOC PFF 4-71 Fed. Register, *36* (175) 18095–18098, (Sept. 9, 1971)..
2. Flammability Standard for Mattresses DOC FF 4-72 Fed. Register, *37* (110), 11362–11367 (June 7, 1972).
3. P. A. Koenig and N. B. Knoepfler, Amer. Dyestuff Reptr. *58* (17), 30–34 (1969).
4. P. A. Koenig and N. B. Knoepfler, Upholstering Industry *37* (8), 10–11, 14–15, 20, 22 (1970).
5. N. B. Knoepfler, *J. Fire and Flammability, 2*, 219–230 (1971).
6. N. B. Knoepfler and P. A. Koenig, U.S. Pat. 3,629,052, Dec. 21, 1971.
7. American Association of Textile Chemists and Colorists Text Method 34-1969, Technical Manual of the American Association of Textile Chemists and Colorists, *46*, 208–209 (1970).
8. N. B. Knoepfler, P. A. Koenig, and W. T. Gentry, Jr., Flame Retardant Multiple Density Cotton Flote Products for Furniture and Mattresses, Agri. Res. Serv. Pub. ARS 72–86, 8 pp. (1970).
9. Motor Vehicle Safety Standard 302, Fed. Register *36* (5), 289–291 (Jan. 8, 1971).
10. N. B. Knoepfler, Agri. Res. Serv. Pub. ARS 72–97, 71–74 (1972).
11. J. P. Neumeyer, P. A. Koenig, and N. B. Knoepfler, Agri. Res. Serv. Pub. ARS 72–98,

55–64 (1972).
12. P. A. Koenig and N. B. Knoepfler, Agri. Res. Serv. Pub. ARS 72–92, 30–32 (1971).
13. P. A. Koenig, J. P. Neumeyer, N. B. Knoepfler, H. L. E. Vix, Organic Coatings and Plastics Chem. ACS 165th Meeting, Dallas, Tex. 33 (1), 476–483 (1973).
14. Lescouer, H. Ann. Chim. Phys. *6* (19), 35–67 (1890); *6* (20) 533–556 (1890); *6* (21) 511–565 (1890).
15. H. Tazaki, J. Sci. Hiroshima Univ., A 10, 76-54, 55–61 (1940); A 10, 109–112, 113–116 (1940).
16. A. Thiel and H. Siebeneck, Z. Anorg Allgem Chem. *220*, 232–246 (1934).
17. L. F. Gilbert and M. Levi, J. Chem. Soc, 527–535 (1929).
18. M. v Stackelberg and F. Quatram, Z. Electrochem. *42*, 551- (1936).
19. M. v Stackelberg, F. Quatram, and Dressel, Z. Electrochem, *43*, 14–28 (1937).
20. L. K. Nash, Anal. Chem., *21*, 1405–1410 (1949).
21. P. Jaulmes and A. Gontard, Bull. Chem. Soc., *5* (4), 139–148 (1937).
22. P. Jaulmes and E. Galhac, Bull. Chem. Soc., *5* (4), 149–157 (1937).
23. J. R. Adams, Jr. Flameproofing Army Clothing, Final Rpt. Sub. Proj., 27 A 8, NRC Quartermaster Corps, (Columbia Univ.) (Aug. 1945).
24. R. W. Little, Flameproofing Textile Fabrics, Amer. Chem. Soc. Monograph Series 104, Reinhold, NY 410 pp. (1947).
25. J. W. Lyons, The Chemistry and Uses of Fire Retardants, Wiley-Interscience, NY 462 pp. (1970).
26. E. F. Hartman, R. Hicks, and F. Hartman, Fire Eng. *92*, 336 (1939).
27. N. B. Knoepfler, H. K. Gardner, Jr., and H. L. E. Vix, U.S. Patent 3,181,225, May 4, 1965.
28. G. S. Buck, Jr., in R. E. Kirk and D. F. Othmer, Encyclopedia of Chemical Technology Ed 1, (6), 543–558 (1951).
29. R. W. Little, J. M. Church, and S. Coppick, Ind. and Eng. Chem., *42* (3), 432–440 (1950).
30. J. L. Gay-Lussac, Anales de Chemie *18* (2), 211–217 (1821.).
31. W. G. Parks, J. G. Erhardt Jr., and D. R. Roberts Amer. Dyestuff Reptr *39* (9), 294–300 (1950).
32. M. M. Sandholtzer, Natl. Bu. Stds. (US) Circular C455 (1946).
33. M. N. Conklin, Color Trade J., *11*, Pt. I, 171–173, Pt. II, 259–260 (1922).
34. F. Versman and A. Oppenheim, J. Franklin Inst. *39*, 354 (1860).
35. A. Kling and D. Florentin, Quat. Natl. Fire Protection Assn. *16*, 134 (1922).
36. National Cotton Council of America (Memphis, Tenn.), Manufacture of Flameproof Cotton Mattresses, 11 pp. processed.
37. J. D. Reid, W. A. Reeves, and J. G. Frick, USDA Leaflet No. 454 (1959).
38. H. R. Richards, Canadian Textile J., *77* (26), 68–76 (1960).
39. F. J. Kilzer and A. Broido, Pyrodynamics, *2*, 151 (1965).
40. A. E. Lipska and F. A. Wodley, Journal of Applied Polymer Science, *13*, 851 (1969).
41. U.S. Borax, Industrial Products Catalogue Technical Data Sheet IP-14.
42. P. L. Walker, *et al*, Chemistry and Physics of Carbon, *4*, 292 (1968).
43. J. M. Smather and H. R. Richards, Amer. Dyestuff Reptr. *59* (3), 21–26 (1970).

L. J. Hillenbrand and J. A. Wray

Battelle
Columbus Laboratories
505 King Avenue
Columbus, Ohio 43201

A FULL-SCALE FIRE PROGRAM TO EVALUATE NEW FURNISHINGS AND TEXTILE MATERIALS*

(Received May 23, 1974)

ABSTRACT: In 1972 Battelle Columbus Laboratories contracted with the National Aeronautics and Space Administration to undertake full-scale fire tests of furnished bedrooms. The series of fires was designed to show the performance of new fire-resistant and fire-retardant materials by providing comparative fire conditions in identical room settings. The measurements of gas mixture compositions, temperature, radiant heat flux, convection, smoke and the visual evidence from photographic and TV coverage led to descriptions of the unsteady-state nature of fire and smoldering conditions during such fires. Fire-retardant items were caused to pyrolyze by the heat of burning of other items in the room and so produced fire conditions not characteristic of any one item individually.

INTRODUCTION

Prior research at Battelle-Columbus was based on the recognition that more deaths occurred in residential fires because of anoxia and the inhalation of toxic gases than because of burns. In an approach to this problem a major source of these toxic gases was assumed to be products formed during smoldering and pyrolysis that occur in room-type fires, especially those involving some types of synthetic materials. Experimental examination of means for modifying the

*In response to numerous requests, this final report was prepared with minimal editing to permit printing and limited distribution on a timely basis. A more extensively edited version is now in preparation and will be distributed as a NASA Special Publication in the next few months.

pyrolytic decomposition of furnishing materials raised the need for a more exact specification of the temperature and environmental conditions to which materials might be exposed during residential-type fires.

With the cooperation and assistance of the Columbus Fire Department, a program was initiated for development of techniques and procedures that would permit observation of the way furnishing materials respond to the period of fire and smoldering in fires set for this purpose. Using a selection of furniture loadings, several room-type fires were set in residential buildings scheduled for destruction at the Fort Hayes military complex located in Columbus, Ohio. These buildings had been occupied in recent times and were in excellent condition for examination of fires under the limited ventilation conditions that would be typical of residences in this part of the United States in winter months.

The plans for a new series of full-scale experimental fires were initiated at the suggestion of NASA following the presentation of a film and discussion illustrating Battelle-Columbus' recent work in this fire research. That film showed bedroom-type fires carried out as a part of a program to determine the influence of the cyclic characteristics of real fires under limited ventilation on the burning and pyrolysis properties of the room furnishings. This film was shown in connection with a presentation on "Fire Resistant Materials", by Mr. Matthew Radnofsky of NASA-Houston. The obvious advantages of the Columbus set-up, in relation to further experimental study of improved fire safety, led to a meeting of Battelle-NASA and others on November 23, 1971. A new series of fires was suggested by NASA designed to show the performance of new fire-resistant and fire-retardant materials by providing comparative fire and smoldering environmental conditions.

More recently, the goal for the new series of fires was written in a meeting with NASA personnel and others at Battelle on May 3 and 4, 1972. The goal was as follows:

> To establish the need for special materials of improved fire safety in domiciliary settings of public concern, and to assess, in a professionally acceptable manner, the potential of materials arising from the new space-age technology for this purpose.

It was anticipated that some new materials arising from the space-age technology and not yet available through conventional commercial channels might provide significant improvements in fire safety if the best of the commercially available materials showed important shortcomings in this area. It was the intent of this program to assess the benefits that could accrue from the use of these new materials.

The fires were intended to provide comparative performance data. The settings, while chosen with buildings of public concern in mind, were selected so as to have general significance for fire safety evaluation.

In planning the program it was recognized that several situations must be considered in planning the comparative fires. Thus, the most popular textile and furnishing materials in present-day use might not include those that are among the best commercially available either for reasons of cost, visual appeal, or conventional practice. Further, the new technology that has been stimulated by NASA and the space-age activities has led to the development of a new generation of "space-age" materials not yet commercially available. For these reasons, the plan of investigation was expected to distinguish between those improvements ordinarily available through use of the best of the commercial materials, and those improvements in fire safety that would require substitution by space-age materials.

It was further anticipated that when a material becomes available for use in interior furnishings, it may be expected to go through stages of piecemeal application by gradual substitution into such furnishings. An exception to this would be the construction of new public buildings where optimum material usage can be achieved through regulatory controls. Thus, the program was expected to make a limited demonstration of both the case of the complete substitution of new materials into the chosen room, and the case of partial substitution where only a few materials that should be most effective in improving fire safety were used to replace the best commercially available materials.

Fire safety is a matter requiring the evaluation of a number of factors. Thus, for example, fire resistance and fire spread, visibility during the fire, toxicity of evolved gases, and the fire-fighting problem that is created, are potentially independent factors that must be evaluated before the relative hazard can be assessed. The plan of the program provided for sampling and instrumentation to evaluate these factors consistent with the goal of technological utilization that has been specified.

OBJECTIVES

To meet the overall goal, the following detailed objectives were set for the investigation.

(1) Obtain comparative performance data for chosen textiles and furnishing materials using fires set in full-scale rooms of identical size and geometry.

(2) Assess fire hazard by evaluation of ignition time, fire spread rate, smoke density, evolution of selected combustibles, toxic gases and combustion products, and heat flux.

(3) Establish baseline data by carrying out a fire in a room equipped with a selection of the popular or typical furnishing materials.

(4) Determine the degree of fire hazard that remains for an improved room when only materials that are among the best commercially available are used under the comparative fire conditions established.

(5) Assess further improvements that can be derived by use of new materials arising from the space-age technology.
(6) Assess the performance of a limited number of new materials selected as most effective in improving fire safety when used for partial substitution in a room otherwise equipped with the typical room furnishings.

BASIS FOR THE EXPERIMENTAL PROGRAM

The previous work at Battelle-Columbus had demonstrated that the early stages of a room fire exhibited stage-wise development of smoldering, flame flareup, oxygen depletion, combustion product accumulation, and oxygen-level recovery through air circulation. The particular degree of stage-wise fire development that was obtained clearly would depend on the choice of room characteristics and the arrangement of furnishings. In the present program these choices were fixed to establish comparative fire data under conditions where only the materials used were varied. This room configuration was chosen partly for convenience to the objectives listed and partly for economy of effort. A bedroom arrangement was chosen for the variety of material combinations that are possible and for the circumstances of realistic fire initiation and development that could be simulated.

The fire load for each room can be described in terms of two quantities: one represents the furnishings and the treatments given to walls, ceiling, and floor, while the other represents the uncontrolled and usually combustible items, brought into the room by the occupant, which have served for ignition of the fire.

The choices available for room furnishings and arrangements were manifold and some basis had to be adopted for deciding among these. On one extreme, all metal furniture might be used such that, if the mattress and bed coverings are suitably fire resistant, very little fire risk exists. This is especially true if the igniter load (brought in by the occupant) is also limited to the degree that fatal conditions cannot arise from this part of the furnishing alone. On the other extreme the use of cheap, highly combustible materials could be carried to such an extreme that severe fire risk could be determined by inspection. Neither extreme was suitable for the present research.

Between these two extremes lie a number of choices, suitable for public occupancy, in which metal is used only where its strength and durability are the primary consideration and where the other materials are chosen with cost, comfort, and decor in mind. For the present program, the furnishings and room arrangements have been based primarily on the materials recommended for trial by NASA-Houston and on the configuration needed to demonstrate their values.

The major factors in fire spread are related to the configuration and layout of the room and the type of furnishings involved. The configuration of the room is important since the relative position and distance of each furnishing item from other items will determine the propagation route and intensity. Also, the relative flammability of each item will also determine the route and the intensity.

Some restraint was exercised in equipping the "Typical room" in that some common fire hazards were avoided; prior experience had shown for example, that ignition of a fire in a polyethylene wastebasket filled with scrap paper aided fire spread because the basket melted and spilled the burning contents into the room. The constraints that have been placed on this "Typical" room arose from the consideration of the reproducibility of the spread conditions and the stated intent of evaluating selected material alternatives in the comparative fires.

The "Improved" room furnishings were intended to show how far it was possible to proceed toward the elimination of fire hazard if the materials were among the best commercially available without regard for cost. Again, this demonstration was to be assessed in terms of the chosen comparative conditions. Further, the performance of these "best" materials would determine the degree to which the new, or "special" materials are needed to achieve suitable fire safety in inhabited buildings.

Two concepts for fire study were considered. In one concept each room ensemble would be observed under conditions of gradually-increasing severity of ignition loading until fire spread was sufficient to cause complete room involvement. In this case more than one set of room furnishings of each type might be necessary to complete the set of trials required depending on the accumulation of fire damage from each trial. For the limited series of trials permitted in this program the above approach was not satisfactory.

The concept for study that was finally chosen for the program involved the choice of a standardized ignition load such that it could be expected to reveal differences in inflammability and fire spread character for the ensembles provided. Since the literature provided very little guidance for the choice of that ignition loading, one preliminary room was required that could be used for ignition trials. The ignition loading chosen in this way was kept the same for all rooms so that differences in fire spread and burning character depended largely on the changes in materials combinations employed. Further, in those cases where the ignition did not cause any appreciable degree of fire spread, a second, and larger, ignition could be tried.

The arrangement of the ignition load in the room was chosen so that at least two paths for fire development were provided. In this way, if special materials (used for an item of furnishings) prevented fire development along one path, room involvement might still occur by another route and in that event the response of the fire-retardant item to the pyrolysis conditions might be observed.

For this program it was anticipated that each room ensemble might be examined from at least two points of view, viz

(a) Its effectiveness in controlling fire spread
(b) The development of other hazards, e.g., smoke, toxic gases, under the conditions of fire spread obtained.

To minimize cost, the series of fires were carried out on as rapid a schedule as possible. This required that each fire in the series be carefully planned in advance so that extensive iterative use of the results of one fire to dictate those of subsequent fire was not possible. Further, to maintain a rapid schedule, the fire rooms were prefabricated in sections for storage and then subsequently used by bolting the sections together. Thus, the manpower and instrumentation were committed to this program for an experimental period of about 4 weeks, during which the four experimental fires and a few trial burns using various igniter loads were tried.

THE EXPERIMENTAL PROGRAM

The Fire Facility

Arrangements were made with the Columbus Fire Department to use an existing six-story concrete building, designed and used as a fire training tower, as the site for the experimental fires. The training tower is a concrete building with unglazed casement window frames on each of three sides. Figure 1 shows floor plan dimensions for this structure. The layout of the fire room and the required personnel area and instrument room in the fire training tower is illustrated in Figure 2. The figure shows the fire room, about 11 ft by 11 ft inside, with an 8 ft ceiling arranged in a space about 18 ft by 19 ft, 6 in, by 11 ft enclosed by three walls of the fire tower and the barricade. The volume of the fire room, 968 ft^3, is about 33 percent of the residual fire tower volume left for circulation of air and gases into and out of the fire room. Communication of such gases between the fire tower volume and the fire room occurred by an open doorway and by a duct opening located in the ceiling of the fire room. All windows were closed during the fire except one, see Figure 2, which was partially opened to provide a continuing supply of air and to relieve pressure fluctuations during the fire. The opening in the window was formed by a wooden panel hinged at the top of the window frame and propped open about 1 foot at the bottom so as to admit air but restrain wind currents that otherwise might affect the fire.

Each fire room was 11 ft by 11 ft square with an 8 ft-high ceiling and one door and one simulated window on opposite sides of the room. The basic construction of 2 X 4 studding topped with conventional 5/8-inch dry wall was used for each fire situation. However, varying wall surface treatments were used for the different room ensembles. The walls, floor, and ceiling were made in sections and assembled in the fire training tower as needed. Ceiling sections were covered with 5/8-inch dry wall material and the floor was covered with 1/2 inch plywood.

ROOM ARRANGEMENT AND FURNISHINGS

Materials Selection

The room arrangement used as shown in Figure 3. The "Typical Room" materials were selected on a basis of frequency of use and were purchased from normal retail outlets. The materials selected for the "Improved Room" and the "Space-age materials" room were based primarily on suggestion from NASA-Houston personnel

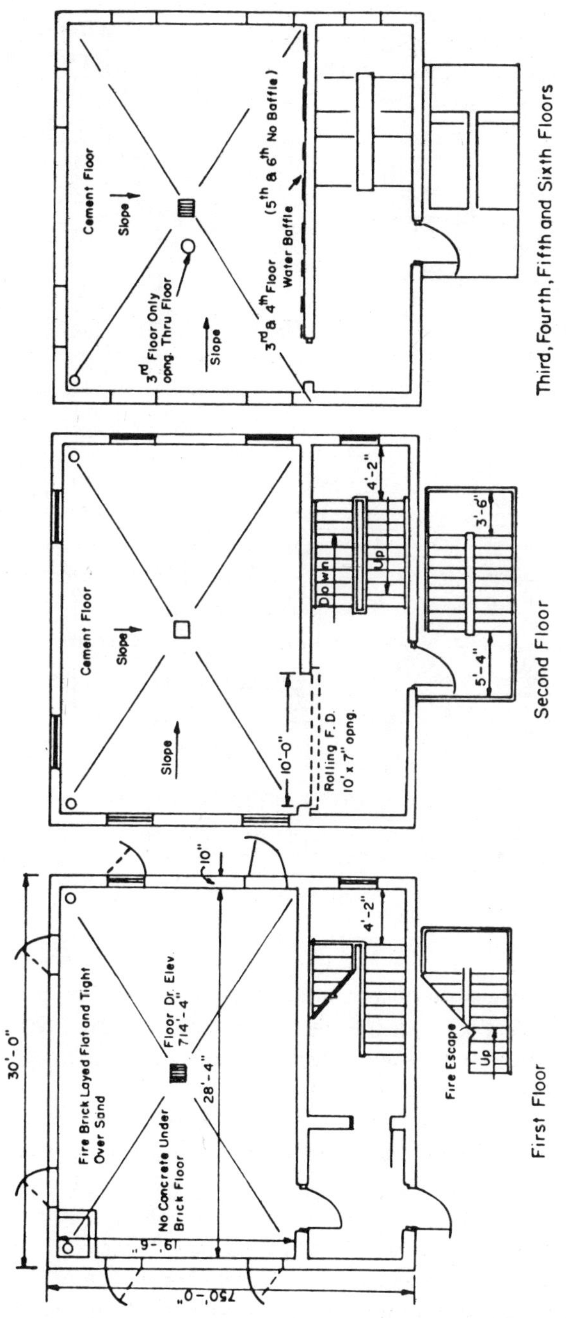

Figure 1. Plan views of the training tower of the Columbus Fire Department.

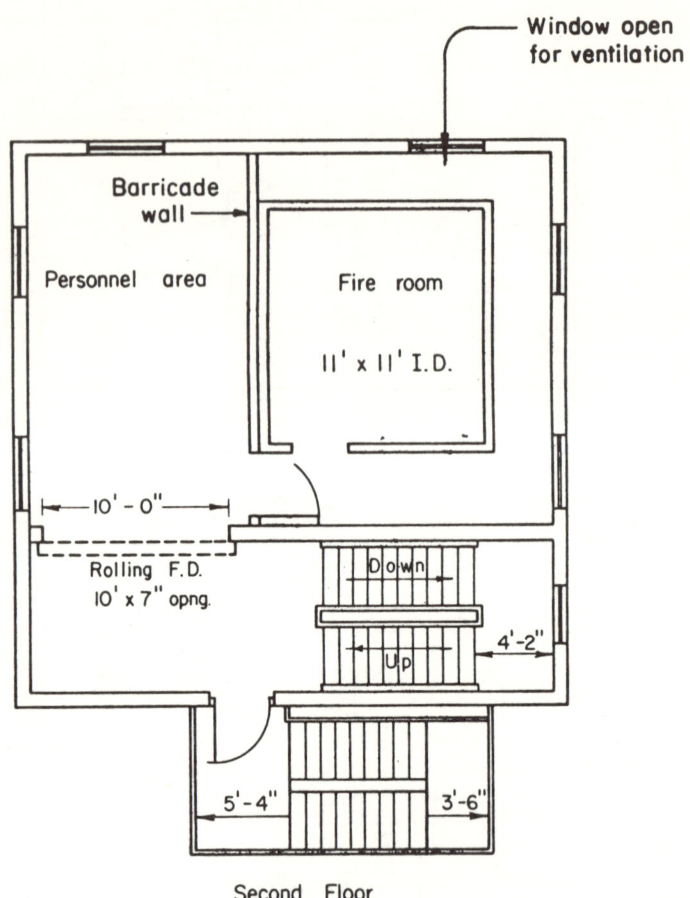

Figure 2. Arrangement of fire room and barricade in the fire training tower.

familiar with both commercial products having a high degree of fire resistance and, of course, the materials developed for space applications. Since a door was not used at the entrance to the fire rooms, a door was added to each fire load by placing it against one wall as shown.

Table 1 lists furnishings and materials included in the four basic fires.* Explanations of material choices specific to each room are given below under the respective room heading. A listing of the weights for each item used is given in Table 2.

Typical Room.—Consultation with Ohio Bedding, Columbus, Ohio, indicated that "typical" bedding construction should be as shown in Table 1. Other choices of materials for this room are based on recommendations from NASA personnel and observations of typical home furnishings available in local outlets. Unpainted

*All Tables at end of article.

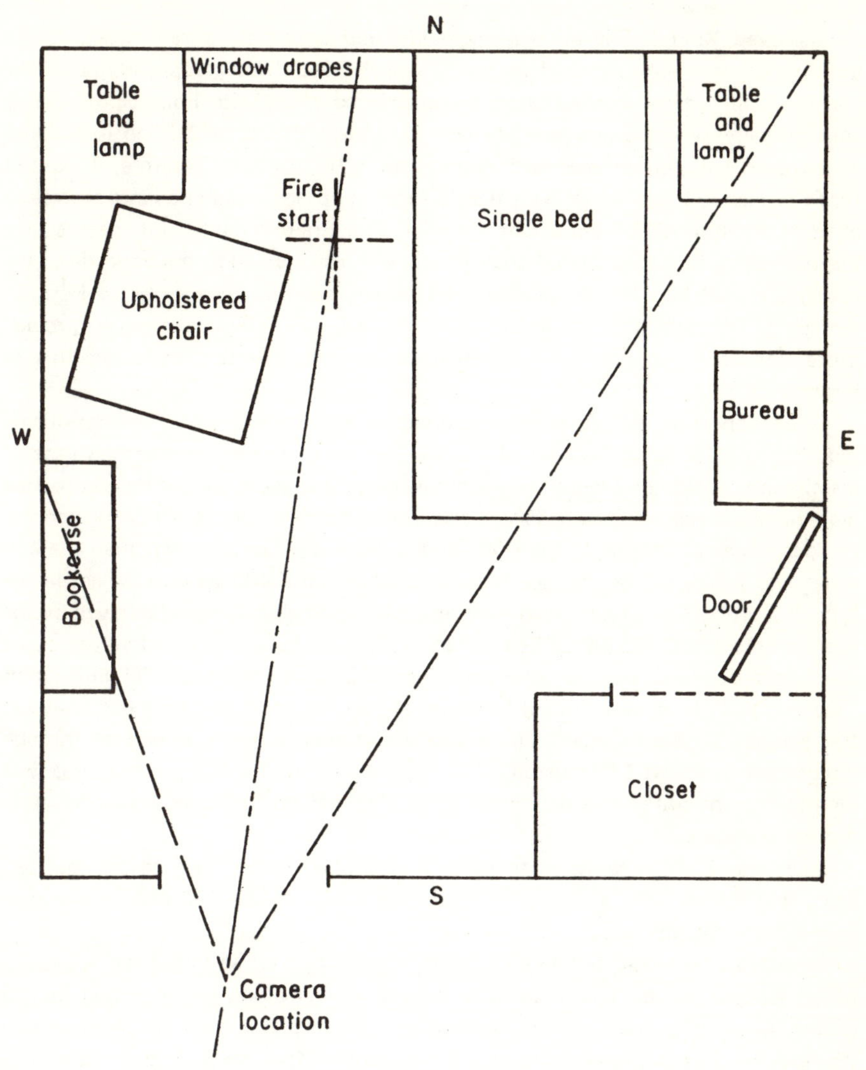

Figure 3. Furnishings arrangement and camera location.

pine end tables, bookcase, and bureau made it possible to use the same solid wood construction for these items in all rooms with changes being made only in the surface finish applied.

Improved Room.—The bureau, end table, and bookcase, were identical to the unfinished pine furniture used in the "Typical Room" for comparison purposes, and the front side of the headboard were surfaced solely with Formica. Norgan is commercially available and was planned for use in this room but no furnishings constructed from or surfaced with Norgan could be located. Therefore, it was felt that Formica, because of its wide distribution and usage, would be more representative of an improved fire-resistant material. The fire-retardant paint specified for the box-spring and upholstered chair frames was also applied to the backside of the headboard and also to the bureau, end tables, and bookcase prior to Formica surfacing. This was done to impart better overall fire retardance and to provide protection to those areas for which lamination with Formica was impossible or impractical.

Discussions with NASA-Houston resulted in the choice of BASF sodium silicate board as wall covering because of its high degree of fire protection and ease of installation. A Nomex blanket was substituted in this room for the Proban-treated wool blanket originally planned since a source of the latter could not be located.

Space-Age.—Selection of the 400 series of Durette fabrics rather than the 420 series was prompted by the greater availability of the 400 series. Changes in the type of Durette fabrics employed have resulted from recommendations given by the supplier. Asbestos was eliminated as recent findings have implicated asbestos as a carcinogen and also because asbestos is not a space-age material. Polyimide plus Fiberglas and Halar were not used because of material and fabrication costs beyond the budget of this program. The substitute structures were arrived at through discussions with NASA-Houston. The base furniture for these structures was identical to the unfinished pine used in the Typical and Improved rooms for comparison purposes.

ADP and Kel-F treatments to have been applied respectively to the mattress foam and ceiling were, as a result of discussions with NASA-Houston, regarded as unnecessary and deleted.

Sheuffelin cellulosic ceiling tiles could not be obtained in time from the designated supplier so the 1/4 in Laminite used as wall covering was also used for the ceiling. In addition, two treated wood panels (foot square) from NASA-Ames and 25 NASA-Houston developed plastic-foam panels (foot square) were selectively positioned on the Laminite.

Materials Preparation

All mattress and box spring construction along with chair upholstering was done by Ohio Bedding of Columbus, Ohio. Construction and preparation for each room is described below under the respective room headings.

Typical Room.—The end tables, bureau, bookcase, and door were stained with Hanna's 4905 Traditional Mahogany Wood Stain and finished with two coats of Hanna's Swan Spar Clear Satin Varnish.

Improved Room.—One heavy coat of Sherwin-Williams Kem Gard Fire Retardant Latex Paint was applied to the box-spring frame, end tables, bureau, bookcase, upholstered chair frame, and back-side of the headboard. Formica 2855 was then selectively laminated over the bureau, end tables, bookcase, and front side of the headboard (that is, Formica was applied to the major portion of the outside surface area of the furniture mentioned), using 3M's Fastbond 30 contact cement. The door in this case was stained and finished as given previously for furnishings in the typical room.

The sodium silicate board wall covering was applied over the dry-wall using nails and staples. The Fiberglas wall covering was adhesively bound to the sodium silicate board using 3M's Scotch Grip 77-N pressure sensitive adhesive.

The ceramic ceiling was of a concealed suspension nature.

Space-Age Room.—One heavy coat of the L-3203-6 Fluorel treatment[1] was applied by spraying to the box-spring frame, mattress and box-spring foam, and upholstered chair frame and foam. Two heavy coats of the L-3203-6 Fluorel treatment were applied by spraying to the end tables, bureau, bookcase, and carpet padding. Sheuffelin paper was then laminated to the end tables, bureau and bookcase using 3M's 4715 contact cement.

The 1/4 inch Laminite wall covering was nailed in place while the Durasan wall covering was adhesively bound to the Laminite using 3M's 4715 contact cement.

Combinations of nails and washers were used to secure the NASA-Ames wood panels and NASA-Houston plastic foam panels to the ceiling.

Typical Room with Space-Age Bed.—The mattress, box spring, and headboard were constructed as described for the Space Room. The end tables, bureau, bookcase, and door were furnished identically as those in the Typical Room.

Igniter Load

As indicated earlier, it was required that the igniter load be large enough to cause complete room involvement in at least some cases, but not so large that it dominated the fire load otherwise arranged. Newspaper seemed to be a reasonable type of material for the fire start and the trials centered about the ways in which such material might be used for the ignition. In addition to the newspaper, other combustibles were arranged about the room as part of the room load that was imagined to be brought in by the occupant. These included

(1) 40 pounds of books arranged in the bookcase located along the west wall of the fire room,
(2) 7 to 8 pounds of clean clothes hung from hangers in the closets located

[1] 550 g Fluorel L-3203-6 dispersed in 2600 g methyl ethyl-ketone.

in the southeast corner of the fire room,

(3) One book open on the floor near the bookcase, one book on the table in the northeast corner, and a few magazines on the tables and bureau,

(4) About five pictures, on illustration board but unframed, hung on the west, north, and east walls of the fire room.

These items were used in each room and no experiments were conducted to evaluate the influence of the individual items. Their location in all portions of the room was expected to contribute to fire spread when irradiation and heat convection in those areas was large enough to bring about ignition.

Initial trials of the newspaper to be used for ignition were made in metal wastebaskets, about 13 inches in diameter and 14 inches high, that held about one pound of crumpled newspaper. Two thermocouple positions were tried, one located along the middle of the side of the basket, and the other 6 inches up and 6 inches out from the top edge. The position above the basket registered a maximum of only 140 F and the maximum temperature of the side of the basket was 300 F. The contents burned rapidly for about 8 minutes and were reduced to smoldering residue in 13 minutes. It was concluded that this ignition source was too small to assure the ignition desired.

As an alternative, a basket formed from wire mesh was tried with the same amount of crumpled newspaper. The maximum temperature registered by the thermocouple above the basket reached 400 F in brief intervals in this trial and the contents were reduced to smoldering ash in 5 minutes. Nevertheless it was concluded that neither of these trials produced enough heat laterally to be good ignition sources for the trial rooms.

Finally, 3 pounds of newspaper in folded sections equivalent to the local Sunday edition was tried. In this trial the sections of newspaper were strewn in an array from the seat of an upholstered chair, to the floor, and up onto a bed adjacent. To ignite this, crumpled newspaper in the wire mesh wastebasket was lit and the basket put on its side so that it was in contact with the newspaper sections on the floor. This arrangement was closely copied in all subsequent rooms and is shown in the fire room photographs included in the Results and Discussion Section. The location of the ignition source with respect to the room arrangement is shown in Figure 3, identified as "Fire Start".

Each room was assembled at least 24 hours prior to the time of ignition, and heaters were placed in the rooms during this interval to assure reasonable and similar temperature and humidity conditions for the ignition. The values measured for each fire start are shown in Table 3.

INSTRUMENTATION AND SAMPLING

Continuous surveillance of selected products of combustion was arranged for the experimental fires. The considerations and choices for each of the instrumentation and sampling problems are discussed in the following paragraphs.

In order to keep sampling probes as short as possible, one wall of the fire room was built adjacent to the barricade. The access opening in the barricade is close to the doorway to the fire room for the same reason. Samples for monitoring purposes were collected at three locations: inside the fire room at about the middle of the wall adjacent to the barricade and about 36 inches from the floor, at the doorway, and at the ventilation duct. The information collected at each of these was as follows

	Inside Room	Doorway	Duct
Smoke density	–	X	X
Ventilation rates	–	X	X
$CO/NO_x/O_2/HC$	X	–	–
Heat Irradiation	–	X	–
Particulate/aldehydes	X	–	–
HCl	X	–	–
HCN	X	–	–
HF	X	–	–

The sampling probe inside the fire room extended about 3 feet into the room and was located in the vicinity of the chair and wastebasket. A pump provided a continuous flow of gases through this probe so as to maintain sufficient flushing of the sampling system.

Temperature Measurements

Chromel-alumel thermocouples, 26 gage, were placed in a number of locations intended to show the time of participation of the individual items of furniture in the fire, and the overall progress of the fire throughout the fire room. The thermocouple locations are summarized in Table 4. Generally, the thermocouples were placed with the bead in contact with the surfaces present at the location, or, in those cases where practical, the bead was inserted through a hole in the upholstery so that it responded to the temperature reached by the textile and not directly to the incident heat irradiation.

Each thermocouple was attached to a channel of a Digitem recorder, Model DAS-1, arranged to scan the channels at the rate of about 2½ channels/second and print out the millivolts found for each. Thus, in each experiment the entire set of thermocouples was scanned and recorded every 14 to 20 seconds depending on the total number of thermocouples used. The data were transcribed to magnetic tape and then plotted by the computer. The entire set of data assembled in this way are presented in Appendix A presented in Appendix A of the NASA Special Publication edition.

Ventilation

The fire room may be ventilated in several different ways and the choice may be anticipated to have considerable bearing on the results obtained in the experimental fires. In our prior experience, air entered the fire room by way of a doorway and, because it was cooler than the room air, flowed along the floor, thus forming a layer through which it was possible to see entirely across the fire room. At the same time the accumulation of smoke in the room had reduced visibility above this layer to only a few feet and the smoke could be seen leaving the room by the same doorway that admitted the fresh air. Thus the door openings served simultaneously to supply air to the fire and to vent the combustion products; in many circumstances, the doorway is the major opening for such transport to occur. For the present program the doorway was planned to be the sole source of air to the fire room and an open window in the training tower, see Figure 3, insured a continued supply of fresh air for that purpose.

One ventilation duct opening was installed in the fire room which was connected to 8 feet of duct terminating in the space above the fire room. This duct was used without blower so that the relative importance of natural convection through this opening could be observed. At the normal blower circulation rate of four room volume changes per hour, the impressed circulation of a blower was expected to be relatively unimportant during the fire periods compared to the circulation through the open doorway, and would serve only to shorten somewhat the smoldering periods that were anticipated.

Transducerized pitot tubes offered the best means for measurements of air flows in the range of temperatures anticipated. The flow measurements at the doorway were of greatest importance and three pitot tubes were located there. Another pitot tube was located in the ventilation duct. The locations of these pitots within the openings must be regarded as compromise choices to provide maximum information with a minimum number of sampling sites. For the doorway, the opening was visualized as being divided into thirds: the middle and top thirds were monitored at the centers to provide data on the efflux of combustion gases, and the bottom third was monitored at its center to provide data of air flow into the room. For the ventilation duct, a single pitot was located at the center of the opening at the outlet end.

For this purpose, 1/8 inch Kiel probes were used with Statham Model PM 197 pressure transducers for readout of the pressure data. This arrangement of apparatus supplied an output of 1 volt at 15 to 20 ft/second flow rates and could be used to estimate flows between about 10 and 70 ft/second under these experimental conditions.

Heat Irradiation

Heat irradiation levels were measured by a Hy-Cal radiometer, Model R-2040-15-072, equipped with a sapphire window as a radiation filter and located in the wall adjacent to the doorway and about 15 inches above the floor. The calibration curve supplied with the instrument shows the radiometer to have an output of 10 millivolts at an incident heat flux of 15.7 Btu/ft^2 · sec and was used to measure flux levels between about 0.1 and 5.6 Btu/ft^2 · sec during the experiments. An enlargement of the calibration curve was used to obtain estimates for low heat flux levels during the initial time periods of the experiments between 0.1 and 1.0 Btu/ft^2 · sec. The time dependent curves obtained in this way are of uncertain precision but serve to characterize the differences in the initial buildup of the experimental fires. The expanded curve, with the values used for its construction, is shown in Figure 4.

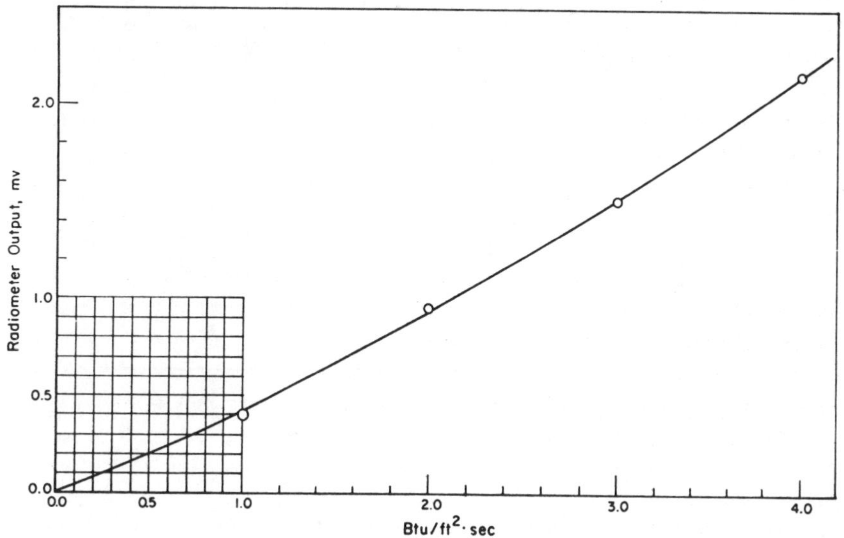

Figure 4. Expanded calibration curve for radiometer.

Since this radiometer could receive radiation over a 150° angle, the irradiation levels obtained are regarded as essentially integrated values of the heat flux at that site.

Smoke Density

Smoke densities were measured vertically across the doorway and the ventilation duct so as to obtain relative values for the smoke being convected through each of these openings. Both the light sources and the photomultiplier tubes were protected from smoke deposition by nitrogen gas that flowed past the optics and then passed through tubulatures extending along the light path. The photomultiplier circuit was based on the 1P21 tube and Signetics operational amplifier No. 741; the wiring diagram for that circuit is shown in Figure 5. Typical plots obtained for the experiment of 10/17/72 are illustrated in Figure 6. Light transmission values between 100 percent and 0.01 percent were recorded continuously at each location.

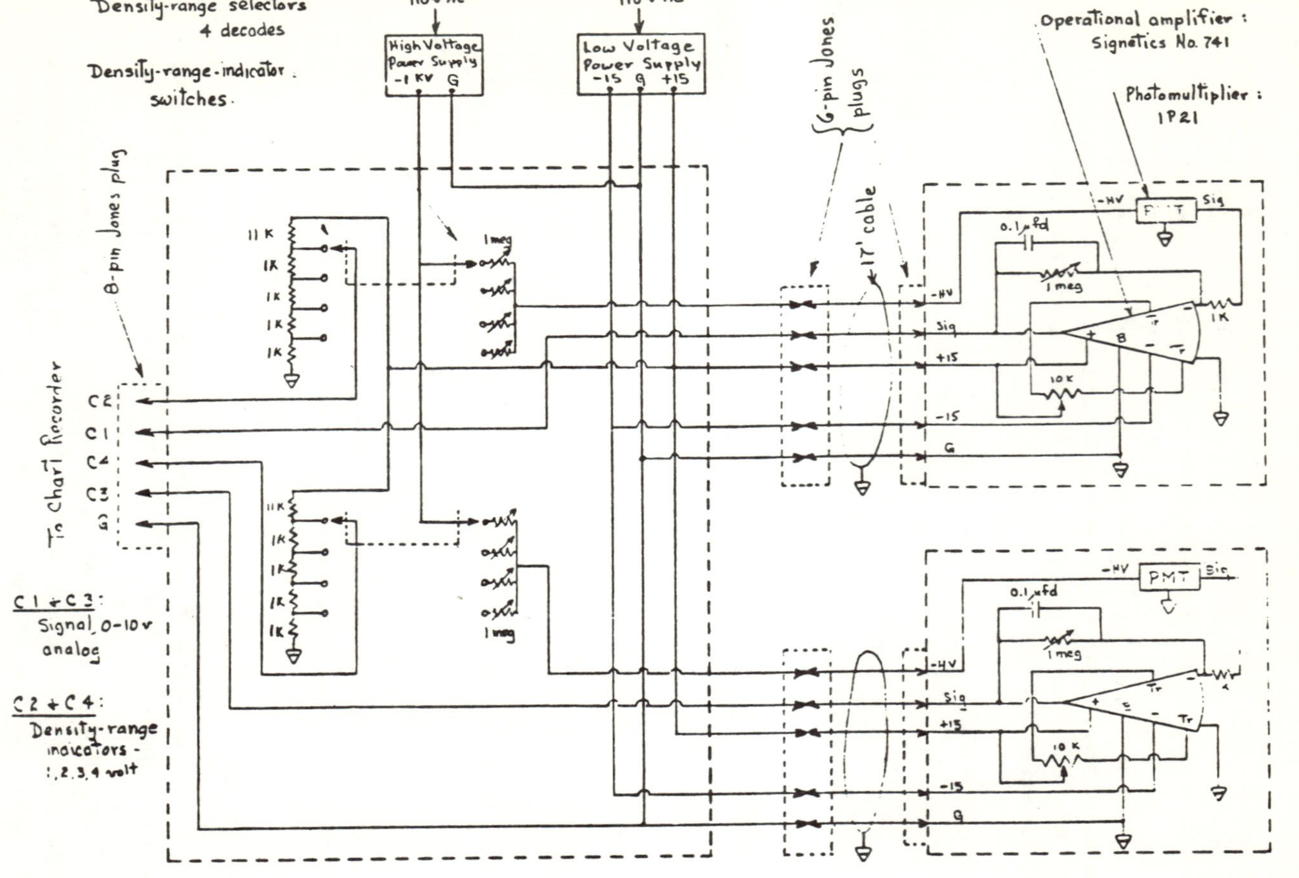

Figure 5. Schematic wiring diagram for smoke density measurement.

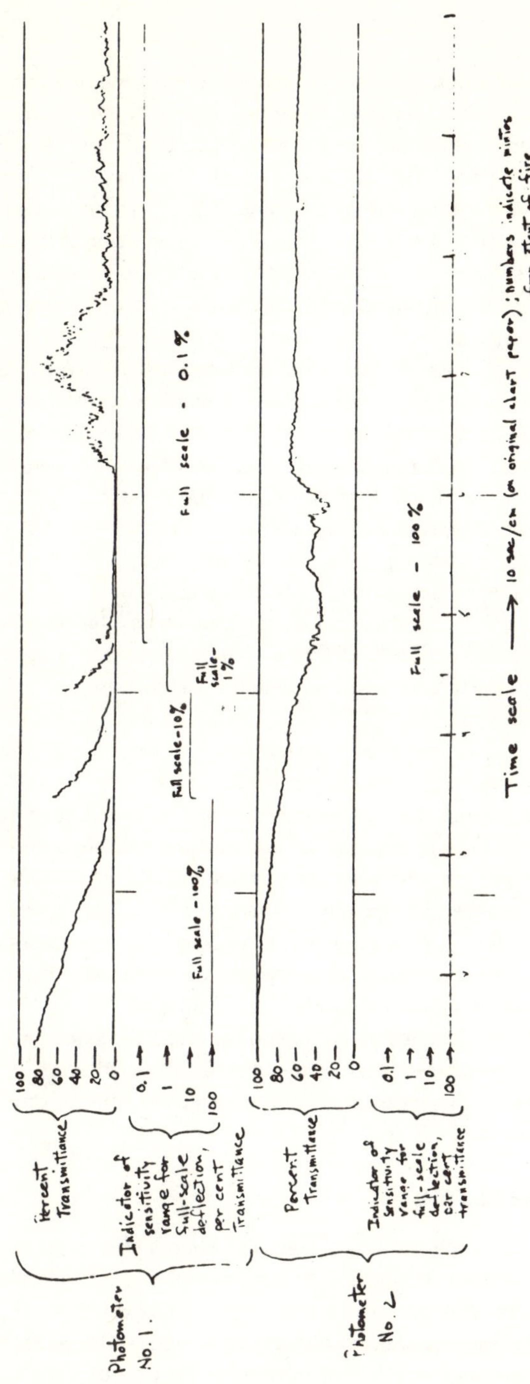

Figure 6. Typical photometer traces.

Gas Analyses

Table 5 lists the instrumentation used for continuous monitoring of a number of gaseous components of the combustion gases. For the hydrocarbon measurements a heated sample line was used to maintain sample temperature at about 300 F until the flame ionization hydrocarbon analyzer was reached. This tended to minimize loss of hydrocarbon by condensation during sampling. All other components were analyzed from a common stainless steel manifold. Prior to entering this manifold the gases passed through a large ice-water trap and particulate trap so as to maintain suitable moisture levels for the analytical instrumentation. An additional dry-ice trap was employed for NO analysis.

In addition, a number of constituents were analyzed as integrated samples representing the entire fire period from ignition to the start of water application. These integrated samples were collected at constant flow rate during the fire period so that the results obtained represent time-averaged samples. The constituents measured in this way were HF, HCl, HCN, and aliphatic aldehydes.

For HF analyses, the sample was drawn through a Vycor probe so that the HF was largely converted to SiF_4 according to the method of Dorsey and Kemnitz [1]. This volatile gas was then absorbed in impinger tubes containing H_2O and the resulting solution analyzed for F^- by the fluoride ion electrode. The only interference reported for this electrode is due to OH^- to which the electrode is about 1/10 as sensitive as for F^-. This interference was eliminated by adjustment of pH of the samples collected.

Cyanide was analyzed by ion-selective electrode which shows interference to sulfide ion (which must be absent), chloride ion (if 10^6 times the CN^- concentration) bromide ion (if 5000 times the CN^- concentration), and iodide ion (if 10 times the CN^- concentration). The absence of sulfide was qualitatively confirmed in these samples and the analyses for chloride plus bromide ion do not indicate significant interference. The CN^- concentrations are believed valid to ± 10 percent.

Chloride ion was determined by titration with $Hg(NO_3)_2$ as a measure of the HCl produced. For this purpose CN^- was first destroyed by peroxide. Any bromide present would be determined along with chloride.

Aliphatic aldehydes are of considerable toxicological importance in combustion processes and have been determined by forming the bisulfite addition product. Ketones are also measured by this method. The methods used for this determination are given in the next section.

Particulate and Condensible Materials

A major source of difficulty in sampling condensible vapors arises from the need to separate such portions from the significant quantities of soot, tars, water, and smoke particles that are usually present. For water soluble vapors such as HF, HCN, and HCl the hot gases were filtered and led directly into water-filled gas scrubbers for collection based on their solubility in that solvent. In the case of organic vapors the basis for separation is much less well-defined. For this program it was decided to try the particulate sampling train designed by the Environmental Protection Agency for collection of particulate and condensible vapors from flue gases of combustion systems. Thus, using this sampling train, a gas sample was drawn through tubes packed with glass wool and then through a glass fiber filter, all at 250 F, where solids and materials condensible at 250 F were removed. The gas stream then flowed through ice-cold sodium bisulfite solution contained in Greenfield-Smith impinger tubes where aliphatic aldehyde vapors and some ketones were retained as the bisulfite addition product. The method of Wilson [2] was used for this analyses and the results were expressed as formaldehyde equivalent to the addition product determined.

The components of the sampling train used for this purpose are shown schematically in Figure 7.

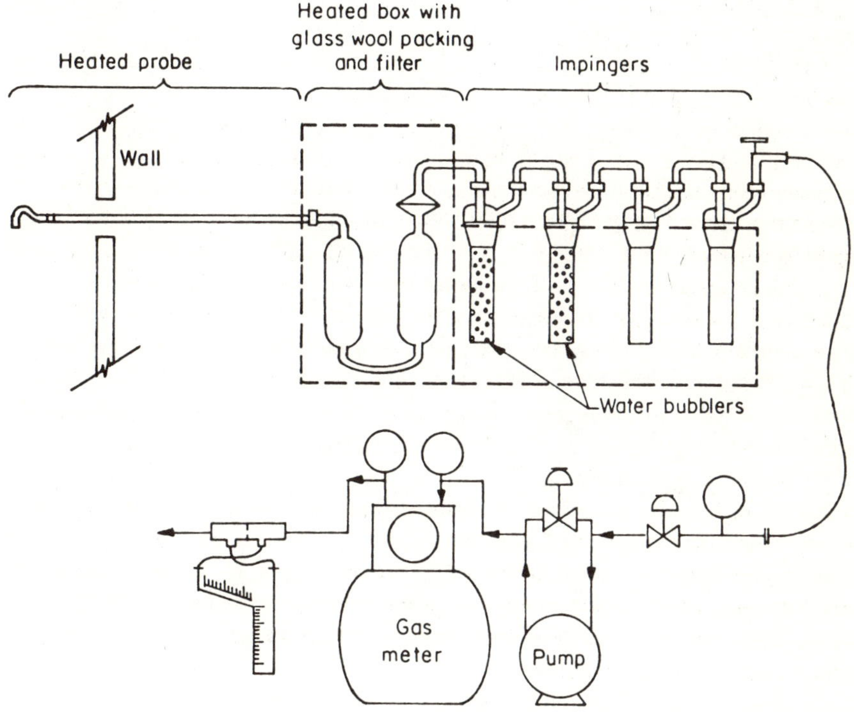

Figure 7. Particulate and aldehyde sampling train.

Photographic and TV Monitoring

A photographic record was made of each experimental fire using 16 mm Ektachrome film. Since the burn times for the rooms could not be predicted, the camera was operated at about four frames per second so that fires of nearly 1 hour duration could be monitored with a 400 foot reel of film. The camera was placed in a box of 1/2 inch Transite having a Vycor window. The box was continuously swept with cylinder nitrogen which vented along the outer surface of the window thereby protecting the window against deposition of smoke during the fire. The location of the camera just outside the doorway of the fire room is shown in Figure 3.

These arrangements for photographic coverage were adequate for all except the Typical Room fire during which the film melted as it threaded through the camera and the photographic record for the final portion of that fire was lost. In subsequent use the box was further protected with a loose-fitting cover of asbestos cloth as an additional heat barrier.

Since the fires could not be observed directly, some additional means was needed to permit the observation needed to guide those recording data of the fire. For this reason a TV camera was also used with a monitor for viewing and with a TV tape unit to provide a record of the pictures obtained. This record, preserved in "real time" was used to provide a time scale for the photographic film record. The TV camera was placed alongside the photographic camera in a similar Transite box.

THE EXPERIMENTAL PROCEDURE

Each room was built in sections and stored so that it and the appropriate furnishings formed a "kit" for the experimental fire arrangement. The walls were made in sections about 3 ft X 8 ft, the ceiling in three pieces about 3.5 to 4 ft wide and 11 ft long, and the floor in two pieces about 5.5 ft X 11 ft arranged for bolting together to form the room. Floor coverage and wall and ceiling surface treatments were generally applied after room assembly was completed, and then the furnishings were put into the room and thermocouples placed and tested. Heaters were placed in the room during construction and heating was continued for at least 24 hours after furnishing was completed so as to adjust the temperature and humidity of the room for the fire trial.

Each instrumental analysis or sampling procedure was calibrated or adjusted on the morning of the fire trial and the record of each measurement was started simultaneously with the signal to ignite the newspaper in the wastebasket. All records except those of temperature were discontinued when sufficient room involvement had occurred to demonstrate the fire response of the ensemble under trial.

At the signal to discontinue, a window of the training tower was opened from the outside and firemen of the Columbus Fire Department entered the corridor area for approach to the fire room. As soon as preliminary fire control activities were completed, they removed the pitot tubes and camera equipment from the doorway and proceeded to put out the fire. As soon as practical after this was completed, the door to the personnel area was opened and photographic records of the remains of the contents of the fire rooms were obtained.

A schedule of one fire per week proved practical and was maintained for the four trials that were made.

RESULTS AND DISCUSSION

The visual evidence provided by TV and photographic coverage of the four room fires shows clearly that these rooms responded very differently to a common ignition condition. In particular

(1) The Typical Room ignited easily and burned rapidly in just the same manner observed for the preliminary room fire used to establish the ignition conditions. See Figures 8 and 9. The fire was stopped after 8.5 minutes because of the excessively large degree of fire development.

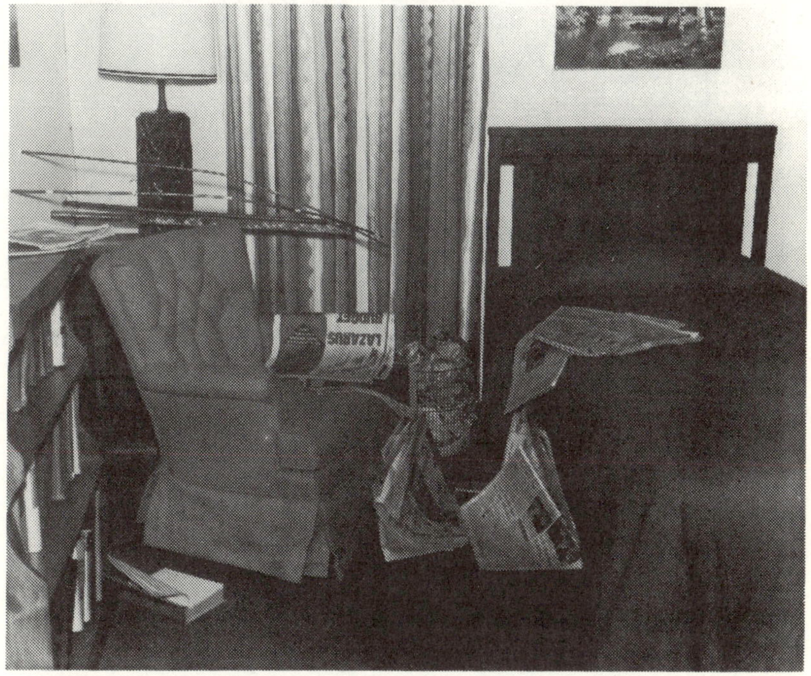

Figure 8. "Typical" room before ignition.

(2) The Improved Room showed substantial improvement over the Typical Room in that there was slower fire spread, but the relatively complete consumption of the room that resulted, and the large amounts of smoke that was generated, made it clear that substantial further improvement in performance was needed. See Figures 10 and 11. The fire was stopped after 29 minutes when complete room involvement had produced the high degree of damage shown.

Figure 9. "Typical" room after fire.

(3) The Space-age Room did not ignite under the common ignition condition and so demonstrated a substantial improvement in fire resistance for those room components closest to the ignition fire, e.g., rug, chair, drapes, and bed. This fire essentially burned out in 12 minutes. The second and larger ignition arrangement used for this room ensemble demonstrated that this room can burn, but the time required for the ignition to take place confirmed the substantial improvement in fire resistance available with this room ensemble. See Figures 12, 13, 14, and 15. Thus, the second fire burned for 27 minutes before room involvement began, apparently because of a hole burned in the floor in the northwest corner of the room. In 6 minutes more the high degree of damage shown in Figure 15 was produced.

Figure 10. "Improved" room before ignition.

Figure 11. "Improved" room after fire.

Figure 12. "Space-age" room before 1st ignition.

Figure 13. "Space-age" room after 1st fire.

Figure 14. "Space-age" room before 2nd ignition.

Figure 15. "Space-age" room after 2nd fire.

(4) The Mixed Room ensemble demonstrated a much moderated fire development obtained by placing the fire-retardant bed from the Space-age Room in a room ensemble otherwise identical with that of the Typical Room. This ensemble illustrated the improvement in fire spread available by careful placement of fire-retardant materials in the important paths for fire spread of an otherwise ordinary room ensemble. See Figures 16 and 17. This room was allowed to burn for 1 hour before the fire was extinguished.

Figure 16. "Mixed-load" room before ignition.

These observations reveal that the ignition conditions chosen have been suitable for demonstration of the differences in fire performance for the four room ensembles. This demonstration was the primary goal of the program.

However, the measurements made during the burning of each room ensemble provide the basis for a much more detailed analysis of performance than the above observations. In order to do this, some basis must be chosen for establishing levels of importance for the numerical information that has been generated. In the following pages a few selected criteria are discussed and used for assessment of room performance and life hazard.

TEMPERATURE PROFILES AND FIRE SPREAD

In all, 126 time-temperature plots were obtained from thermocouples placed in various locations in the four fire rooms. The locations of the thermocouples are

Figure 17. "Mixed-load" room after fire.

listed in Table 4, the 126 time-temperature curves are shown in Appendix A of the NASA Special Report, see note p 115.

An approximate description of the time-temperature plots obtained from this research program might be made by noting that active fire periods usually showed bulk temperatures above about 1000 F whereas in-between these fire periods the smoldering conditions usually appeared in the 200 to 500 F range. Intermediate temperatures in the range of 700 F usually were observed in times of rapid transition between active fire and smoldering conditions.

Temperatures of about 700 F also appear to lie in the middle of the range where cellulosic and organic polymers are pyrolyzed to yield combustible gases and vapors. Thus in a review of some earlier publications, Wagner [13] noted that four distinct temperature zones have been given for the thermal decomposition of wood. Two of these zones, covering the range up to 536 F were characterized by the appearance of noncombustible gases, primarily H_2O, arising by endothermic processes. The third zone, 536 F to 932 F was described as that of active pyrolysis under exothermic conditions that led to formation of combustible products such as CO, CH_4, tars, and smoke. These gases and vapors were ignitable under piloted conditions. Above 932 F the charcoal residue glowed and was consumed.

Similarly, for organic polymers, Wall [3] lists kinetic data from laboratory studies describing the thermal decomposition of a variety of polymers in the temperature range of about 425 F to 1022 F, in which the organic monomer is one of the decomposition products. This range of pyrolytic conditions can be assumed to include the formation of combustible decomposition products. Thus, it may safely be concluded that pyrolysis of these materials occurs generally in the same temperature range as the active pyrolysis range given previously for wood. These considerations have led in this work to the choice of 700 F as an approximate representation of the boundary between active fire zones and smoldering. In Figures 18, 19, 20, and 21, the apparent boundary between fire and smoldering is shown for selected times during each fire experiment.

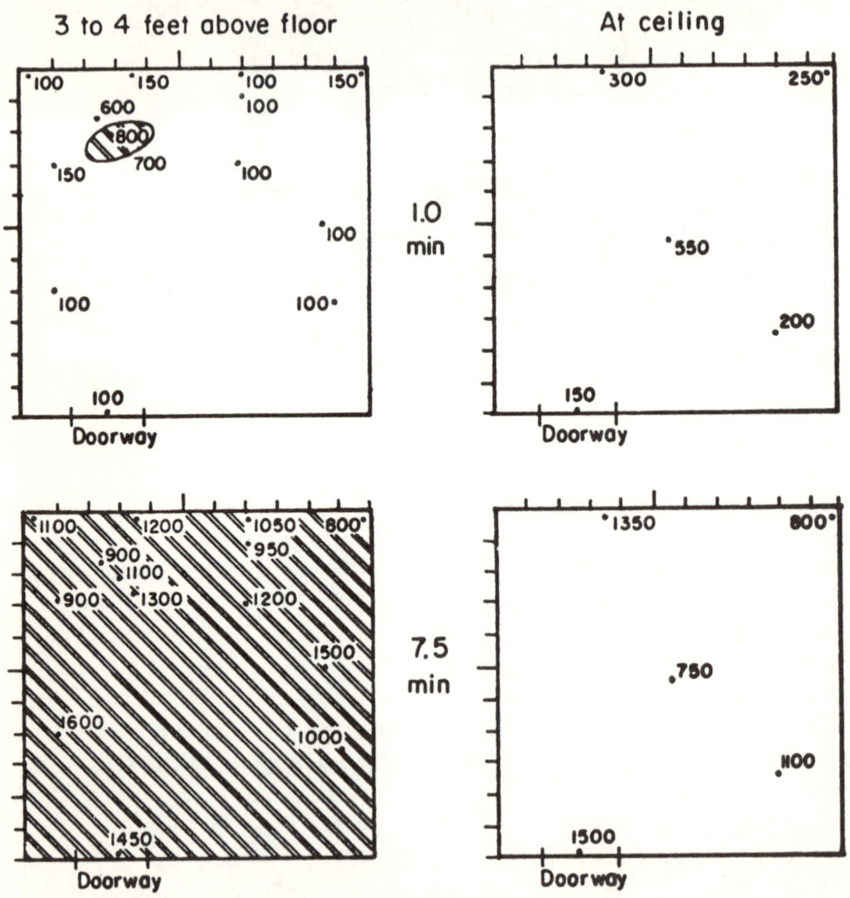

Figure 18. Fire spread diagrams for typical room. (700 F contour shown only at 3 to 4 ft level)

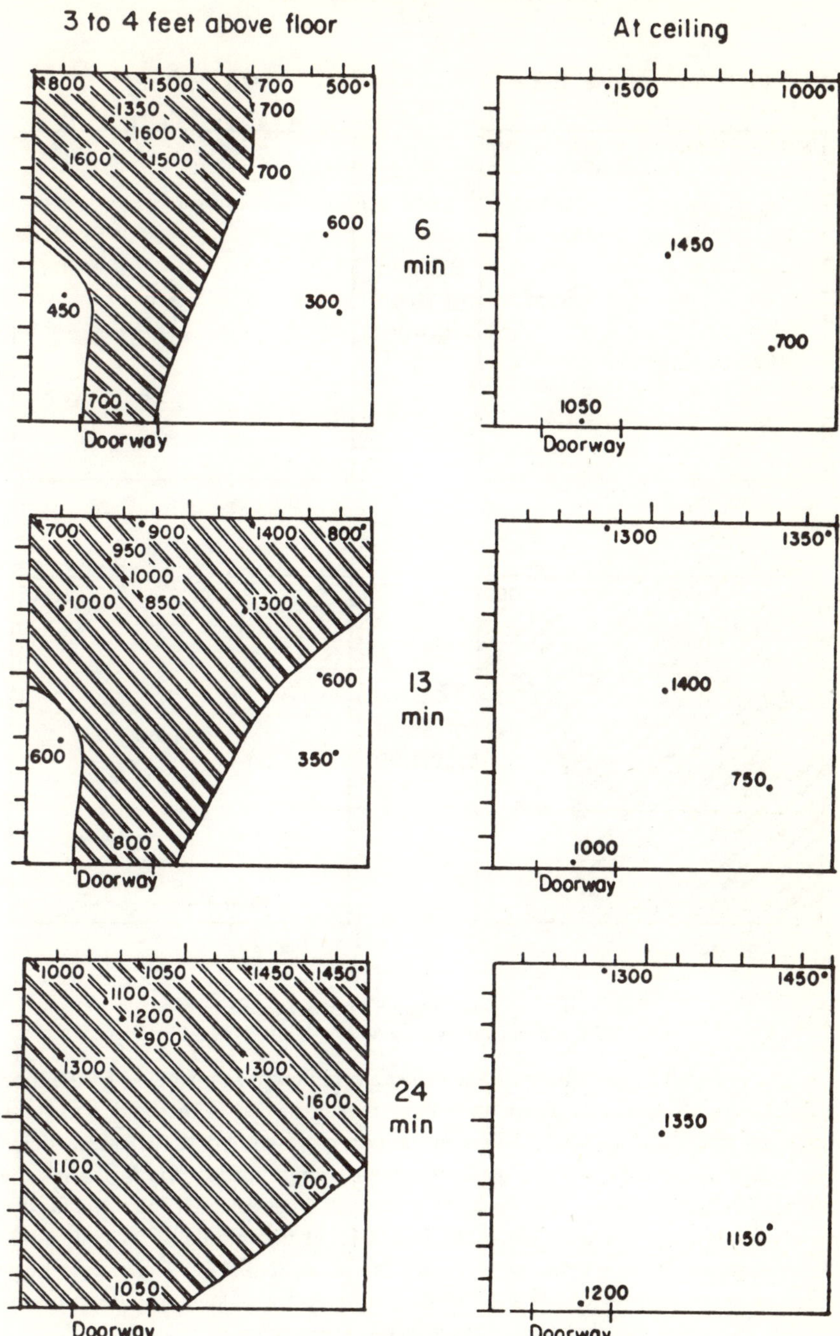

Figure 19. Fire spread diagram for improved room. (700 F contour shown only at 3 to 4 foot level)

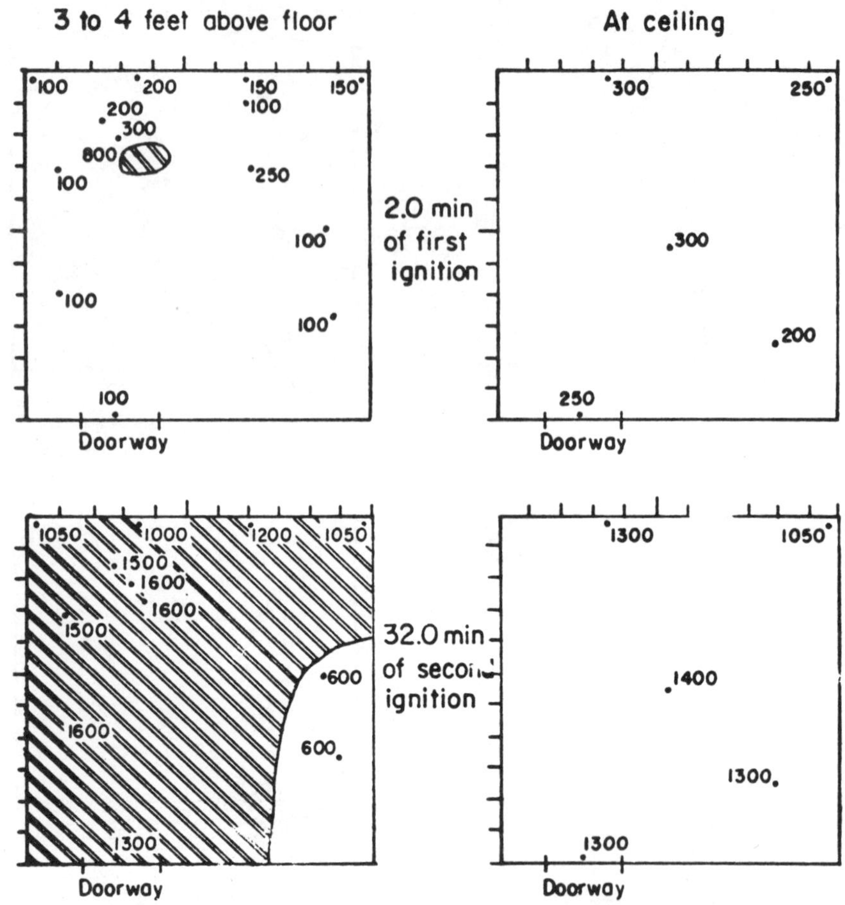

Figure 20. Fire spread diagram for space-age room.
(700 F contour shown only at 3 to 4 foot level)

In those Figures, the square of each diagram represents a plan view of the room at the specified elevations above the floor, that is, at 3 to 4 feet and 7 to 8 feet above the floor (at ceiling). These elevations represent zones where sufficient thermocouples were located to give some approximate idea of the isotherms that existed. The temperature conditions are shown at times selected from the plots of gas temperatures at the sampling probe (No. 21), at the middle of the ceiling (No. 14), and above the open doorway, inside the room (No. 16). The times chosen represent times during which gas temperatures at these locations were passing through maximum values and are taken to represent times of peak fire activity. In each diagram, the shaded areas represent those in which temperatures are everywhere above 700 F and which are interpreted to represent the active flame area.

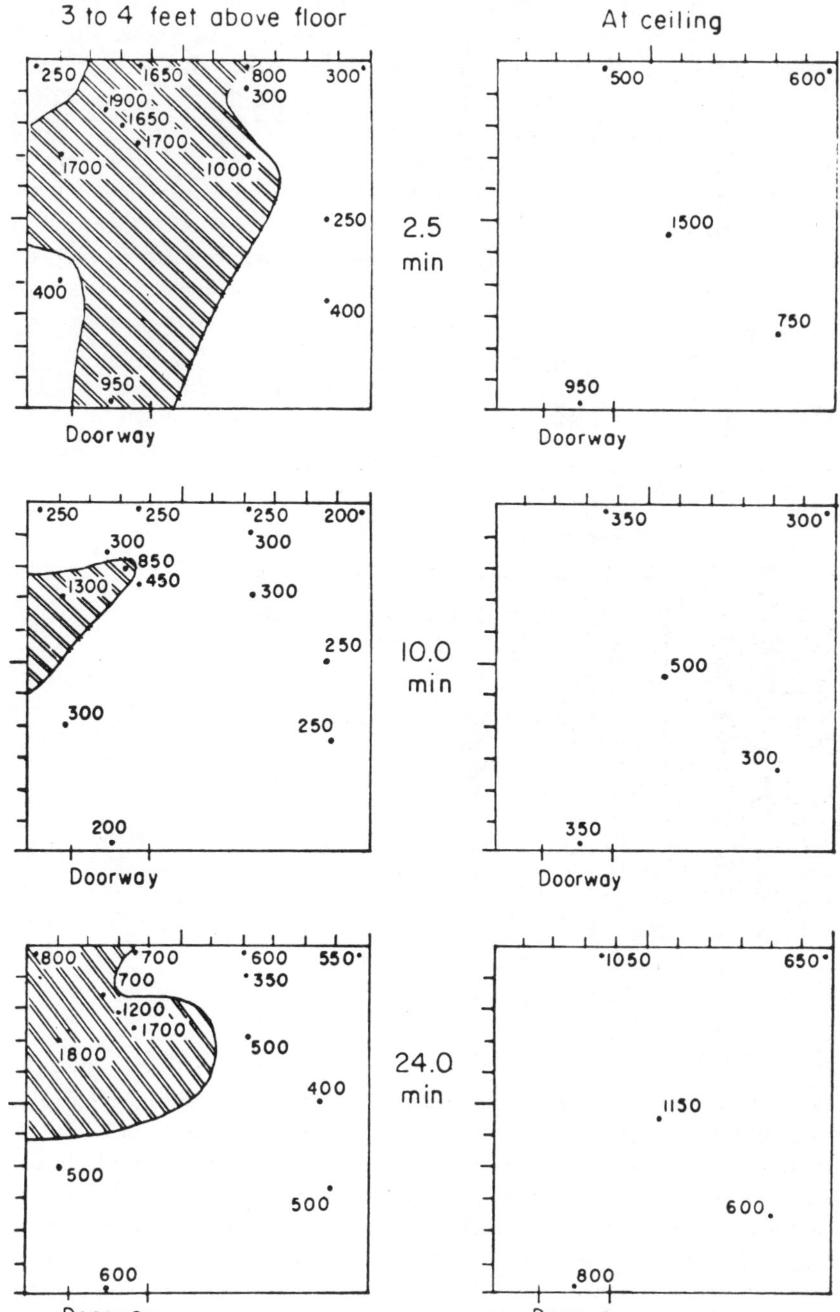

Figure 21. Fire spread diagram for mixed-load room. (700 F contour shown only at 3 to 4 foot level)

Although the boundary curves cannot be drawn with any great precision, and then only in the 3 to 4 foot elevation data, comparison of the four rooms reveals the differences in fire spread noted from visual evidence.

It is of interest to compare these times of peak fire activity with another estimate of fire activity based on the variations in oxygen concentration in the fire rooms. Thus Huggett [4] has presented an approximate correlation of flame spread rates over solid fuels with the heat capacity of the gas environment per mole of O_2. This correlation, shown in Figure 22 for paper, has been examined for other fuels and Huggett concludes [5] that organic fuels cease to burn when the heat capacity of the atmosphere reaches the range of 40 to 50 cal/°C per mole O_2 at atmospheric pressure. Indeed, the correlations for other fuels[2] frequently show zero fire spread at even lower heat capacities per mole O_2 than the correlation shown for paper. Thus this correlation for paper may be assumed to represent the greatest tolerance that will be met for continued fire spread with oxygen depletion.

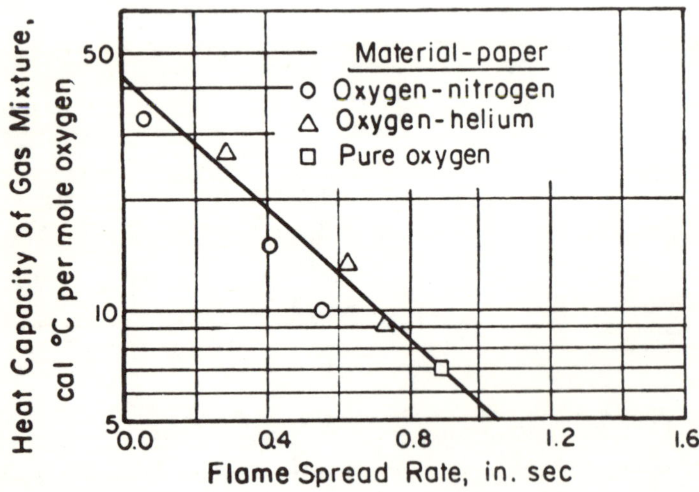

Figure 22. Rate of flame spread for paper as a function of heat capacity [4].

For the present purpose, the implications of the above correlation can be visualized if we note that the gas mixtures of interest are largely N_2, CO_2, and O_2 and that the molar heat capacities of such mixtures are relatively insensitive to variations in composition in the range of interest experimentally. Thus estimated values of $C'p$, the heat capacity in calories per degree C per mole O_2, appear as follows

20.9 percent O_2 + 79.1 percent N_2 = 33 cal/°C mole O_2
16 percent O_2 + 84 percent N_2 = 43
16 percent O_2 + 4.9 percent CO_2 + 79.1 percent N_2 = 44

[2] Data presented at NFPA meeting in St. Louis, Missouri (1973).

Based on the correlation for paper shown in Figure 22, C′p values of about 44 represent almost a zero fire spread rate. Compared to the fire spread rate at C′p = 33 for normal air this represents perhaps a ten-fold decrease. On this basis, the active fire periods in our fire experiments might be expected to occur only when ambient oxygen concentrations are above about 16 percent. All other time periods may be anticipated to represent weak flame and smoldering conditions when smoke and toxic organic vapors are expected to evolve. The estimate of periods for active fire spread based on the experimental oxygen concentration curves are compared in Table 6 with the times selected to represent peak fire activity based on heat accumulations in the room. These estimates for fire spread and peak fire activity will be compared with the time variations in heat irradiation and smoke accumulation in later sections.

In summary, temperature distributions in the fire rooms can be used to describe areas of active burning at selected times, and it has been concluded that 700 F can be used as a reasonable estimate of the boundary between active fire and smoldering zones. In this way the distributions of fire in each room have been plotted for selected times corresponding to maximum accumulations of heat in the room. These times for maximum heat accumulation are taken to represent times of maximum fire activity. Published data, showing the dependence of fire spread rates on ambient O_2 concentrations, lead to the conclusion that such active fire periods probably would be confined to those times when ambient O_2 is $>$ about 16, so that the experimental O_2 concentration curves represent an estimate of fire and smoldering periods in each fire.

IRRADIATION EFFECTS

The irradiation curves shown in Figures 23 and 24 graphically illustrate the differences in fire intensity obtained for the four Fire rooms. The low-range heat irradiation curves shown in Figure 25 were obtained using the expanded calibration curve discussed on page 129, and have superimposed on them two irradiation levels, marked A and B, that serve to help in interpretation of these irradiation curves.

The level B, represents 0.18 Btu/ft^2 sec and is said [6, 7] to be the irradiation level for exposed skin that will produce pain after 1 minute exposure. The other level, A, represents 1.1 Btu/ft^2 sec and is the minimum level found [8, 9] to produce piloted ignition of wood. Level B can be taken to represent a threshold condition of discomfort, or a stimulus to awaken an occupant of the fire room in question.

With this basis for specification of critical environmental conditions for an occupant, the four room fires can be compared on the basis of irradiation hazard. Thus in Figure 25, the irradiation level in the Typical Room rises so rapidly to the piloted ignition level that severe burning of skin might be anticipated almost as soon as the pain is sensed. If we imagine an occupant of the room about as far from the ignition site as is the radiometer, about 9 feet, the hazard due to irradiation

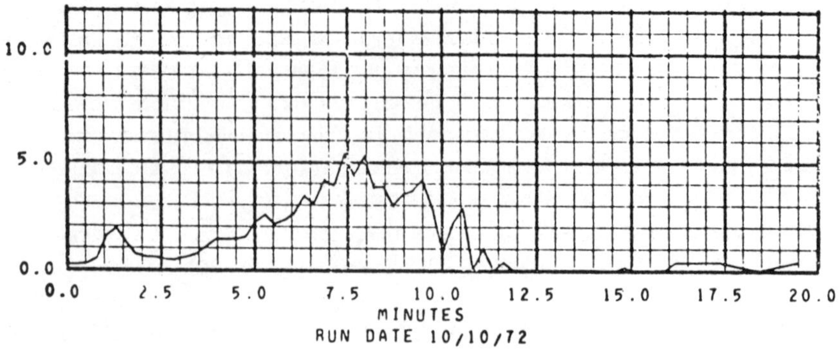

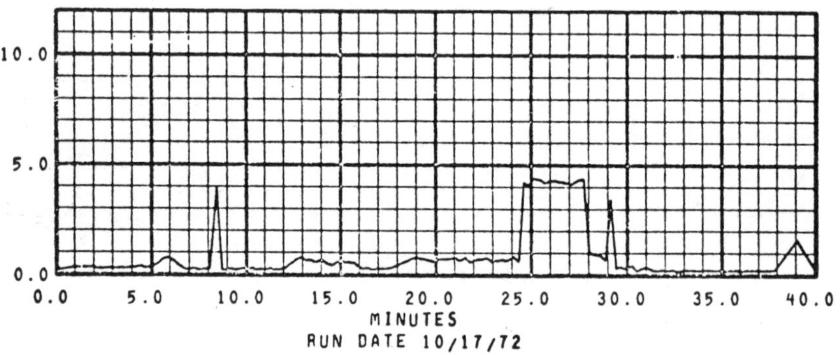

Figure 23. Heat irradiation curves for the "typical" and "improved" room fires.

becomes appreciable in the Typical Room after about 50 seconds and remains so for the duration of the fire. Comparison of this irradiation curve with the curves for the other experimental room fires indicates a substantial decrease in irradiation due to the slower burning of the Improved Room, and an appreciable delay in the pulse of heat irradiation for the Mixed-load Room which is similar to the Typical Room except for the insertion of a much improved bed from the Space-age Room. The Space-age Room curve represents essentially the heat from the igniter load and shows no appreciable hazard to the occupant at the assumed distance.

It is of interest to note that the times of peak fire activity selected in Table 6 for the Typical Room compare well with the peaks in the heat irradiation curve for this room in Figure 23 and 25.

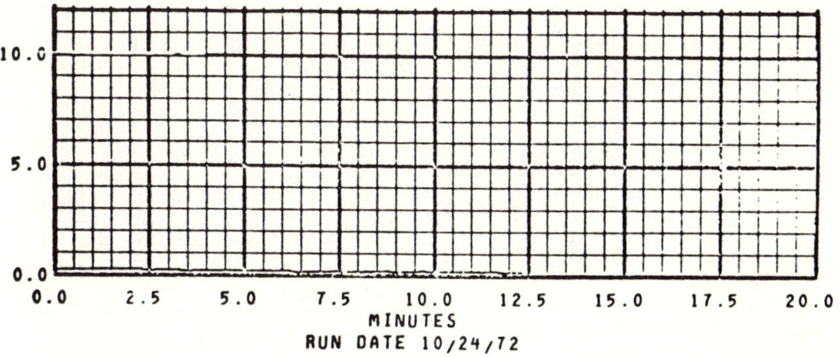

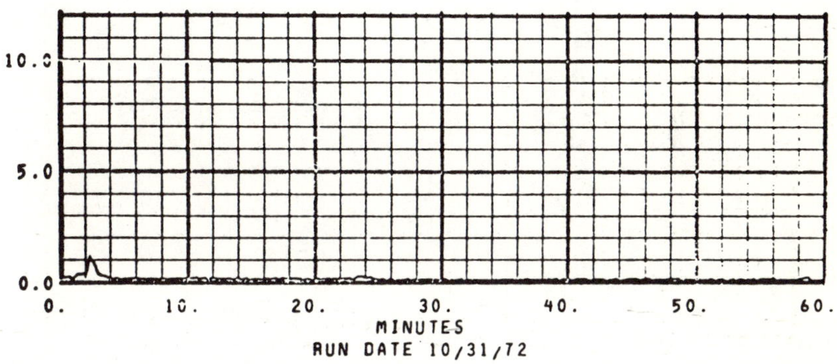

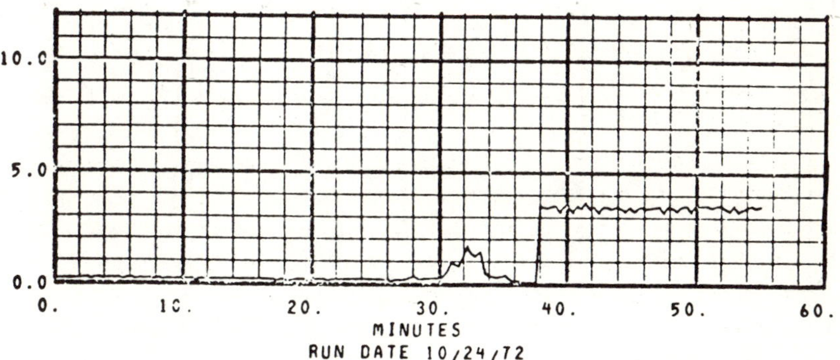

Figure 24. Heat irradiation curves for the "space-age" and "mixed-load" room fires.

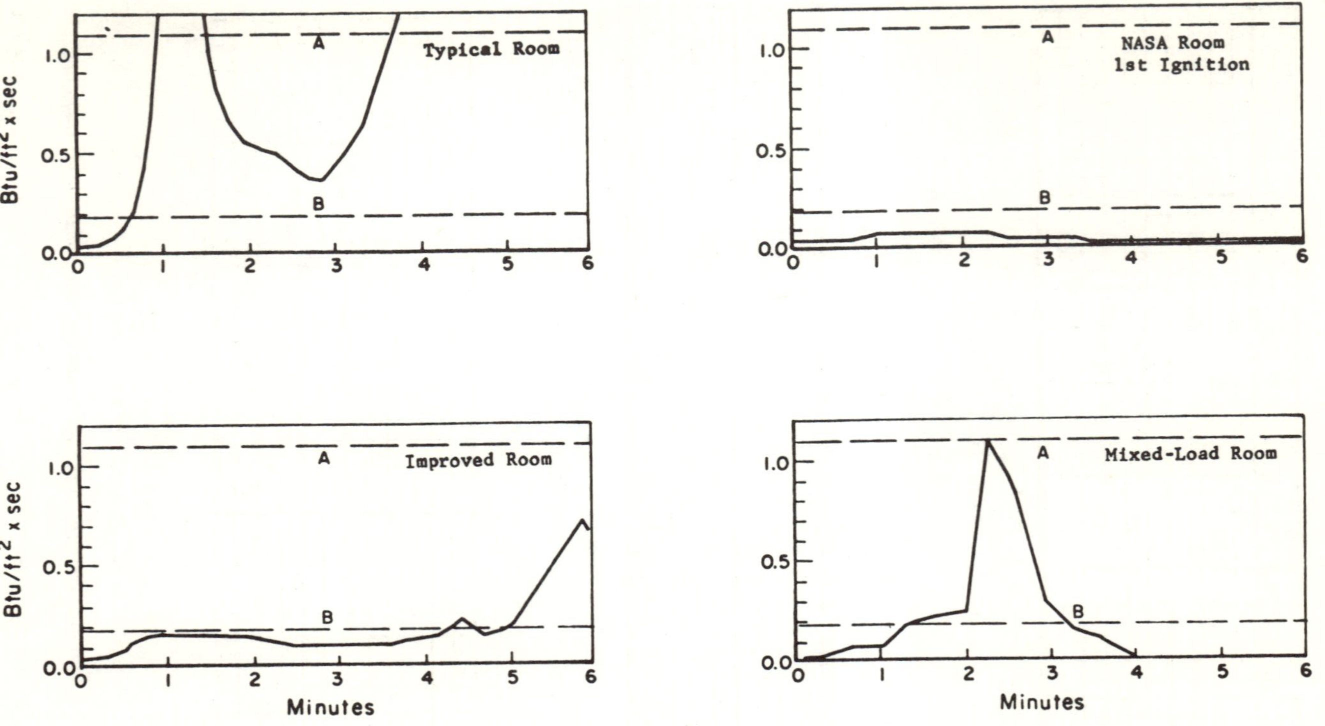

Figure 25. Comparison of heat irradiation levels in the initial period of each room fire.

SMOKE ACCUMULATION

The optical densities of the smoke measured in the doorway and in the ventilation duct of each fire room are shown in Figure 26 and 27. In each case where substantial fire spread occurred, the initial flareup of the fire is followed by oxygen depletion and consequent evolution of large amounts of smoke which temporarily reach light transmission levels of 0.01 percent or less (optical densities > 4). If we are to assess the hazard due to smoke by estimation of the changes in visibility that occur, it is necessary to specify the distance above the floor at which viewing is to be done.

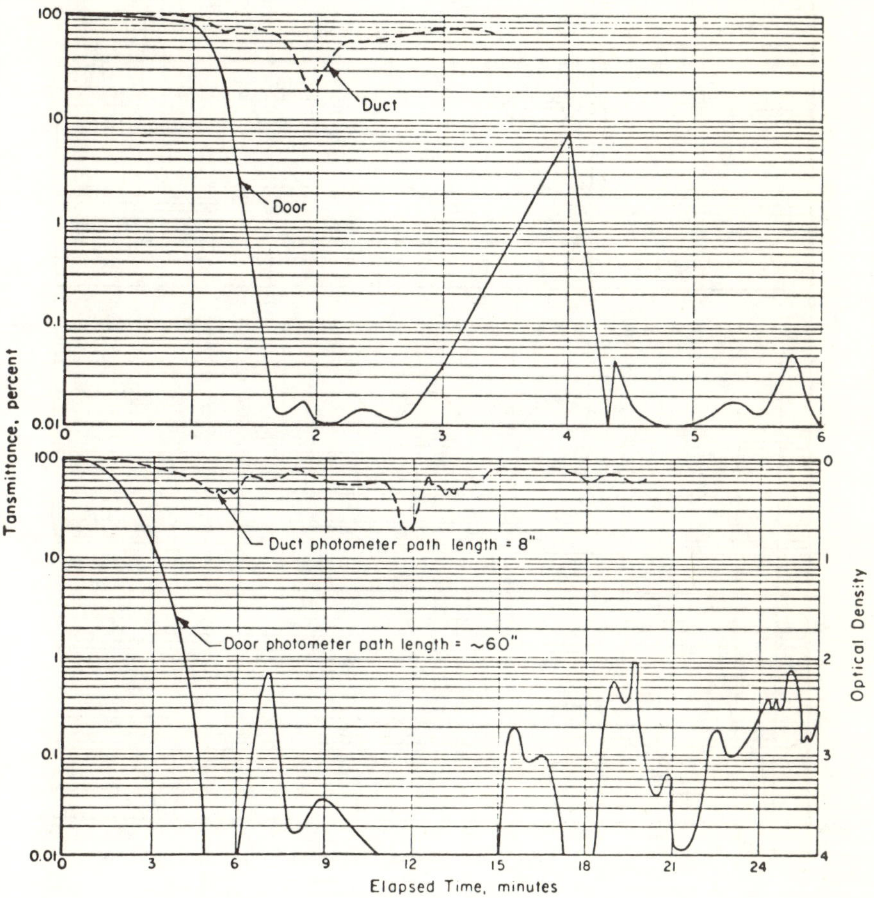

Figure 26. Smoke density curves for typical and improved rooms.
Upper Figure - Typical room
Lower Figure - Improved room

Because of the circulation pattern for air into the fire room and the convection of hot gasses out, the smoke accumulation is layered horizontally, usually with

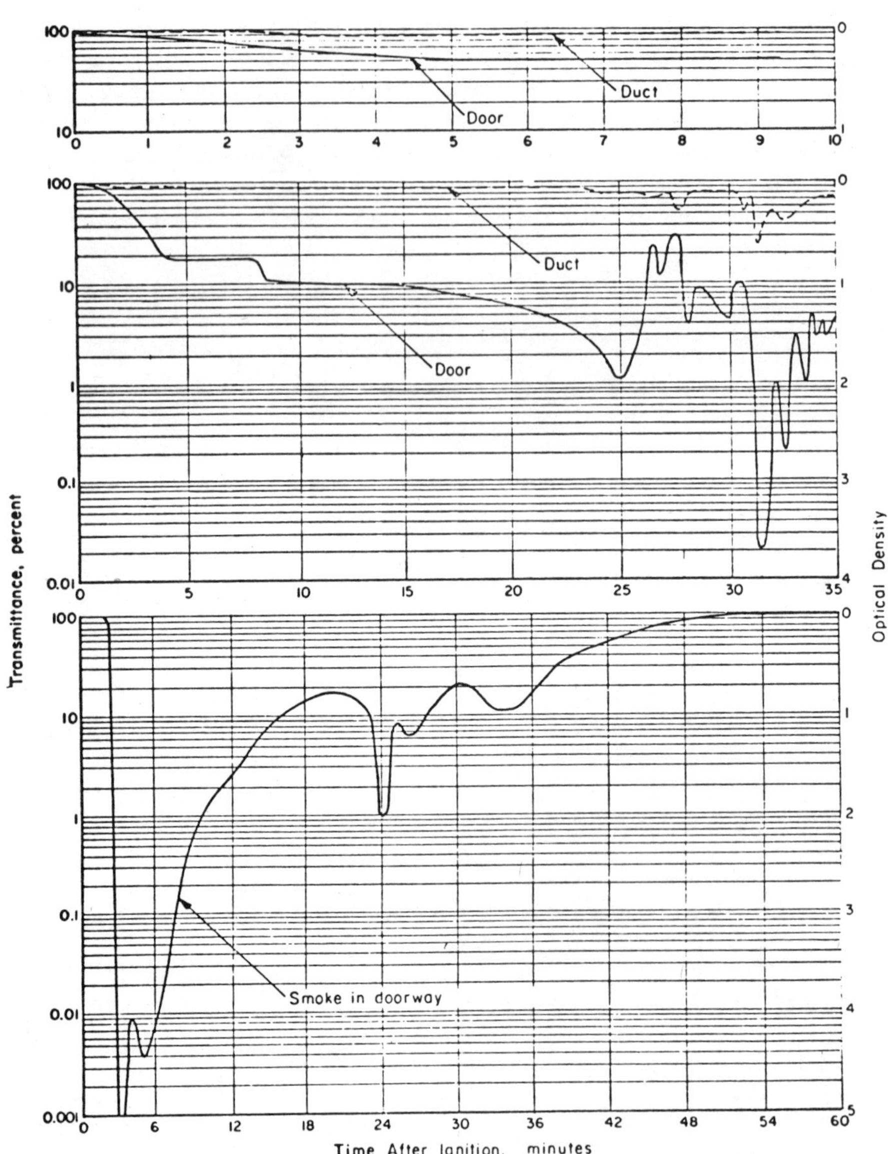

Figure 27. Smoke density curves for space-age and mixed-load rooms.
Top Figure - Space age room, first fire
Middle Figure - Space age room, second fire
Bottom Figure - Mixed-load room

maximum density near the ceiling and minimum near the floor. In the present research the smoke density was measured vertically so that the results represent an averaged value. Assuming that this averaged value can be taken as typical of normal viewing elevations, somewhere between 3 and 6 feet above the floor, the experimental curves can be used in assessment of hazard along with the heat and toxic gas data. The curves of Figures 26 and 27 show that in each fire the smoke density increases very rapidly soon after the first large flare-up of fire and then oscillates considerably with fire activity and convection.

In an assessment of the relationship between visibility and smoke density, Silversides [10] showed that at an optical density of 0.4/meter a standard light source is just visible at a distance of about 4 feet. With this relationship, the time at which light transmission falls below 10 percent in the present study corresponds to the time when an assumed occupant of each fire room would be unable to see the doorway of the room (through which he might escape) if it is about 4 to 5 feet distant from him. The times estimated in this way are of the same order as the heat irradiation hazard times and these two quantities along with toxic gas accumulation represent an approximate assessment of the initial hazard. Since there are no available data on the reaction of an occupant to the combined effect of heat, smoke, and toxic gases at the hazard-threshold level, these assessments of hazard are of more value as a means to compare the individual rooms than as a realistic expression of human survivability. A summary comparison of these estimates of hazards for each room is undertaken in the discussion of toxic gas accumulations during the same periods.

A comparison of these smoke accumulation curves with the time estimates of active and peak fire activity in Table 6 indicates a complex relationship probably because in some cases the fire activity results mainly in consumption of previous smoke accumulations and in others the generation of smoke by the fire activity predominates.

TOXIC GAS ACCUMULATION

The results of the continuous gas analyses carried out during each fire are summarized in Figures 28, 29, 30, 31 and 32; the data for the other components collected during the fires and analyzed as single integrated samples are presented in Table 7, and summarized in Table 8. Finally, an attempt has been made to express the relative hazard represented by these gas concentrations.

Some systematic interrelationships are noticeable, but in general the concentration changes with time show rather complex interdependences that require further experimental investigation. It can be seen that the analytical results for O_2 CO_2, and CO vary in consistent manner such that the total oxygen equivalent is reasonably constant. Further, the accumulations of hydrocarbons and CO tend to occur simultaneously as might be expected if they are taken as an expression of oxygen deficiencies. Usually, CO values reached levels above one percent as O_2 concentrations fell below about 15 percent, and spikes of extreme CO concentrations accompanied O_2 levels below about 10 percent. These observations are broadly

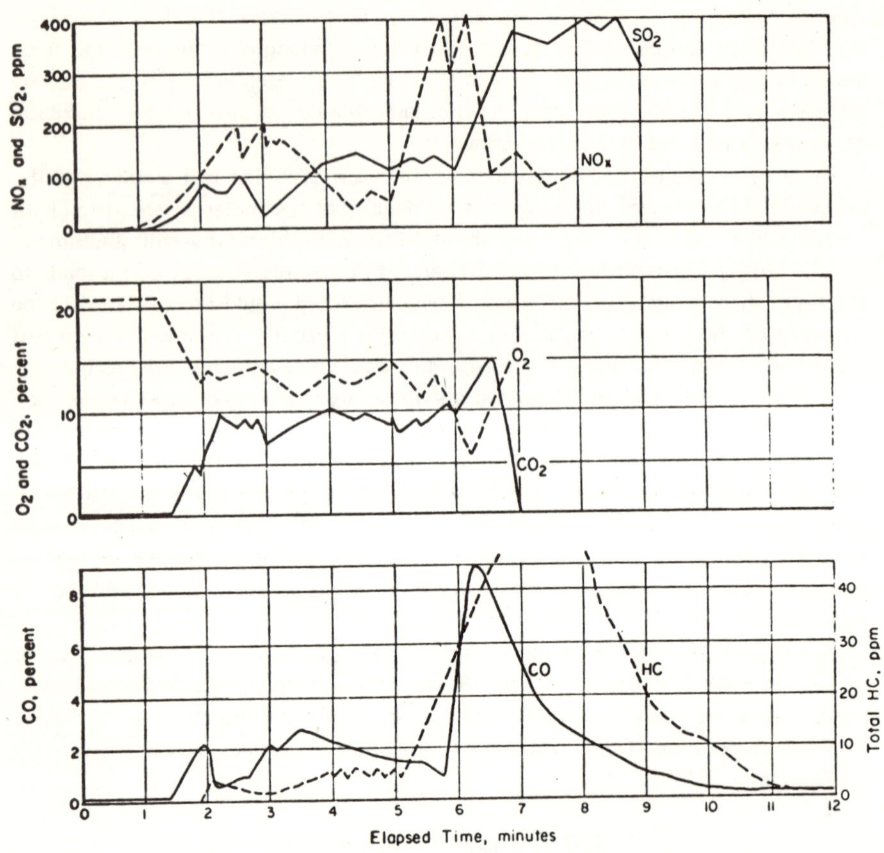

Figure 28. Gas accumulation curves for the typical room experiment.

consistent with the conclusion reached earlier that smoldering and pyrolysis are expected to predominate at O_2 levels below 16 percent.

The SO_2 and NO_x concentration curves show considerable variation and do not necessarily respond consistently to the same variations in oxygen level. Undoubtedly some changes seen in these and other concentration curves are related to the times when the fire spreads to different components of the room, but the identification of such gaseous products with individual furnishing items in the rooms cannot be made in these experiments.

An assessment of the hazard represented by each of these gases individually can be attempted by comparing the measured concentrations with published data on the response of man and animals to the gases when separately present in air. However, the complexities of the combined physiological response expected for mixtures such as found in these fires are such that the conclusions serve more as a

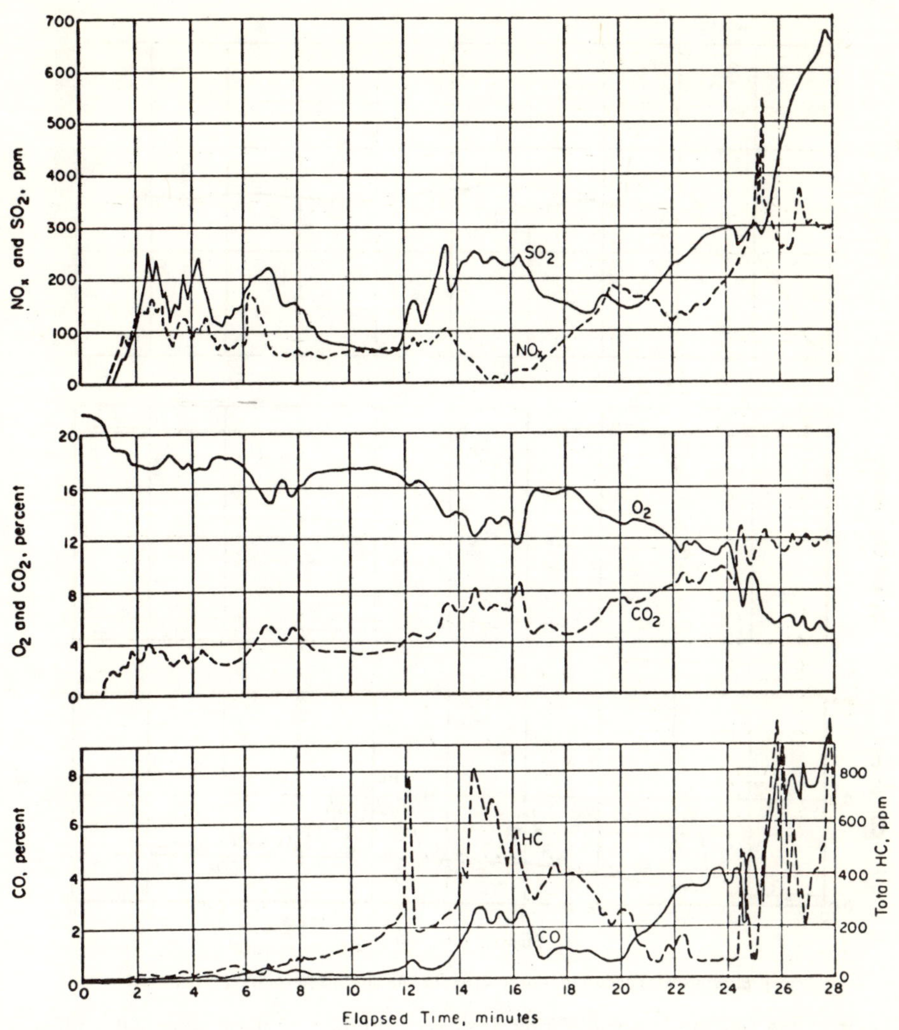

Figure 29. Gas accumulation curves for the improved room experiment.

basis of comparison of the results for these experimental fires than as a statement of human survivability.

Insofar as an occupant of the room is concerned, the hazard due to smoke, heat, and flames has become severe in all but the Space-age Room after about 3 to 5 minutes, so that interest in the hazard due to toxic gases might be expected to center on the additional risk arising from such gases during an early fire period, say the first 10 minutes. Occupants of areas adjacent to the fire room would be exposed to this risk over longer periods of time as a result of convection processes

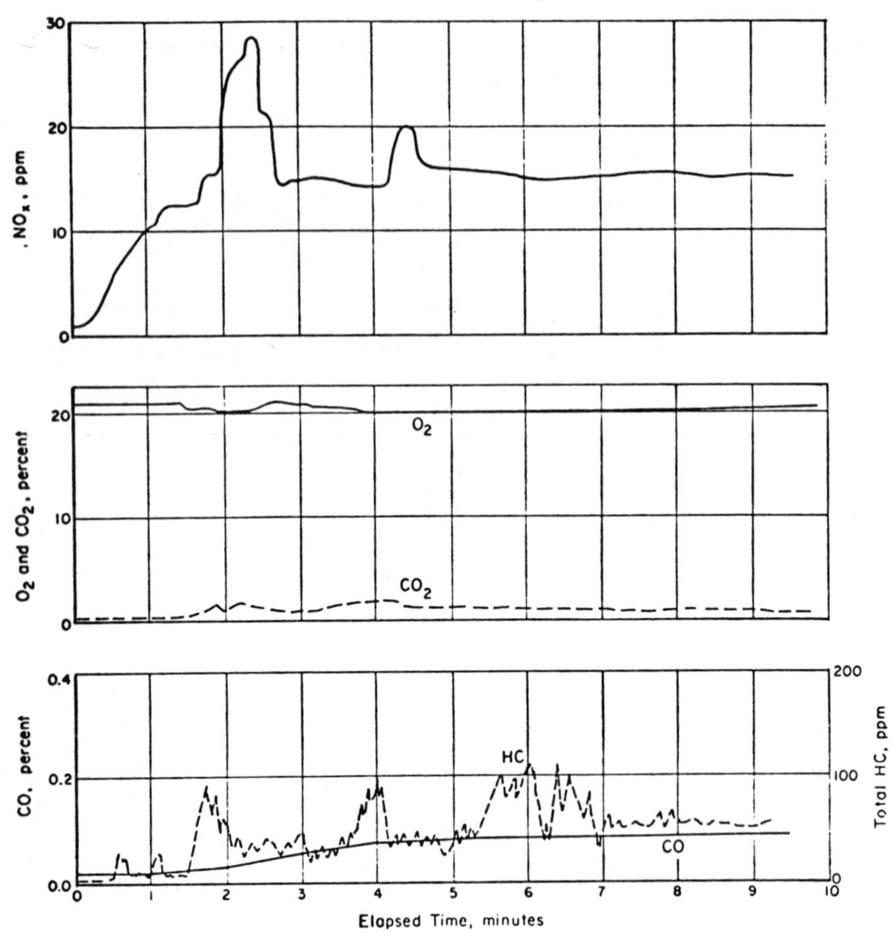

Figure 30. Gas accumulation curves for the space-age room; 1st ignition.

out of the fire room, but differing degrees of dilution of these gases with ambient air make that assessment uncertain. For this latter purpose detailed analyses of convecting gases at the doorway and ventilation duct would be required.

With these considerations in mind, the published data for the appropriate toxic gases have been assembled in Table 9, and the measured average concentration levels found in each fire room during the first 10 minutes have been listed for comparison. In most instances, the published data that are reported represent the hazard for about 1/2 hour exposure, so that a direct comparison of concentrations shown in the Table tends to overemphasize the estimate of risk. From that table, the conditions of major importance would appear to be the HCN, CO, and SO_2 concentration levels since they equal or exceed the identified danger levels in Table

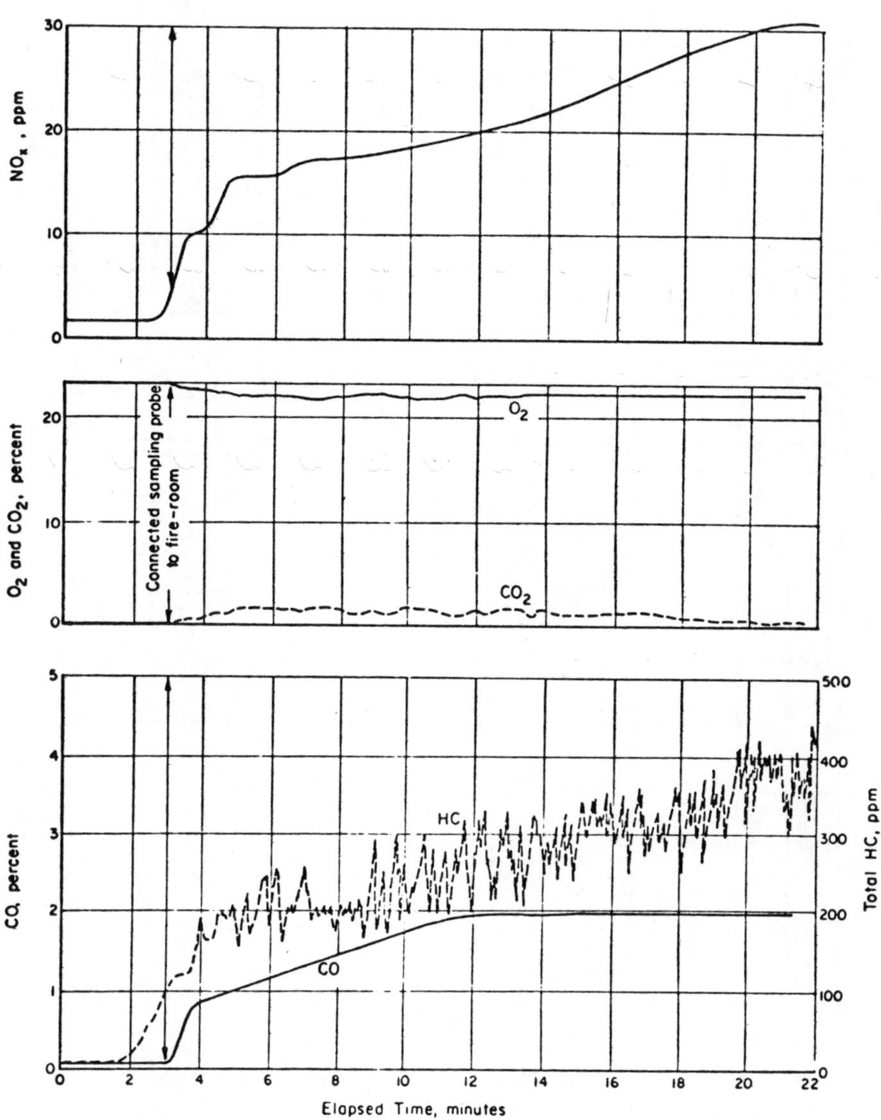

Figure 31. Gas accumulation curves for the space-age room; 2nd ignition.

9. Other conditions that can be recognized as significant hazards are the high aldehyde concentration in the Improved Room fire, the peak NO_x concentrations in all but the Space-age Room fire, and the temporary O_2 depletion conditions seen during active fire periods. These are only the most easily recognizable hazards and ultimately some other conditions such as the peak hydrocarbon concentrations may

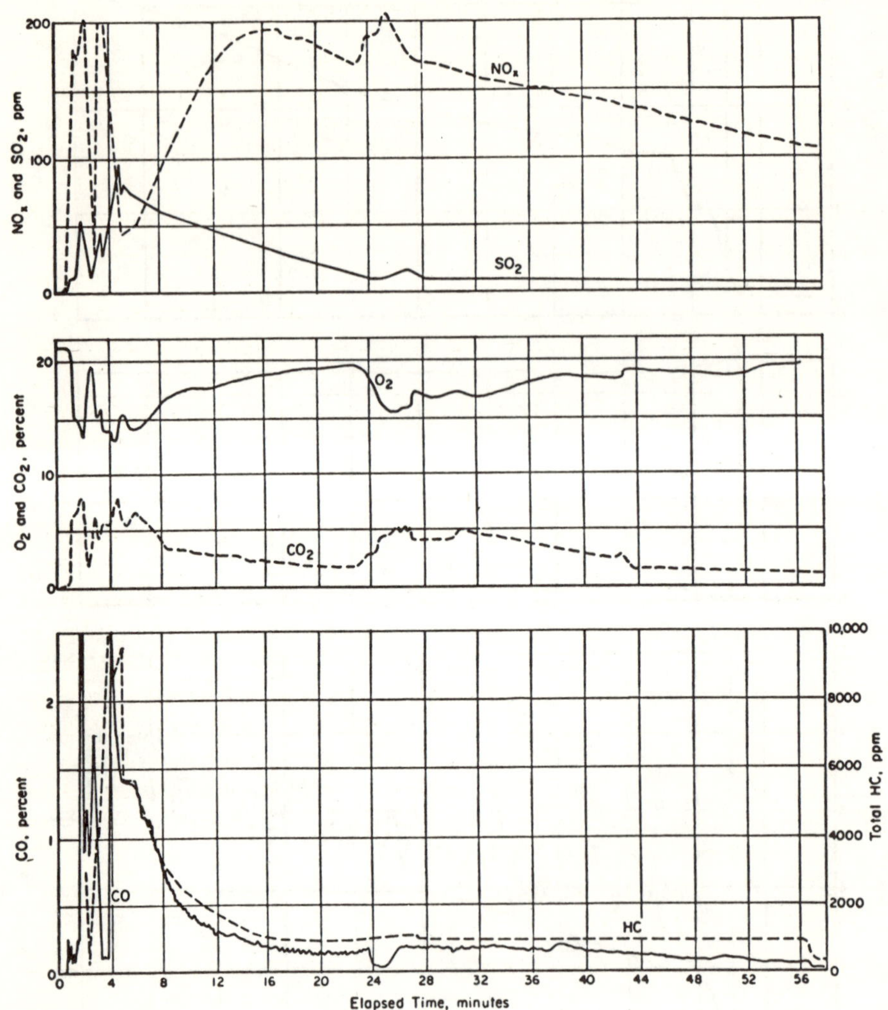

Figure 32. Gas accumulation for the mixed-load room experiment.

prove of equal importance when the identities of individual compounds are obtained, or when the synergistic effects of the combined action is demonstrated.

The Space-age Room shows only minor accumulations of these combustion and pyrolysis products and those seen probably represent mostly the results of the newspaper ignition-fire itself. Both the Typical and Improved Rooms show significant hazard due to toxic gas formation, and the moderation of hazard obtained from the controlled fire of the Mixed-load Room is clearly noticeable.

BURNING VERSUS PYROLYSIS

Since, in each case, the flame represents the combustion of vapors produced by thermal decomposition of the furnishings, the burning of the various items used in these fire rooms is related in some way to the pyrolysis processes for each. Initially, as the flames spread from the fire-start, the ignition and burning of the fuel items represents the burning of those vapors in normal air. However, as the flames build in size the rate of consumption of oxygen in the fire chamber can temporarily exceed the convection of oxygen into the fire room and the resulting depletion of O_2 has a large effect on the speed of the burning processes in those flames. At this point in the fire-time the rapid buildup of heat in the fuel items is causing a rapid increase in the rate of pyrolysis so that combustible vapors begin to accumulate faster than the processes of burning can consume them. Increases in flame size along vertical surfaces lead to impingement of large turbulent flames on cool walls and ceiling which, together with the weakening of flame structure by O_2 depletion, leads to accumulations of smoke and combustible vapors. Eventually these processes can bring the fire chamber to the state where flame speeds are so small that the large flames become extinguished by heat lost through impingement and convection, and the room passes into a smoldering condition while the continuing convection processes start to bring O_2 levels back to the normal ambient value. When this recovery has been achieved, the increase in flame speed that results can cause local reignition of flame and a new cycle occurs. This cyclic, unsteady-state burning and pyrolysis cycle has been observed to be typical of room fires in which the enclosure represents some significant limitation on the accessibility of normal air to the site of the flame. It is not surprising that the rate of burning that is found, and the flammability of fuel items under these conditions, differs markedly from the observations and conclusions reached for the same items under steady-state burning conditions such as practiced in most laboratory tests of flammability.

The Ensemble Effect

A further distinction arises between chamber fires and laboratory fire tests through the use of fire retardants. Most such retardants are intended to inhibit the flame processes occurring at the surface of the treated item so that, once a source of flame is removed, the flame goes out. In fires of the sort used in the present work, the substantial buildup of heat in a treated fuel item, as a result of uncontrolled burning of other items, causes it to continue to generate combustible gases by pyrolysis and these circulate into other parts of the chamber since burning of these at the generation site has been inhibited. These gas mixtures now can burn in the free-space of the chamber without necessarily being attached to any individual items of the fuel load. A free-standing gas-phase fire of this sort is visible in the film made of the Typical Room fire in this study. The result is that a fire can be supported by both the burning of an untreated item and by the resulting pyrolysis of a treated item, whereas individually only the former would support the continuation of the burning process. This is clearly documented by the rapid burning ultimately obtained in the Space-age Room during the second ignition trial when, after 27 minutes, the ignition of one corner of the room brings about complete involvement of the fire-resistant furnishing of the room in less than 6 minutes more. This effect of a mixture of combustible materials to produce burning not characteristic of any one item individually we have chosen to call the ensemble effect. The serious consequences of the ensemble effect are suggested to be a major reason why current laboratory trials of individual textiles and furnishing items fail to predict actual performance.

It should be noted that the major effects produced on a fuel item by the burning of another item in a chamber of restricted circulation are

(1) Change in convection
(2) Depletion of O_2 and generation of CO and CO_2 in a somewhat cyclic manner
(3) Generation of heat for pyrolysis.

so that, if the time-fluctuations of these were known and imitated, a reasonable reproduction of the response of that fuel item to ensemble burning conditions might be made. Further full-scale fire trials may be anticipated to show that only certain combinations of variations in these time-dependent quantities are significant for fire testing purposes and in that event a programmed fire chamber might be developed to yield realistic laboratory results.

AIR-FLOW MEASUREMENTS

Figures 33, 34, 35, and 36 show the measured convection velocities found during each fire trial for air and combustion gases passing through the doorway and the ventilation duct. In that figure the flow data for "top", "middle", and "bottom" refer to the pitot tubes located in those sections of the doorway. The pitot tubes in the ventilation duct and in the top and middle sections of the doorway were oriented to measure flow velocities out of the fire room, while that located in the bottom section was oriented to measure flow into the room. Inspection of the curves shows that there were a number of occasions when flow reversal was indicated. During these periods substantial negative flow heads were developed but these could not be measured conveniently and are shown in Figure 37 by the line drawn below zero flow. These periods of negative flow head have been examined and are believed to correspond to true reversal of flow; it is especially noteworthy that the reversal of flow is a complete reversal of flow pattern in that during the reversal, flow is outward in the bottom section of the doorway and inward in the middle and top sections. The following observations support this surprising evidence for flow reversal

(1) In separate trials the pitot tubes were shown to produce negative heads when the reversed flow direction was tried, so that the negative head seen

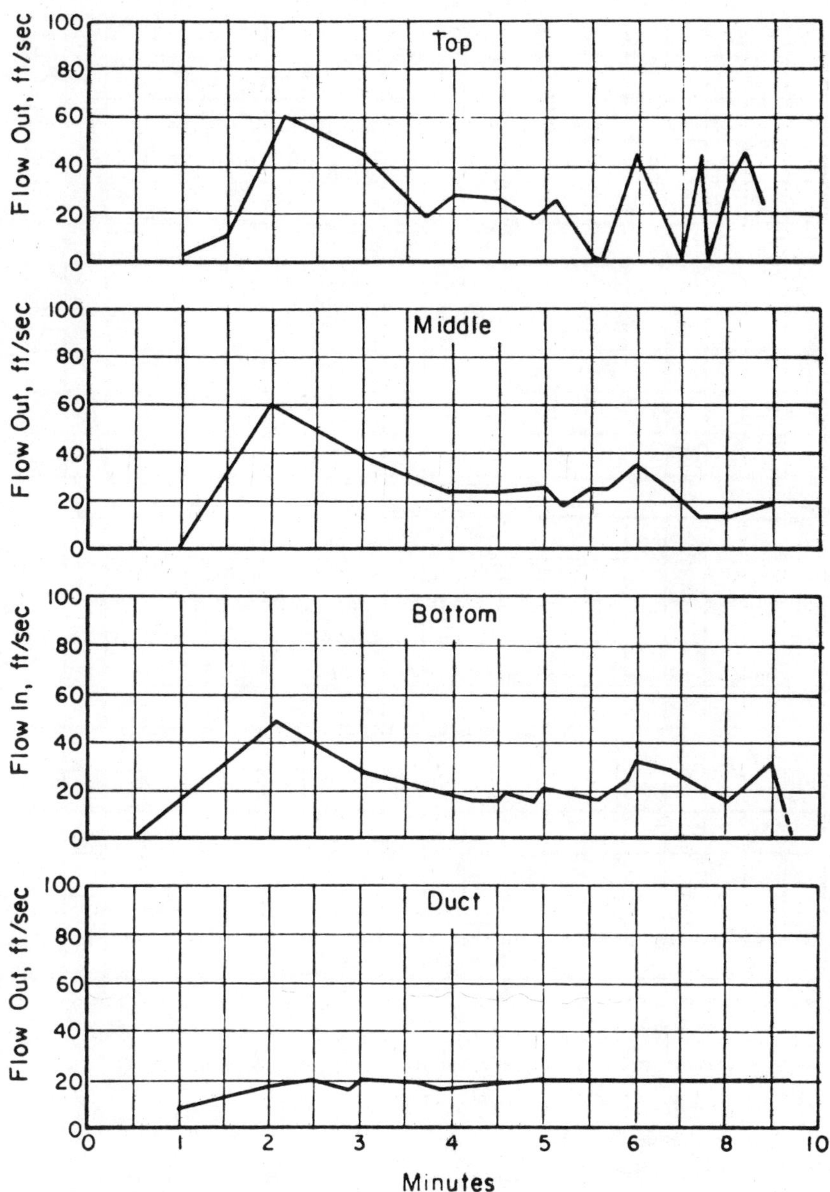

Figure 33. Ventilation flow rates for the typical room fire.

in the fire experiments would indeed represent a reversed flow.

(2) The negative output of the pressure transducer of the pitot tube was confirmed in one of the fire-room experiments by having a liquid

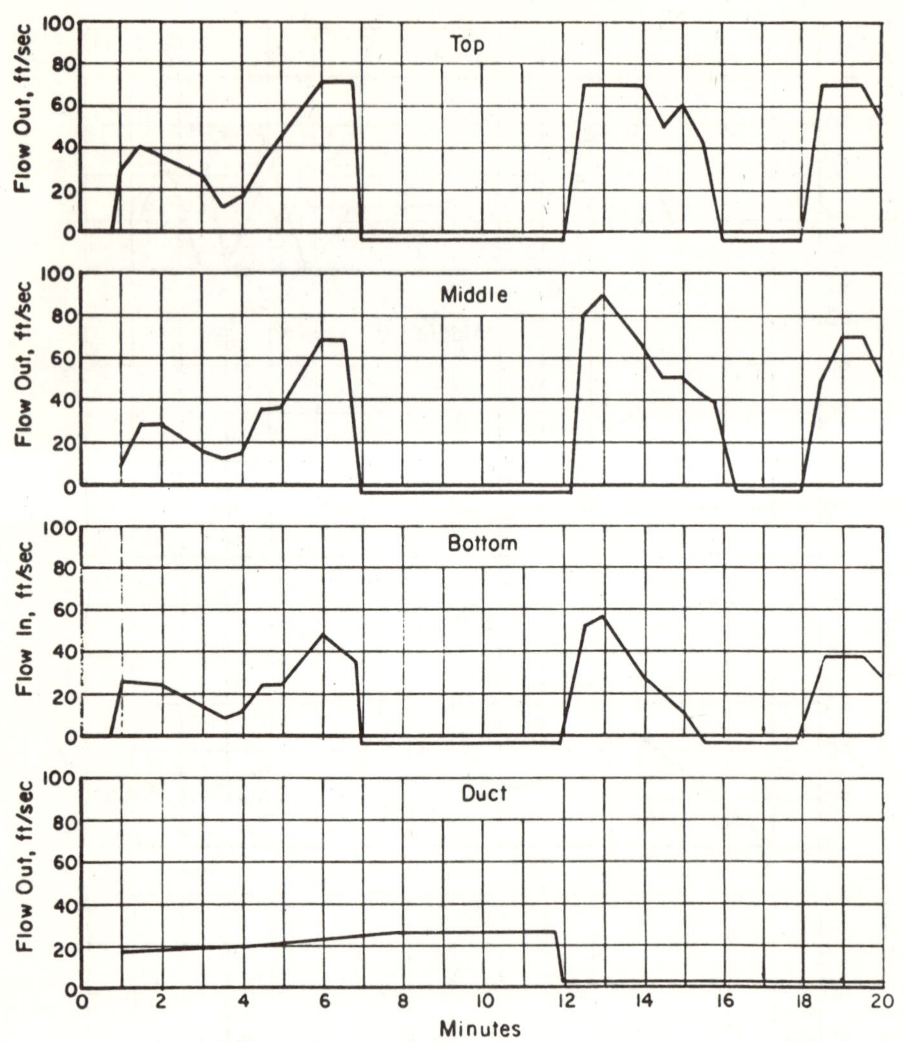

Figure 34. Ventilation flow rates for the improved room fire.

manometer in parallel with the transducer so that the reversal could be observed separately on that manometer.

(3) The temperature record obtained at each pitot tube was compared with the corresponding flow curve, see for example the data for the Improved Room in Figure 37, and this comparison showed that the temperature in the top and middle sections dropped sharply when flow reversal occurred probably because at the time the hot combustion gases ceased to flow out of those sections and relatively cool air from the corridor began to

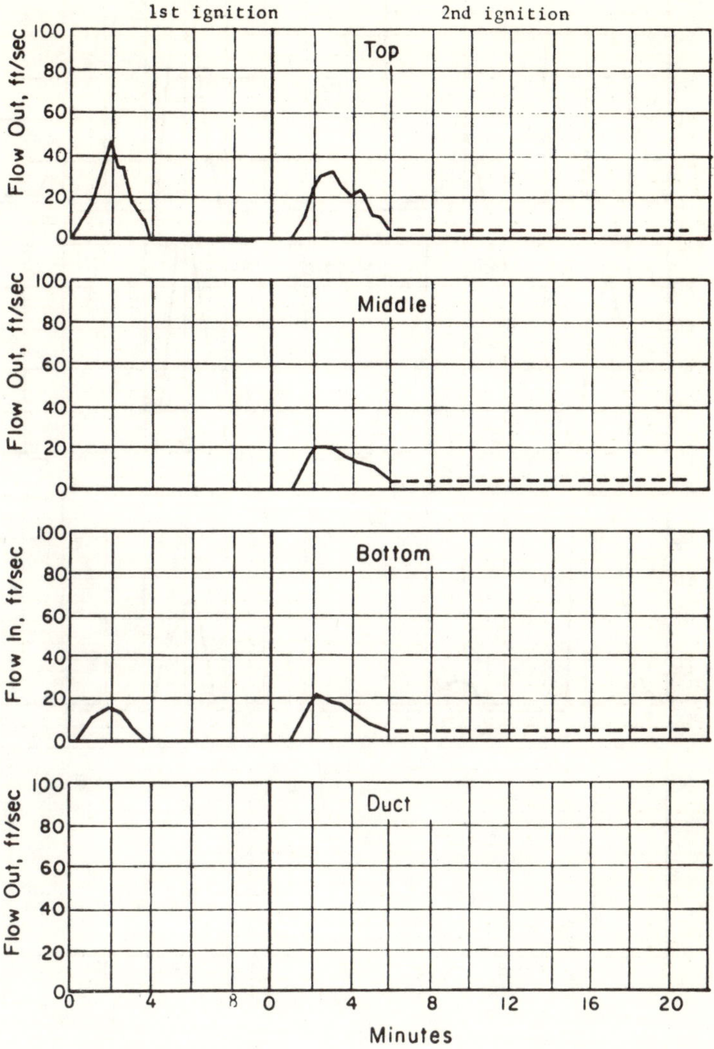

Figure 35. Ventilation flow rates for the space-age room fire.

flow in; it should be noted that the direction of the thermal gradient in the doorway does not reverse although its magnitude is substantially reduced.

(4) The temperature at the ventilation duct grill, in the center of the ceiling, dropped sharply at the same time that the flow reversal indicated that cooler air was entering by way of the top and middle sections of the doorway, see Figure 37. Because of the flow impedance of the duct, reversal did not occur in the duct, and the gases reaching the outer end,

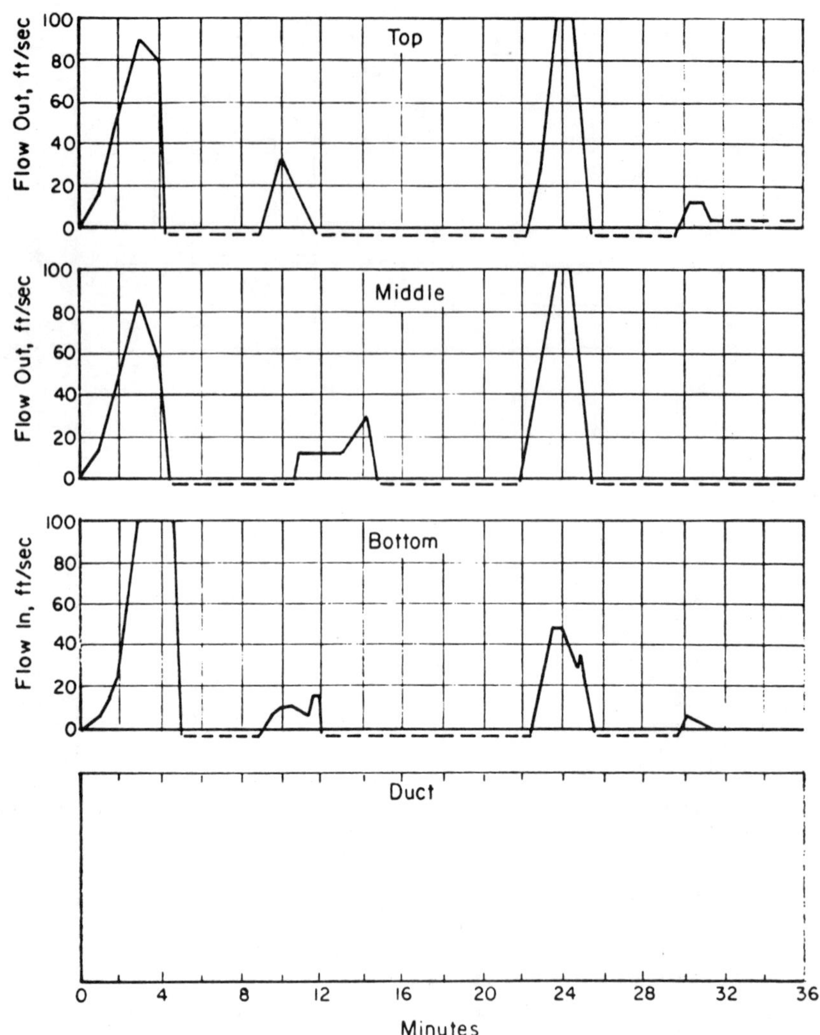

Figure 36. Ventilation flow rates for the mixed-load room fire.

where the duct flow was measured, were relatively cool throughout.

This complex pattern of flow was not anticipated and its relationship to other characteristics of the fires is not clear. It would not be surprising to find that the reversal is related to the restrictions to flow posed by the ventilation arrangements used and the phenomenon requires further investigation. The occurrence of the flow reversal could be taken to imply temporary hazardous conditions along the floor of the fire room and the implications of this should be examined from the

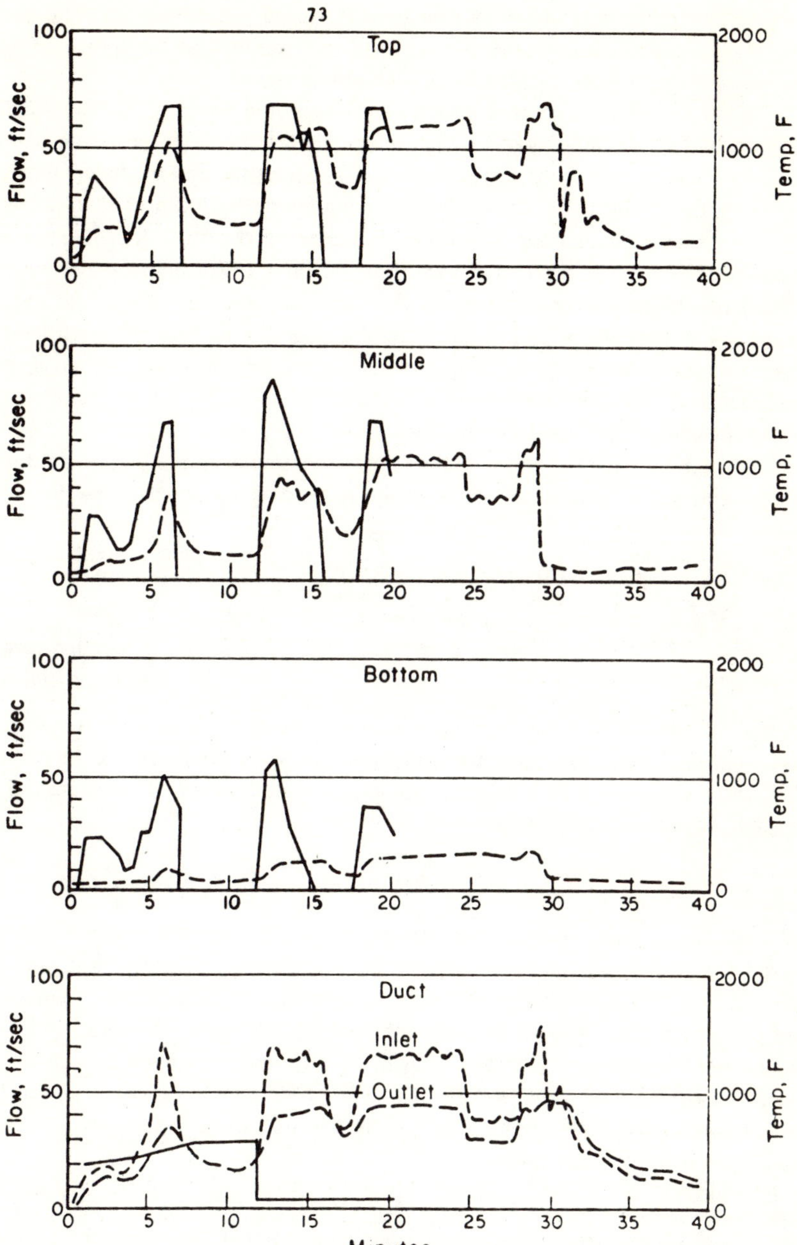

Figure 37. Comparison of flow and temperature conditions at the doorway and ventilation duct in the improved room fire. Solid line represents flow data and the dotted line represents temperature data.

standpoint of occupants who might be seeking escape along the floor. The occurrence of the reversal should also be examined for its effect on the spread of toxic gases and smoke through adjacent areas of the building.

Since the ventilation occurred entirely by way of the doorway, which was 30 in X 84 in, the commonly used ventilation parameter, $A(H)^{1/2}$, could be applied readily to these experiments. In this parameter A equals the area of the opening and H is its vertical height. In other work, this parameter has been used to correlate values of other characteristic quantities for chamber fires using proportionality constants derived from a series of experimental fires. Unfortunately such proportionality constants have been found to vary with chamber size, the size of the fuel elements, and the degree of insulation of the chamber so that the usefulness of these correlations in prediction of quantities for the present study is questionable. Correlations of this sort have been discussed by a number of workers, see for example Friedman [14], and Tsuchiya and Sumi [16].

No prediction value has been found for this parameter in the present study; for the convenience of others the appropriate quantities are listed below in metric terms:

$A(H^{1/2}) = (0.76 \times 2.14)\ (2.14)\qquad (2.14)^{1/2} = 2.37$ values in meters

W = weight of fire load	Typical room	=	1000 kg	
	Improved room	=	1090 kg	excluding special wall and ceiling cover

The values of W for the other two rooms require some interpretation of the term "fire load" since certain items were not readily combustible. The values can be estimated as needed from Table 2.

SUMMARY

A program of experimental fires has been carried out to establish the advantages offered by new space-age materials for the improved fire safety that might be provided. For this purpose four full-scale bedrooms, differing only in the materials used to furnish them, have been built and burned to provide comparative data on the fire hazards produced. Cost and availability differences were not considered.

The visual evidence provided by TV and photographic coverage of the four experimental room fires showed clearly that the rooms responded very differently to a common ignition condition.

In particular:

(1) The Typical Room, furnished from conventional retail sources, ignited easily and burned rapidly so that after 8 minutes the contents of the room were nearly destroyed.

(2) The Improved Room, furnished with materials selected as being among

the best commercially available, showed substantial improvement over the Typical Room in that there was slower fire spread. However, the relatively complete destruction of the room contents that resulted, and the large amounts of smoke, made it clear that substantial further improvements were needed. This fire was stopped after 29 minutes.

(3) The Space-Age Room, furnished completely with new materials that were not yet commercially available, did not ignite under the common ignition condition and so demonstrated the substantial improvement in fire resistance available for those components close to the ignition source. A second and larger ignition arrangement showed that this room can burn, but the difficulty with which this was brought about confirmed the improved fire resistance available with use of these materials.

(4) The Mixed Room ensemble, furnished with materials identical to the Typical Room except for the substitution of the bed from the Space-Age Room, illustrated the improvement in control of fire spread available by careful placement of fire-resistant materials in the important paths of fire development of an otherwise ordinary room.

In addition to these visual observations of fire development, the resulting temperatures, ventilation and convection velocities, heat irradiation levels, smoke accumulation, and a large variety of chemical combustion and pyrolysis products were monitored during each experimental fire. As might be expected, a considerable degree of interrelation was noted between the variations in ambient oxygen concentration produced in the fire room and the other fire-related quantities that were monitored.

The published data on fire spread rates have been used to deduce that active fire spread periods would be expected to occur when ambient O_2 concentrations are at about 16 percent or above. The temperature and heat irradiation measurements made in the present study are reasonably well interpreted on this basis and it is concluded that smoldering and pyrolytic processes characterized the times when O_2 levels fell below 16 percent.

This being the case, some correlation of toxic gas accumulations with O_2 concentration would be expected. Taken broadly, the CO, and hydrocarbon accumulations showed the expected correlation in that sharp rises in concentration accompanied the important dips in ambient O_2 levels that were observed. Thus, usually, CO values reached 1 percent or more as O_2 concentrations fell below 15 percent, and spikes of extreme CO concentrations accompanied O_2 levels below about 10 percent. These variations appear to be broadly consistent with the conclusion that smoldering and pyrolysis are expected to predominate at O_2 levels below about 16 percent.

The most significant hazards at early times in each fire were due to the rapid rise in heat flux and the abrupt obscuration of vision by smoke. The most consistent toxicity hazard was due to CO and its importance would depend on the ability of the occupant to survive the initial heat and smoke menace which characterized each fire room. Other gases and vapors were shown to reach hazardous levels in certain fire rooms and, again, their significance to an occupant would relate to the times in which such hazards occurred, and probably to the synergistic nature of the hazard arising from mixtures of such gases. The individual estimates of hazard are summarized in Table 9.

Since, in each case, the flame represents the combustion of vapors produced by thermal decomposition of the fuel item, the burning of the various items used in these fire rooms is related in some way to the pyrolysis processes for each. During periods of rapid fire spread, the rate of consumption of O_2 can temporarily exceed the convection of O_2 into the fire room and the resulting depletion of O_2 decreases the speed of the burning processes in those flames. Accumulated heat causes pyrolysis to continue at a rapid rate while the burning subsides to a smoldering condition, and meanwhile convection begins the process of replenishment of the O_2 level in the fire room. This cyclic unsteady-state burning and pyrolyses cycle is typical of room fires in which the room enclosure significantly limits ventilation.

Fire-retardant items in the room are caused to pyrolyze by the heat of burning from other items in the room and so contribute to combustible and toxic vapor accumulations even though they may not have entered into the burning process. This effect of a mixture of combustible materials to produce burning and pyrolyses not characteristic of any one item individually we have chosen to call the "ensemble effect".

The major interactions of a fuel item with the burning of another item in a chamber arise through changes in O_2 level, accumulations of CO and CO_2, and radiant and convected heat reaching the fuel. Thus, if the time-fluctuations of these quantities are known and imitated, a reasonable reproduction of the response of that fuel item to ensemble burning conditions might be made. Further full-scale fire trials may be anticipated to show that only certain combinations of such fluctuations are significant for fire testing purposes and in that event a programmed fire chamber might be developed to yield realistic laboratory results.

RECOMMENDATIONS FOR FUTURE WORK

The interdependencies that have been noted among the many fire-related variables that were monitored have revealed a glimpse of the unsteady-state relationships that result between burning and pyrolytic processes. These processes in turn describe the basis for estimation of fire hazard and the role that new materials should serve. It is clear that present-day laboratory tests based on steady-state burning conditions do not necessarily reveal the manner in which a furnishing item reacts to the actual fire conditions nor the degree to which a particular use is important because of fire spread or pyrolytic decomposition effects. The following recommendations for future work are suggested to offer a proper bases for continuation of the work that has been started. While they are thought of as extensions of the present study they could be applied to new fire hazard configurations if necessary.

(1) This phase of the recommended work is concerned mainly with fire spread and ventilation conditions that are of importance to extend the scope of the relationships already established. The immediate objectives of the work would be

(a) To examine the reproducibility of burn results and thereby determine the degree of reliability that can be placed on the observations

(b) To examine the important differences that can be anticipated by changes in ventilation arrangements, especially through windows and forced convection

(c) To examine the importance of igniter location and especially the fire spread control provided by materials used for wall and ceiling construction

(d) To examine the importance of convected gases and heat in detail by monitoring compositions at the doorway (or window) and by carrying out each fire to essentially complete burnout so that the maximum rates of evolution can be described as well as the total amounts of each convected quantity. In addition an especial effort should be made to show how the hazard arising in the first minute or two is related to the choice of materials and conditions.

For this program, full-scale bedroom arrangements are suggested similar to those used in the present study so that the results can be related to the present study. In some cases an arrangement that imitates the Mixed-Load Room would be of value. For this purpose the Space-Age Bed could be replaced by a nonflammable barrier of similar height so that the convection and fire-spread from one part of the room to another can be studied for the influence of wall, ceiling, and floor choices.

(2) This phase of the recommended work is concerned with the gas composition and heat flux variations and their effect on the performance of selected furnishing items. The objectives of this phase would be

(a) To establish the performance of individual materials under the variations of atmosphere and heat flux that occur in a real fire. Both rapid fire buildup as in the Typical Room, and also slow development as in the Improved Room would be examined

(b) To identify as far as possible the minimum set of variations that must be described in order to predict the material performances that are observed.

For this program a material would be placed on a suitable frame or furniture item and located at some distance from the furnishings where fire start occurs. Monitoring of heat and gas compositions would be made at the test material so that its performance could be related to the local conditions that arise at its site. The

Table 1. List of Furnishing and Materials Included in Four Basic Fires[a, b]

Furnished Item	Typical Room	Improved Room
Mattress Components[c]		
Mattress cover	Cotton	Nomex[6]
Mattress padding	Cotton felt + sisal pad[1.1]	HR foam – 3 lb density[7]
Box Spring		
Cover	Cotton	Nomex[6]
Padding	Cotton felt + sisal pad[1.2]	HR foam – 3 lb[7]
Frame	Wood	Fire-retardant paint on wood[4.2]
Hollywood Frame	Metal[2]	Metal[2]
Headboard	Wood – pecan[3.1]	Formica[8] on wood – white oak[2]
Bedspread	Cotton[2]	Nomex[6]
2 Sheets and 1 pillow cover	50% Cotton + 50% polyester[2]	Nomex[6]
Blanket	Acrylic[2]	Nomex[6]
Pillow	Polyester fiber and 50% cotton + 50% polyester cover[2]	HR foam – 3 lb[7] and Nomex ticking[6]
Auxiliary Furniture		
End tables	Wood – pine[5.1]	Formica[8] on wood – pine[5.1] treated with fire-retardant paint[4.2]
Lamps	Ceramic[3.2]	Ceramic[3.2]
Lamp shades	Paper parchment[3.2]	Fiberglas[9]
Upholstered Chair		
Upholstery	Polypropylene	Proban-treated wool[10.1]
Frame	Wood – oak[2.1]	FR paint[4.2] on wood – oak
Padding	Urethane	HR foam – 3 lb[7]
Bureau	Wood – pine[2.2]	Formica[8] on wood – pine[2.2] treated with fire-retardant paint[4.2]
Bookcase	Wood – pine[2.3]	Formica[8] on wood – pine[2.2] treated with fire-retardant paint[4.2]
Floor Covering		
Carpet	Nylon – jute backing[2.4]	Untreated wool[10.2]
Pad	Fiber pad[2.5]	HR foam – 3 lb[7]
Wall Covering	Paint[4.1]	Sodium silicate board[11] + Fiberglas wall cover[9]
Ceiling	Paint[4.1]	Ceramic ceiling tile[12]
Drapes	Polyester-cotton[2]	Fiberglas[2]
Door	Hollow core – birch[5.2]	Solid wood[5.2]

[a]Numbers in parenthesis respectively refer to material suppliers and significant trademark or designation given in Table 1-A.

[b]A listing of furnishing weights is given in Table 2.

[c]Single bed mattress – 39 in × 74½ in.

[d]Cellulose – urea-formaldehyde formulation developed by NASA Houston and prepared by Battelle.

Space-Age Room	Typical Room with Space-Age Bed
Durette 400-1[13] HR foam – 2-lb density[7] + L-3203-6 Fluorel treatment[14]	Durette 400-1[13] HR foam – 2-lb density[7] + L-3203-6 Fluorel treatment[14]
Durette 400-1[13] HR foam – 2 lb[7] + L-3203-6[14] L-3203-6[14] on wood	Durette 400-1[13] HR foam – 2 lb[7] + L-3203-6[14] L-3203-[14] on wood
Metal[2]	Metal[2]
Norgan laminated with FX impregnated fiber glass[15]	Norgan laminated with FX impregnated fiber glass[15]
Beta glass[9]	Beta glass[9]
Durette 400-6[13]	Durette 400-6[13]
Needle batt Durette 400-11[13] between 2 plies of Durette 400-2[13]	Needle batt Durette 400-11[13] between 2 plies of Durette 400-2[13]
HR foam – 2 lb[7] + L-3203-6[14] Cover – 2 plies Durette 400-6[13]	HR foam – 2 lb[7] + L-3203-6[14] Cover – 2 plies Durette 400-6[13]
L-3203-6[14] + Scheuffelin paper[18] on wood – pine[5.1]	Wood – pine[5.1]
Ceramic[3.2]	Ceramic[3.2]
Durasan – 5 mil[16]	Paper parchment[3.2]
Durette 400-6[13]	Polypropylene
Wood (oak) + L-3203-6[14]	Wood – oak[2.1]
HR foam – 2 lb[7] + L-3203-6[14]	Urethane
L-3203-6[14] + Scheuffelin paper[18] on wood – pine[5.1]	Wood – pine[2.2]
L-3203-6[14] + Scheuffelin paper[18] on wood – pine[5.1]	Wood – pine[2.2]
Fiberglas[17]	Nylon – Jute backings[2.4]
HR foam – 2 lb[7] + L-320306[14]	Fiber pad[2.5]
1/4 inch Laminites[18] + Durasan – 5 mil wall covering[16]	Paint[4.1]
1/4 inch Laminites[18] + Ames wood panels[19] + Plastic foam panels[d, 20]	Paint-cotton[4.1]
Beta-fiber[9]	Polyester-cotton[2]
Norgan laminated with FX impregnated fiber glass[15]	Hollow core – birch[5.2]

Table 1-A. Footnotes to Table 1.

Code No.	Supplier	Location	Material Trade Name or Design
1.1	Ohio Bedding	Columbus, Ohio	Youkon
1.2			Youkon
2	Lazarus Department Store	Columbus, Ohio	--
2.1			Model No. 851-230
2.2			Bailey - 1439
2.3			Bailey - 1202
2.4			Lee - 7503-711
2.5			Startex
3.1	Furniture Warehouse	Columbus, Ohio	
3.2			
4.1	Sherwin-Williams	Columbus, Ohio	Rogers Celeste White Latex
4.2			Kem Gard Latex Paing
5.1	Hilliard Lumber Company	Hilliard, Ohio	Mastercraft - 1731
5.2			--
6	DuPont	Wilmington, Delaware	Nomex
7	Scott Paper Company	Chester, Pennsylvania	HR Foam
8	Arco Linoleum and Rug Company	Columbus, Ohio	Flat Cut Regency Walnut (Formica 3855)
9	Owens-Corning Fiberglass Corp. Textile Development Laboratories	Ashton, Rhode Island	--
10.1	Collins & Aihman	Albermarle, North Carolina	Fontant
10.2			Alpine
11	BASF-Wyandotte Chemical Cellasto Division	Ypsilanti, Michigan	Palusol
12	Armstrong Cork Company	Lancaster, Pennsylvania	Ceramaguard
13	Monsanto	St. Louis, Missouri	Durette
14	Raybestas-Manhattan	North Charleston, S. Carolina	Refset L-3203-6
15	Air-Transmission Systems Inc.	San Francisco, California	Norgan
16	Allied Chemical	Morristown, New Jersey	Duratan
17	Carolina Narrow Faline Company	Winston-Salem, North Carolina	U.S. Coast Guard Quality
18	NASA (Houston)	Houston, Texas	
19	NASA (Ames)		
20	Battelle-Columbus NASA (Houston)	Columbus, Ohio Houston, Texas	

overall fire test configuration would again be the same as for the bedroom used in the present study.

(3) This phase is concerned with the development of a programmed laboratory test procedure that suitably imitates the performances detailed in Phases 1 and 2. Work on this phase could be undertaken simultaneously with 1 and/or 2 but should continue in time beyond those phases so that the results of each can be utilized fully. Most of this phase would be carried out in laboratory chambers of suitable simplicity in which heat fluxes and atmosphere variations are artificially produced.

ACKNOWLEDGMENT

The authors gratefully acknowledge the invaluable advice and assistance received in the planning of this project from numerous individuals within NASA, including the Headquarters and the Houston Ames and Lewis Research Centers; the Department of Housing and Urban Development; the National Bureau of Standards; the National Fire Protection Association; The National Research Council of Canada; and the American Society for Testing and Materials.

The authors are also indepted to the Columbus Ohio Fire Department, under Chief Raymond R. Fadley, for their cooperation and for the use of its facilities which was essential to the conduct of the program.

Table 2. Weights of Furnishings Used in Each Fireroom[a]

Furnished Item	Typical Room	Improved Room	Space-Age Room	Mixed-load Room
Mattress	53	32	27	27
Box Spring	44	42	39	39
Hollywood Frame	20	20	20	20
Headboard	18	21	46	46
Bedspread	2.5	2.5	4.5	4.5
2 Sheets and 1 Pillow cover	2.5	2.5	3	3
Blanket	2.8	2.5	5	5
Pillow	1.5	3	1	1
Auxiliary Furniture				
End Tables (2)	23 each	28 each	23 each	23 each
Lamps and Lamp Shades (2)	7.5 each	7.5 each	6 each	7.5 each
Upholstered Chair	49	45	39	49
Bureau	43	55	46	43
Bookcase	23	29	24	23
Floor Covering				
Carpet	61	66	68	61
Pad	29	21	11	29
Wall Covering	--	226[b]	156[c]	--
Ceiling	--	148	54[d]	--
Drapes	2.5	2.5	3	2.5
Door	37	80	110	37

(a) Weight in pounds.
(b) 210 lb sodium silicate board + 16 lb fiberglass.
(c) 138 lb Laminite + 18 lb 5-mil Durasan.
(d) 47 lb Laminite + 2 lb NASA Ames panels + 5 lb NASA Houston panels.

Table 3. Temperature and Humidity Conditions for the Experimental Fires

Date	Fire Room	In Room		Outside Air	
		Temperature °F	Humidity %	Temperature °F	Humidity %
10/3	Preliminary Room	72	50		
10/10	Typical Room	63	33	56	34
10/17	Improved Room	63	42	50	55
10/24	Space-age Room	60	58	50	74
10/31	Mixed load Room	60	48	49	61

Table 4. Thermocouple Locations

Thermocouple Locations	Channel Number 10/10	10/17	10/24	10/31
Ventilation duct, outer end	0	0	0	0
NW corner, 41 in. above floor	1	1	1	1
Bed headboard, ~ 41 in. above floor	3	3	3*	3
NE corner, 41 in. above floor	4	4	4	4
At ceiling above drapes	5	5	5	5
In carpet, below drapes	6	6	6	6
In carpet, 30 in. S, 51 in. E	7	7	7	7
Under pad, 30 in. S, 51 in. E	8	8	8	8
In center of bed, under spread	9	9	9*	9
In pillow	10	10	10*	10
In chair, corner near ignition	11	11	11	11
In bureau, middle of top front edge	12	12	12	12
In bookcase, middle of top front edge	13	13	13	13
In ventilation diffuser, in ceiling center	14	14	14	14
Above center of doorway, inside room	15	15	15	15
At pitot tubes, top 1/3 doorway	16	16	16	16
At pitot tubes, middle 1/3 doorway	17	17	17	17
At pitot tubes, bottom 1/3 doorway	18	18	18	18
At radiometer, 15 in. above floor	19	19	19	19
In movie camera box	20	20	20	20
At gas analyses sample probe	21	21	21	21
At HX, aldehyde sample probe	22	22	22	22
Above center closet doorway, in room	23	23	23	23
Inside TV camera box	24	24	24	24
At ceiling, NE corner	25	25	25	25
Center of door, east side	26	27	26	26
Center of door, west side	27	26	27	27
In chair, topcenter of back	28	28	28	28
Above ceiling, 30 in. S, 41 inc. E		36		36
At top nightstand, NW corner				37
NASA tile A			37	
NASA tile B			36	
At ceiling, 30 in. S, 41 in. E				39
Under drapes, 41 in. above floor	2	38	38	38
Horizontally arranged across doorway, 2/3 up from floor West to East				41 42 43 44 45
Above ceiling near closet		37		
Tile made at BMI			39	
Tile from Dow			40	

(*)Not pertinent, 2nd trial 10/24/72

Table 5. Gaseous Emission Instrumentation Used in Fire Study

Pollutant	Instrument	Range (a)	Principle of Operation	Comments
Total HC	Beckman Model 402	0-120,000	Flame Ionization	Continuous, fast response, portable with selected elevated temperature sampling line and oven
NO_x	Faristor	0-5000	Electrochemical (dry)	Continuous, fast response, portable; SO_2 interference can be accommodated
NO	Beckman Model 315	0-1000	NDIR	Continuous, portable; water must be removed (dry ice trap)
SO_2	Faristor	0-5000	Electrochemical (dry)	Continuous, fast response, portable; no NO_2 interference
CO	Beckman Model 215A Olsen-Horabi	0-1250 0-10%	NDIR NDIR	Continuous, portable; water and CO_2 interference can be accommodated
CO_2	Beckman Model 215A	0-20%	NDIR	Continuous, portable; water interference can be accommodated
O_2	Beckman Model 715	0-25%	Amperometric	Continuous, portable

(a) ppm except as noted.

Sample line for total HC measurements connected directly to fire-room (No trapping) using Heated sample line (~ 300 F)

All other pollutants were analyzed via a common stainless steel manifold. Up stream from manifold, the sample gas passed through a large ice-water trap and particulate trap. An additional dry-ice trap was employed for NO analysis.

Table 6. Comparison of Estimates of Fire Activity from Fire Room Performance Data

	Time from Ignition, Minutes	
	showing > 16% Oxygen Concentration[a]	for gas temperatures in rooms[b]
Typical Room	0 to 1.7	1.0
		7.5
Improved Room	0 to 6.5	6
	8 to 12.8	13
		24
Space-Age Room	all	2.0 after first ignition
	all	32.0 after second ignition
Mixed-Load Room	0 to 1.2	–
	2.2 to 3.0	2.5
	8 to 25	10 and 24

[a]Assumed necessary for active burning and fire spread. Data show time span in which O_2 exceeded 16 volume percent at sampling probe

[b]Peak gas temperature times assumed to represent maximum burning condition

Table 7. Average Concentrations[1] of some Species in the Atmospheres of Full-scale Room Fires: Details

Fire No.	1		2		3		4	
Sample Type	Aldehyde	HX	Aldehyde	HX	Aldehyde	HX	Aldehyde	HX
Date of Fire	10/10/72	10/10/72	10/13/72	10/17/72	10/24/72	10/24/72	10/31/72	10/31/72
Sample								
Collection time, min	(2)	(2)	28	28	31	31	58	58
Volume, cf	0.18	2.1	2.43	19.3	2.85	21.8	5.7	40.0
Volume, SCF[3]	0.18	2.1	2.45	18.9	2.86	21.6	5.8	39.6
m^3[3]	0.0051	0.059	0.069	0.534	0.081	0.610	0.161	1.12
Meter temp., F	56	60	56	70.6	58	66.3	55	68.7
Meter pressure, in. Hg	29.80	29.80	29.34	29.34	29.40	29.40	29.56	29.56
Moisture, percent[4]		20[4]		9.25		1.29		5.01
Particulate,[5]								
On glass wool, g		0.104		1.54		0.031		0.406
On filter, g		0.191		0.20		0.0038		0.0122
Total, g		0.295		1.74		0.035		0.418
(g/m^3)		(4.95)		(3.25)		(0.057)		(0.372)
Aldehydes,[6] mg	0.78		122.2		4.2		43.2	
(mg/m^3)	(153)		(1760)		(52)		(263)	
Cl,[7] mg		8.0		205		6.3		37.3
(mg/m^3)		(135)		(383)		(10.3)		(33.2)
F,[7] mg		0.13		0.30		0.38		0.13
(mg/m^3)		(2.2)		(0.56)		(0.62)		(0.12)
CN,[7] mg		46.0		60.3		1.50		6.7
(mg/m^3)		(770)		(113)		(2.5)		(6.0)

(1) Samples collected at constant rate during entire duration of fire experiments
(2) Sampling systems plugged after about 2 to 3 minutes operation
(3) SFC corrected to 70 F, 1 atmosphere, m^3 calculated at 68 F (20 C) and 1 atmosphere
(4) Water content of hot inlet gases from sampling probe. Run 1 uncertain
(5) Includes solids and condensed material, 250 F
(6) Separate sampling train, expressed as HCHO
(7) Collected after separation of particulate, see Note 5

Table 8. Summary Table: Average Concentrations[1] of some Species in the Atmospheres of Full-scale Room Fires

Component	1 (typical)[2]		2 (improved)		3 (space-age)		4 (mixed)	
	$\frac{mg}{m^3}$[3]	ppm[3]	$\frac{mg}{m^3}$[3]	ppm[3]	$\frac{mg}{m^3}$[3]	ppm[3]	$\frac{mg}{m^3}$[3]	ppm[3]
Particulate[4]	4960	–	3250	–	57	–	372	–
Chloride ion	135	92	383	260	10.3	7.0	33.2	22.5
Fluoride ion	2.2	2.8	0.56	0.71	0.62	0.79	0.12	0.15
Cyanide ion	770	710	113	105	2.5	2.3	6.0	5.6
Aldehyde (as HCHO)	150	120	1760	1410	52	42	263	210

(1) Samples collected at constant rate for entire duration of fire experiment
(2) Gas sampling systems plugged after 2 to 3 minutes and so sample represents the initial period of substantial smoke evolution
(3) mg/m^3 at 68 F (20 C) and 1 atmosphere; ppm on volume basis
(4) Total particulate and condensate at 250 F

Table 9. Summary Comparison of Gaseous Component Concentrations of Fire Room Trials with Published Values for Hazardous Concentrations

	Danger Level[1]	Fire Experiment[2]			
		1	2	3	4
CO_2	>8%[1a]	8	3–4	1	3(7)
O_2	<14%[1b]	12	17	20	16
CO	Ca0.1–0.2%[1c]	1.5	0.5	small	1
Hydrocarbons	(ppm)	50,000	150		5000
SO_2	50–100 ppm[1d]	100(400)	200	small	50(100)
NO_x	100–150 ppm[1e]	150(300)	100	15	100(200)
HCl	1000–2000 ppm[1f]	92	260	7	22
HF	50–250 ppm[1g]	3	0.7	0.8	0.2
HCN	100–200 ppm[1h]	710	105	2.3	5.6
Particulate	(g/m^3)	5.0	3.3	0.057	0.37
Aldehydes	ppm[1c]	120	1410	42	210

[1] Based on the following estimates for hazard

- a CO_2, unconsciousness after 4 hrs, see Robison, et al, [11]
- b O_2, dizziness, fatigue, poor coordination, <14%, see Robison, et al, [11]
- c CO, time to collapse (minutes) $= \frac{4.5}{\%CO}$, see Robison, et al, [11]
- d SO_2, danger after 30/60 minutes, see Hafer & Yuill [12]
- e NO_x, danger after 30/60 minutes for NO = 100–150 ppm, see Wagner [13]
  danger after 30/60 minutes for NO_2 = 100 ppm, see Hafer & Yuill [13]
- f HCl, danger after brief exposure, see Wagner [13]
- g HF, danger after brief exposure, see Wagner [13]
- h HCN, fatal after 30/60 minutes, see both Wagner [13] and Hafer & Yuill [12]
- i May be quite hazardous: formaldehyde considered unsafe at 5 ppm for 8 hr day, see Friedman [14], others such as acrolein are very poisonous, see [15]

[2] Units are the same as indicated under Danger level for each component. Values in parentheses indicate peak values, others are average values, during 1st 10 minutes, for those measured continuously, and for entire fire for those measured as single integrated sample.

REFERENCES

1. J. A. Dorsey and D. A. Kemnitz, "A Source Sampling Technique for Particulate and Gaseous Fluorides," J.A.P.C.A., *18*, 1244 (1968).
2. K. W. Wilson, "Fixation of Atmospheric Carbonyl Compounds by Sodium Bisulfite, " Analytical Chem., *30*, 1127–29 (1958).
3. L. A. Wall, "Condensed Phase Combustion Chemistry," Fire Research Abstracts and Reviews, *13*, 204–19 (1971).
4. C. Huggett, "Combustion Processes in the Aerospace Environment," Aerospace Medicine, *40*, 1176–80 (1969).
5. C. Huggett, "Habitable Atmospheres Which Do Not Support Combustion," Combustion and Flame, *20*, 140–2 (1973).
6. Konrad Huettner, "Effects of Extreme Cold on Human Skin. II. Surface Temperature, Pain and Heat Conductivity in Experiments with Radiant Heat," J. of Applied Physiology, *3* (12), 703–713 (June 1951).

7. S. Atallah and D. S. Allan, "Safe Separation Distances from Liquid Fuel Spill Fires," Presented at the Central States Section of the Combustion Institute Meeting of Disaster Hazards, April 1970, Houston, Texas.
8. D. I. Lawson and D. L. Simms, "The Ignition of Wood by Radiation," British J. of Applied Physics, *3*, 288–292 (1953).
9. G. H. Tryon and G. P. McKinnon, Eds., "Fire Protection Handbook, Thirteenth Edition, 1969," National Fire Protection Association, pp. 8–210.
10. R. G. Silversides, "Measurement and Control of Smoke in Building Fires," Symposium on Fire Test Methods-Restraint and Smoke, 1966, ASTM STP 422, Am. Soc. Testing Materials, 125–165 (1967).
11. M. M. Robison, P. E. Wagner, R. M. Fristrom, and A. G. Schulz, "The Accumulation of Gases on an Upper Floor during Fire Buildup," Fire Technology, *8*, 278–290 (1972).
12. C. A. Hafer and C. H. Yuill, "Characterization of Bedding and Upholstery Fires," Final Report NBS Contract No. CST-792-5-69, March 31, 1970.
13. J. P. Wagner, "Survey of Toxic Species Evolved in the Pyrolysis and Combustion of Polymers," Fire Research Abstracts and Reviews, *14* (1), 1–23 (1972).
14. R. Friedman, "The Role of Chemistry in Fire Problems: Gas-Phase Combustion Kinetics," Fire Research Abstracts and Reviews, *13*, 187–203 (1971).
15. H. E. Christensen, "Topic Substances," Annual List, 1971, U.S. Department of Health, Education, and Welfare (1971). U.S. Government Printing Office Stock No. 1716-0004.
16. Y. Tsuchiya and K. Sumi, "Evaluation of the Toxicity of Combustion Products," *J. Fire & Flammability, 3*, 46–50 (1972).